Marine Biotechnology: Applications in Food, Drugs and Energy

Muhammad Dawood Shah •
Julian Ransangan • Balu Alagar Venmathi
Maran
Editors

Marine Biotechnology: Applications in Food, Drugs and Energy

Springer

Editors
Muhammad Dawood Shah
Borneo Marine Research Institute,
Universiti Malaysia Sabah
Kota Kinabalu, Sabah, Malaysia

Julian Ransangan
Borneo Marine Research Institute, Universiti
Malaysia Sabah
Kota Kinabalu, Sabah, Malaysia

Balu Alagar Venmathi Maran
Borneo Marine Research Institute,
Universiti Malaysia Sabah
Kota Kinabalu, Sabah, Malaysia

ISBN 978-981-99-0623-9 ISBN 978-981-99-0624-6 (eBook)
https://doi.org/10.1007/978-981-99-0624-6

This Springer imprint is published by the registered company Springer Nature Singapore Pte Ltd.
The registered company address is: 152 Beach Road, #21-01/04 Gateway East, Singapore 189721, Singapore

Preface

Food security, drug development, and energy production are among the most important issues currently globe is facing. Marine biotechnology, a fast-expanding field, may be key to solving many of these issues. This field is becoming more and more significant due to the rising need for clean and renewable resources. This book offers an in-depth analysis of the most recent advancements in this fascinating field, ranging from basic science to applied research.

The book is designed for students, researchers, and professionals in the field of marine biotechnology as well as those interested in using these tools for solving the main issues. It addresses major areas of marine biotechnology including food, drugs, and energy. The chapters of the book focus on various topics including the discovery of nutraceuticals, the hybridization of fish, the application of molecular markers to fish breeding, seaweed utilization, the potential and challenges of sea cucumber mariculture, and the application of jellyfish-derived collagen. This book also offers insightful analyses of the state-of-the-art tools and methods employed in marine biotechnology, including the use of bacteriophage as a therapeutic approach, the identification and characterization of the marine microbiome, and the application of artificial intelligence techniques.

We are very thankful to the staff of the Borneo Marine Research Institute, Universiti Malaysia Sabah, for their full support and cooperation in the preparation of the book. We also thank the management team of Universiti Malaysia Sabah for helping us to make this book a reality by enabling us to gather a wealth of expertise and information from across the world of marine biotechnology using their technologies.

Kota Kinabalu, Sabah, Malaysia

Muhammad Dawood Shah
Julian Ransangan
Balu Alagar Venmathi Maran

Contents

About the Editors

Muhammad Dawood Shah is a Senior Lecturer at the Borneo Marine Research Institute, Universiti Malaysia Sabah (UMS), Malaysia. Dr. Dawood earned his MSc and PhD in Biotechnology from the Biotechnology Research Institute, UMS in 2011 and 2016, respectively. In 2011, he received the excellence award for his MSc outstanding research work and best postgraduate student. From 2014 to 2022, he received four gold and three silver medals for his research work. In 2021, he was the winner of the Malaysian-Croatia Technology Exchange and Malaysia Technology Expo 2021 awards for his innovation. He has more than 40 peer-reviewed international publications and has delivered oral and poster presentations at numerous international conferences and symposia. Dr. Dawood has filed two patent applications and established a research group network both nationally and internationally. Besides, his research work has also been published in the printing media.

Julian Ransangan is an Associate Professor at the Borneo Marine Research Institute, Universiti Malaysia Sabah (UMS), Malaysia. He obtained his MSc and PhD degrees in Marine Biology and Aquaculture from the Borneo Marine Research Institute, UMS in 1999 and 2009, respectively. He has authored more than 55 peer-reviewed articles in international Scopus-indexed journals. He was the Bronze Medallist in the Salon International Des Inventions De Geneva in 2009 and the Seoul International Invention Fair in 2010. In recognition of his excellence in research and publication, he was awarded the Top 5 Prolific Author by the Universiti

Malaysia Sabah in 2018. Currently, Dr. Ransangan is actively conducting research and publishing high-impact articles in the areas of bivalve aquaculture and fisheries.

Balu Alagar Venmathi Maran is an Associate Professor at the Borneo Marine Research Institute, Universiti Malaysia Sabah (UMS), Malaysia. He received his BSc, MSc in Zoology from The American College, Madurai, India; Ph.D. in Marine Biology from Hiroshima University, Japan and CAS in Marine Biology, Annamalai University, India as well. He currently works in the field of Aquaculture and Marine Science and has about 25 years of teaching and research experience. He is specialized in the field of "Taxonomy/Systematics of Parasitic Copepods and Application of Natural Products to Control Fish Parasites in Aquaculture". In addition, he is intensively doing research on the "Biodiversity and Ecology of Jellyfish in Malaysia with a focus on Harmful Jellyfish, Toxins and Jellyfish collagen & its applications". He received several best researcher awards and gold medals from UMS for his research and innovation. He has authored more than 80 peer-reviewed research articles and ten book chapters. He has edited two books. He is the project leader for several research projects including an international project on marine biodiversity. Currently, he is an Academic Editor of a journal and Guest Editor for several other journals.

List of Contributors

Khodzori Fikri Akmal Borneo Marine Research Institute, Universiti Malaysia Sabah, Kota Kinabalu, Sabah, Malaysia

Sajid Ali Department of Biotechnology, Abdul Wali Khan University Mardan, Khyber Pakhtunkhwa, Pakistan

Thyviaah Ananthan Borneo Marine Research Institute, Universiti Malaysia Sabah, Kota Kinabalu, Sabah, Malaysia

Sumithra Pasumalai Arasu Department of Biotechnology, Srimad Andavan Arts and Science College, Bharathidasan University, Trichy, India

Sujjat Al Azad Borneo Marine Research Institute, Universiti Malaysia Sabah, Kota Kinabalu, Sabah, Malaysia

Helena Biun Likas Fisheries Complex, Department of Fisheries Sabah, Kota Kinabalu, Sabah, Malaysia

Grace Joy Wei Lie Chin Biotechnology Research Institute, Universiti Malaysia Sabah, Kota Kinabalu, Sabah, Malaysia

Fui Fui Ching Borneo Marine Research Institute, Universiti Malaysia Sabah, Kota Kinabalu, Sabah, Malaysia

Wei Sheng Chong Borneo Marine Research Institute, Universiti Malaysia Sabah, Kota Kinabalu, Sabah, Malaysia

Norfazreena Mohd Faudzi Borneo Marine Research Institute, Universiti Malaysia Sabah, Kota Kinabalu, Sabah, Malaysia

Aurelia Jackson Borneo Marine Research Institute, Universiti Malaysia Sabah, Kota Kinabalu, Sabah, Malaysia

Bhuvaneshwari Jothirajan Department of Microbiology, Srimad Andavan Arts and Science College, Bharathidasan University, Trichy, India

Nazia Binti Abdul Kadar Borneo Marine Research Institute, Universiti Malaysia Sabah, Kota Kinabalu, Sabah, Malaysia

Muthu Subash Kavitha School of Information and Data Sciences, Nagasaki University, Nagasaki, Japan

Imran Khan Department of Biotechnology, Abdul Wali Khan University Mardan, Khyber Pakhtunkhwa, Pakistan

Mogana Kumar Borneo Marine Research Institute, Universiti Malaysia Sabah, Kota Kinabalu, Sabah, Malaysia

Mohammad Tamrin Bin Mohamad Lal Borneo Marine Research Institute, Universiti Malaysia Sabah, Kota Kinabalu, Sabah, Malaysia

Balu Alagar Venmathi Maran Borneo Marine Research Institute, Universiti Malaysia Sabah, Kota Kinabalu, Sabah, Malaysia

Muralindran Mariappan Faculty of Engineering, Universiti Malaysia Sabah, Kota Kinabalu, Sabah, Malaysia

Nurzafirah Binti Mazlan Borneo Marine Research Institute, Universiti Malaysia Sabah, Kota Kinabalu, Sabah, Malaysia

Zepri Abdul Muhaimin Department of Marine Science, Kulliyyah of Science, International Islamic University Malaysia, Jalan Sultan Ahmad Shah, Kuantan, Pahang, Malaysia

Siti Syahirah Binti Muzli Borneo Marine Research Institute, Universiti Malaysia Sabah, Kota Kinabalu, Sabah, Malaysia

Muhammad Nor Afdall Bin Nazahuddin Borneo Marine Research Institute, Universiti Malaysia Sabah, Kota Kinabalu, Sabah, Malaysia

Rafidah Othman Borneo Marine Research Institute, Universiti Malaysia Sabah, Kota Kinabalu, Sabah, Malaysia

Anand Paul The School of Computer Science and Engineering, Kyungpook National University, Daegu, South Korea

Nur Nashrah Binti Abd Rahman Borneo Marine Research Institute, Universiti Malaysia Sabah, Kota Kinabalu, Sabah, Malaysia

V. Manikandarajan Kinabalu International School, Kota Kinabalu, Sabah, Malaysia

Julian Ransangan Borneo Marine Research Institute, Universiti Malaysia Sabah, Kota Kinabalu, Sabah, Malaysia

Kenneth Francis Rodrigues Biotechnology Research Institute, Universiti Malaysia Sabah, Kota Kinabalu, Sabah, Malaysia

Nur Suhaila Binti Rosli Borneo Marine Research Institute, Universiti Malaysia Sabah, Kota Kinabalu, Sabah, Malaysia

Vidya Sankarapandian Faculty of Biomedical Science, Department of Microbiology and Immunology, Kampala International University, Kampala, Uganda
Department of Microbiology, Srimad Andavan Arts and Science College, Bharathidasan University, Trichy, India

Shigeharu Senoo Borneo Marine Research Institute, Universiti Malaysia Sabah, Kota Kinabalu, Sabah, Malaysia
UMS-Kindai Aquaculture Development Centre, Kindai University, Wakayama, Japan

Muhammad Dawood Shah Borneo Marine Research Institute, Universiti Malaysia Sabah, Jalan UMS, Kota Kinabalu, Sabah, Malaysia

Saad Shahbudin Department of Marine Science, Kulliyyah of Science, International Islamic University Malaysia, Jalan Sultan Ahmad Shah, Kuantan, Pahang, Malaysia

Sitti Raehanah Muhamad Shaleh Borneo Marine Research Institute, Universiti Malaysia Sabah, Kota Kinabalu, Sabah, Malaysia

Sarah Syazwani Binti Shukhairi Borneo Marine Research Institute, Universiti Malaysia Sabah, Kota Kinabalu, Sabah, Malaysia

Sumaliny Subramaniam Borneo Marine Research Institute, Universiti Malaysia Sabah, Kota Kinabalu, Sabah, Malaysia

Riaz Ullah Department of Biological Sciences, National University of Medical Sciences, Rawalpindi, Pakistan
Department of Biological Sciences, National University of Medical Sciences, Islamabad, Pakistan
Department of Food Science and Technology, The University of Haripur, Pakistan

Wilson Thau Lym Yong Biotechnology Research Institute, Universiti Malaysia Sabah, Kota Kinabalu, Sabah, Malaysia

Wahidatul Husna Zuldin Borneo Marine Research Institute, Universiti Malaysia Sabah, Kota Kinabalu, Sabah, Malaysia

Marine Biotechnology: A Frontier for the Discovery of Nutraceuticals, Energy, and Its Role in Meeting Twenty-First Century Food Demands

1

Imran Khan, Khodzori Fikri Akmal, Wei Sheng Chong, Balu Alagar Venmathi Maran, and Muhammad Dawood Shah

Abstract

One of the most underutilized biological resources in the world is the marine environment, which makes up nearly three-quarters of the Earth's surface. A variety of organisms with unique biological systems and features can be found in the marine environment. They have evolved special characteristics that allow them to survive in a variety of hostile environments. By applying a wide variety of screening tools, extracts and purified compounds of these organisms can be studied for food processing, biological activities, and bioenergy production. Biomolecules derived from marine organisms have a wide range of applications in the food industry, including colorants, preservatives, and flavor enhancers. Some of the most useful marine-derived food ingredients are pigments, polyunsaturated fatty acids, sterols, polysaccharides, proteins, and enzymes. Among the therapeutics, more than 60 % of the active pharmaceutical formulations come from natural products or their derivatives, which have been reported to possess biological activities (anticancer, anti-inflammatory, antioxidant, antimicrobial, etc.). Using marine resources to produce biodiesel is one of the hottest areas for renewable energy. International cooperation, novel biotechnological tools, mass production of marine organisms, integration of biotechnology with other sectors, etc., will be necessary to fully explore the potential of marine sources.

I. Khan
Department of Biotechnology, Abdul Wali Khan University Mardan, Khyber Pakhtunkhwa, Pakistan

K. F. Akmal · W. S. Chong · B. A. Venmathi Maran · M. D. Shah (✉)
Borneo Marine Research Institute, Universiti Malaysia Sabah, Kota Kinabalu, Sabah, Malaysia
e-mail: dawoodshah@ums.edu.my

M. D. Shah et al. (eds.), *Marine Biotechnology: Applications in Food, Drugs and Energy*, https://doi.org/10.1007/978-981-99-0624-6_1

1.1 Introduction

Despite covering around three-quarters of the Earth's surface, the marine environment is one of the world's least utilized biological resources. The marine environment is home to a wide variety of organisms (Fig. 1.1), each with its own set of biological systems and characteristics. For example, marine algae, sponges, corals, fish, and microbes have evolved specific characteristics that allow them to flourish in a variety of hostile environments such as salinity, pressure, temperature, and darkness (Rasmussen and Morrissey 2007).

Marine organisms hold useful industrial products. About 230,000 species have been estimated to be dwelling in the marine environment. However, many species

Fig. 1.1 Some of the marine invertebrates. (**a**) Sponge, (**b**) hard coral, (**c**) soft coral, (**d**) cushion star, (**e**) ascidian, (**f**) tunicate, (**g**) star fish, (**h**) feather star, (**i**) nudibranch

have yet to be discovered and characterized. Marine organisms produce a wide variety of metabolites (primary and secondary) which have remarkable biological applications. The integration of modern tools into marine research is unmasking the hidden potential and fast-tracking marine research for the discovery of novel products, characterization of marine resources, and exploitation of these resources for human welfare. According to the Dictionary of Marine Natural Products and MarinLit database (http://pubs.rsc.org/marinlit/), over 39,000 compounds have been identified so far in the marine environment, and approximately 38,700 articles have been published on marine products. According to the record of the PubMed database, there has been a consistent and steady increase in the research work carried out on marine products and marine biotechnology (Fig. 1.2).

To address the current and future challenges from the perspective of the marine environment, marine (blue) biotechnology has been developed to modernize the established tools and engineer new technologies for the efficient utilization of marine resources. Marine biotechnology thrives to discover, exploit, and utilize the potential of marine resources (including organisms and the environment) for the prosperity of humanity while maintaining the natural ecosystem of the marine environment. Not only for humanity's benefit, but marine technologies are also innovating new ways and harnessing modern technological tools for the welfare of marine life. From a wider perspective, marine biotechnology involves the use of marine organisms or their components to produce goods or services and exploit marine resources for ubiquitous applications in the fields of medicine (particularly drug discovery), cosmetics, environmental remediation, food, feed supply and processing, and energy production (Freitas et al. 2012; Baerga-Ortiz 2009; Tramper et al. 2003). Like other types of biotechnology, marine biotechnology also utilizes and innovates the tools of molecular and cellular biology, genetics, chemistry, OMICS, and bioinformatics. Advances in these fields have facilitated the application of marine biotechnology, in which marine organisms and their compounds are explored and useful components are identified, obtained, and characterized for use in a variety of fields such as food and feed, pharmaceutical, and biomedical industries (Rotter et al. 2021).

By the year 2025, it is predicted that the global market for marine biotechnology will reach $6.4 billion, covering a wide range of commercial objectives for the pharmaceutical, chemical, and biofuel industries (Hurst et al. 2016; Vierros et al. 2016). Another report has recently projected the marine biotechnology market value at $7.3 billion by 2026 (IndustryARC 2020). Due to the potential of blue growth, marine biotechnology is progressively making its way into economic-driven policies. Biotechnology holds the potential to play at the forefront and apply the capabilities of marine organisms in the economic, food, and therapeutic sectors. Marine organisms are a major source of a variety of molecules (of biomedical and industrial value) as they have evolved to dwell in the extreme conditions of chemistry, pressure, temperature, and darkness (Poli et al. 2010). One of the many ways to get exclusive benefits from these resources is through the "marine genetic resources" (MGRs) which offer genetic materials of potential value and economic benefits (United Nations 1992). We presume that strong cooperation between academics could enhance the utilization of marine resources in the form of detecting, isolating,

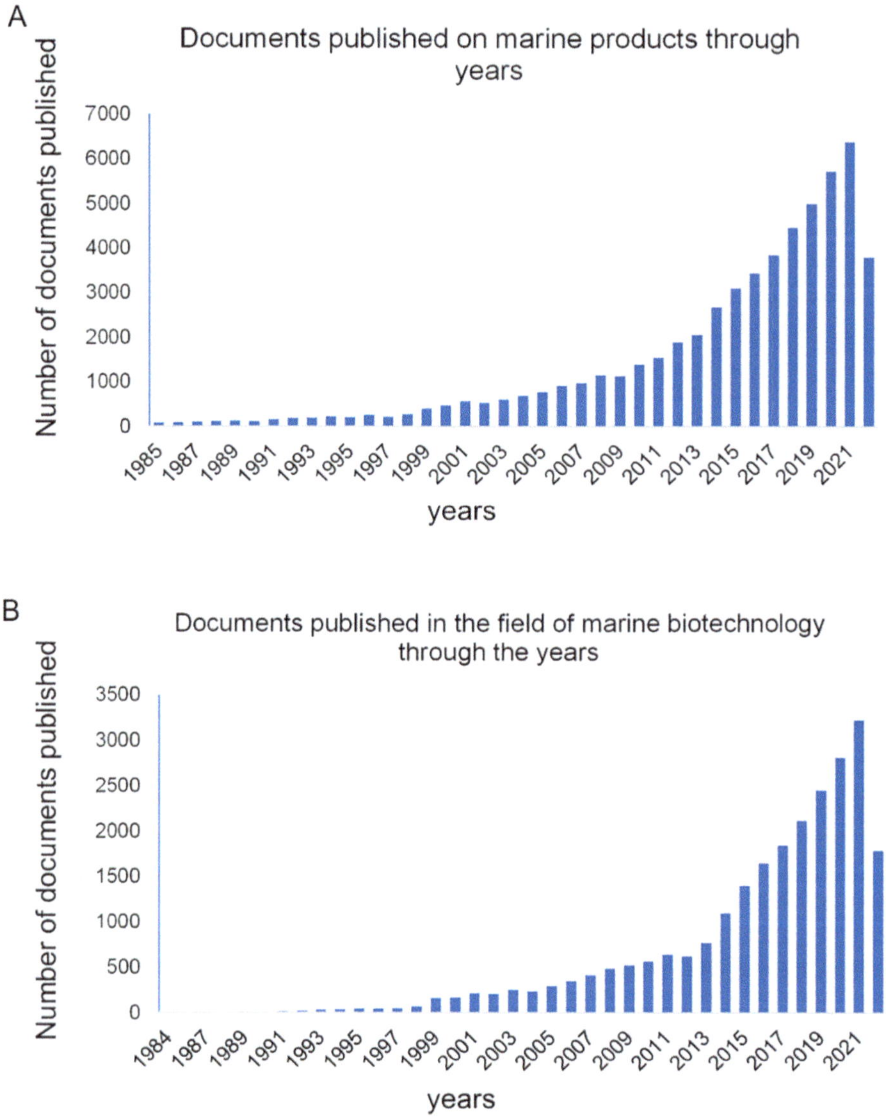

Fig. 1.2 Number of documents published on marine products (**a**) and marine biotechnology (**b**). The PubMed database was accessed for this information on the 15th of July 2022

and classifying marine organisms (such as bacteria, marine invertebrates, fungi, and microalgae) and characterizing them for bringing their products to industries.

By applying a wide variety of screening tools, extracts and purified compounds of these organisms can be studied for food processing, therapeutically, and industrially significant biological activities, including anticancer, anti-inflammatory, antiviral, antibacterial, and anticoagulant activities, as well as for ion channel/receptor

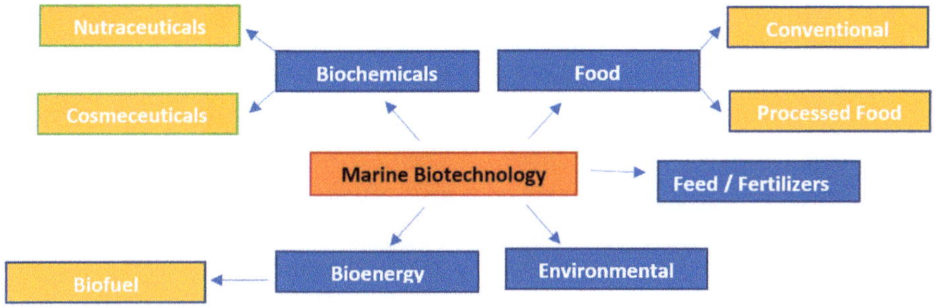

Fig. 1.3 Marine biotechnology's role in food, energy, and nutraceutical production

modulation and plant growth regulation (Shah et al. 2022; Kijjoa and Sawangwong 2004; Abdelnasser et al. 2017; Pech-Puch et al. 2020).

Marine biotechnology's important role in food, nutraceuticals, and energy production has been discussed below in Fig. 1.3.

1.2 Food

Biomolecules derived from marine organisms have a wide range of applications in the food industry, including food production at high temperatures and pressures, coloring agents, preservatives, and flavor enhancers. Some of the most useful marine-derived food ingredients include photosynthetic pigments, polyunsaturated fatty acids (PUFAs), sterols, polysaccharides, proteins, and enzymes (Rasmussen and Morrissey 2007).

Marine-based food ingredients obtained from marine algae (macroalgae and microalgae) (Figs. 1.4 and 1.5) are an important source of nutrients. Algae are inhibited all over the world, a rich source of bioactive compounds and nutritional compounds including calcium, sodium, magnesium, iodine, phosphorus, potassium, iron, and zinc (Ścieszka and Klewicka 2019). The usage of algae in the biotechnology industry has extensively grown since the chemical composition and bioactive substances identified in algae are suitable to be used in various fields, especially in the food industry. Polysaccharides originating from algae, such as algins, carrageenans, and agar, are widely utilized in a range of foods for their capacity to form gels and function as thickeners and stabilizers (Rasmussen and Morrissey 2007). Algae have largely been used in meat and bakery products to improve their quality and safety. The presence of *Porphyra umbilicalis*, *Undaria pinnatifida*, *Enteromorpha*, and *Himanthalia elongata* algae altered the antioxidative capacity of meat and cereal-based products (Gupta and Abu-Ghannam 2011).

Macroalgae, according to their color are divided into three groups: brown algae from the family Phaeophyceae (which gets its brown or yellow-brown color from fucoxanthin), red algae from the family Rhodophyceae (which contains phycoerythrin and phycocyanin), and green algae from the family Chlorophyceae (dominating

Fig. 1.4 Macroalgae, (**a**) *Pyrodinium bahamense*, (**b**) *Akashiwo sanguinea*

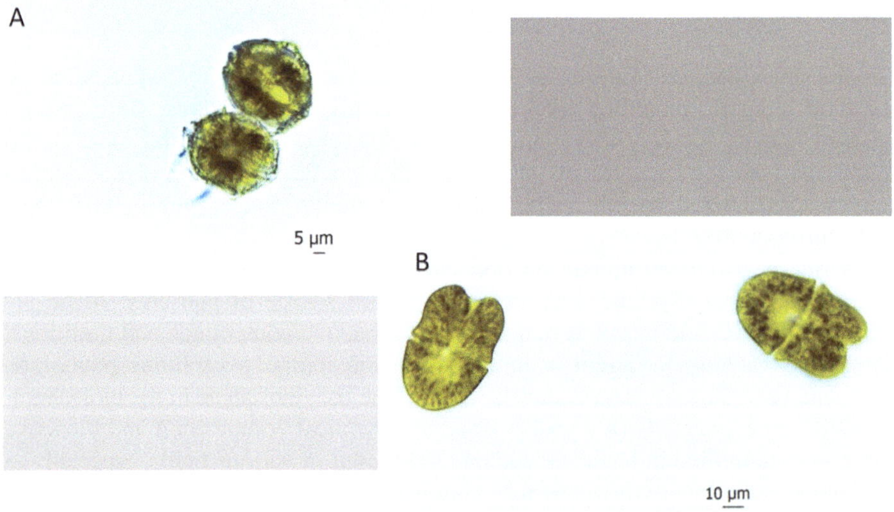

Fig. 1.5 Microalgae, (**a**) *Pyrodinium bahamense*, (**b**) *Akashiwo sanguinea*

chlorophyll a and chlorophyll b) (Shah et al. 2022; Domínguez 2013). Macroalgae are predominant producers in the sea and coastal areas. A major amount of their biomass is driven away to the deep sea and sediment (Ortega et al. 2019). Macroalgae is added to dairy products such as milk desserts, cheese, ice cream, yoghurt, cottage cheese, and processed cheese to improve their nutritional values. Brown algae, for example, *Laminaria*, is added to dairy products to make them iodine-rich. Green algae *Chlorella* and brown algae *Undaria pinnatifida* are added to

Fig. 1.6 Sea cucumber *Pearsonothuria graeffei*

increase the quality of the cheese during storage. Fermented foods with a high concentration of lactic acid bacteria incorporated with algae, which have biologically active metabolites of natural origin, improve product quality and develop a completely new category of fermented foods. For instance, calcium is trapped in casein in cheese, which prevents those without the necessary enzymes from absorbing calcium from those products. As a result, the inclusion of calcium-rich algae might boost the amount of the element in dairy products and aid in the treatment of hypocalcemia (Ścieszka and Klewicka 2019).

Microalgae can carry out photosynthesis and they are classified according to their cytological and morphological characteristics, pigments, type of reserve metabolites, and components of the cell wall. Marine diatoms are golden-brown due to xanthophyll pigments, and blue-green algae possess chlorophyll a, and blue phycocyanins (Domínguez 2013). Microalgae are fast-growing algae species that can double their biomass more than once in 24 h. It is estimated that the microalgae can yield 20 kg/m^2/year (Varshney et al. 2015). Microalgal biomass (defatted) holds application in the feed industry for carnivorous fish. It is reported that defatted *Nannochloropsis oceania* biomass could be used as a replacement for fishmeal as it has shown a positive impact on the growth, feed intake, and health of Atlantic salmon (Sørensen et al. 2017). Further, the addition of *Nannochloropsis* spp. biomass as a feed additive to the diet of Pacific white shrimp enhanced its resilience toward temperature change and improved its level of reactive oxygen species (Guimarães et al. 2021).

Sea cucumbers (Fig. 1.6) are marine invertebrates and are characterized by leathery skin, a soft body, and a single-branched gonad. So far, about 1716 species have been identified and classified. These organisms live in a hostile marine environment (Pangestuti and Arifin 2018). Sea cucumbers have been traditionally used

Fig. 1.7 Seagrass *Halophila ovalis*

as food and medicine in Asia. They could be used in soups, pickled food items, or stir-fried foods. In Indonesia, sea cucumbers are known as "teripang" or "trepang" and "beche-de-mer" in France. According to the Ming dynasty literature, sea cucumbers possessed similar therapeutic abilities as herbal ginseng and they are known as "haishen", which means "ocean ginseng" (Bahrami et al. 2014). Sea cucumbers are an ideal tonic food, as these organisms are rich in protein and interestingly, the lipid level is lower. They are collagen-rich and contain an elevated level of gelatin content (Pangestuti and Arifin 2018). *Holothuria poli* is a type of sea cucumber and is ubiquitous in the Mediterranean Sea, Canary Islands Sea, and the northern Red Sea. *Holothuria poli* has been extensively examined for the presence of secondary metabolites (Ismail et al. 2008). *Pearsonothuria graeffei* is another sea cucumber with a good source of triterpene glycosides, which can act as a functional food (Zhao et al. 2012).

Seagrasses are types of plants (also known as angiosperms) that live in marine environments and contribute to the sustainability of coastal ecosystems (Fig. 1.7) (Grignon-Dubois and Rezzonico 2013). In addition to being a source of animal feed, seagrasses have been utilized for centuries as food, medicine, and fertilizer. Seagrasses may be used as a functional food since it is an important source of protein, carbohydrate, lipids, fiber, phenol, flavonoids, and tannin. Seagrasses also contain essential elements (carbon, hydrogen, and nitrogen) as well as photosynthetic pigments (chlorophyll a, chlorophyll b, total chlorophyll, and carotenoid) (Rengasamy et al. 2013).

Fish is a vital source of protein and most people on earth still largely catch fish from the wild, with significant negative effects (Fig. 1.8). Aquaculture development is one of the most significant biotechnology applications in marine research. Producing fish species that have a faster growth rate, higher adaptability and

Fig. 1.8 Aquaculture facilities

survivability, and better fish yield has always been the major goal of the aquaculture sector (Uddin and Islam 2019).

The ocean's fisheries are becoming more stressed due to heavy expenditures on fishing fleets and technology as well as constantly rising yields. Numerous fishing areas are currently so overfished that their further viability is in jeopardy (Shakouri et al. 2010). Food security can be greatly aided by using biotechnology in sustainable aquaculture and fisheries. In this regard, fish that have been genetically modified (GM) have the potential to significantly boost fish farm yields, but they have also raised significant environmental concerns in the US and Europe over potential effects on wild species. To allay these worries and address public resistance to biotechnology, it is crucial to establish a solid, trustworthy, and widely recognized technique for calculating the potential for harm caused by GM fish escaping into the wild. To explore the impact of the transgenic itself on wild populations, a novel technique based on population genetics has just been devised; however, it has limitations (Muir 2004). Gene transfer technology was employed to promote fish growth. In China, a developed gene containing the promoter gene of antifreeze protein and salmon growth hormone cDNA was introduced into the red sea bream fish genome by the technique of electroporation and significant results were obtained in terms of cold tolerance and an increment in body weight (Zhang et al. 1998). In Malaysia, a new type of grouper known as the hybrid grouper (TGGG) has been produced by researchers by crossing sperm of *Epinephelus lanceolatus* with eggs of the *Epinephelus fuscoguttaus* grouper species (Fig. 1.9). *Epinephelus lanceolatus*, also known as Giant grouper, lives in marine, brackish, and reef-associated environments, whereas *E. fuscoguttaus*, also known as Tiger grouper, is an Indo-specific species that live primarily on coral reefs and lagoon pinnacles (Ching and Senoo 2008; Pears et al. 2006; Othman et al. 2015; Shapawi et al. 2019). Hybrid

Fig. 1.9 Hybrid grouper (*Epinephelus lanceolatus x Epinephelus fuscoguttaus*)

groupers reach maturity more quickly and consume feed more efficiently, implying a decrease in the cost of feed for commercial farming compared to nonhybrid parents. It can also survive a wide range of climatic conditions due to its genetic improvement (Ching and Senoo 2008; Shapawi et al. 2019).

1.3 Nutraceuticals

Many potently active compounds worthy of therapeutic use have been found via research into the pharmacological characteristics of marine natural materials. The marine environment is a remarkable source of bioactive natural products, many of which have chemical and structural characteristics that are not present in terrestrial natural products (Kong et al. 2010). Among the therapeutics, more than 60 % of active pharmaceutical formulations come from natural products or their derivatives (Cragg and Newman 2013). Secondary metabolites from marine resources have vast and profound applications in the pharmaceutical industry. These compounds have evolved from millions of years of natural selection. Due to the intrinsic abilities of marine natural products, these molecules can identify and attach to macromolecules, disturb their function, and affect their biological activities (Mayer et al. 2010).

The horizon of marine bioactive compound discovery has exponentially expanded. For instance, in the 1960s, researchers could access and study shallow-water subtidal creatures down to a depth of around 40 m. Whereas scientists have recently gained access to an array of undiscovered marine settings and habitats through the acquisition of advanced tools, such as manned submersibles and remotely operated vehicles, which are currently making it possible to visit depths of 5000 m and deeper (Miyake et al. 2011).

For functions like communication, reproduction, and defense against predation, competition, and infection, marine species have evolved biochemical and physiological systems that involve the creation of bioactive compounds. According to a comparison study by Kong et al. (2010), marine natural products have a higher level of chemical novelty than terrestrial natural products. Nearly every class of marine organisms shows a diversity of molecules with distinctive structural characteristics because of the physical and chemical circumstances in the marine environment. But in addition to its incredible chemical diversity, marine water also offers a remarkable variety of life. About 32 of the 34 basic phyla of life are found in marine waters, whereas 17 are found on land (with some overlap). The ocean is far more diversified from a basic point of view, making it the ideal area to start the development of a natural pharmacy (Kijjoa and Sawangwong 2004). Moreover, the utilization of modern computational tools could also help to find effective synergies of two or more compounds for enhanced therapeutic properties (Harakeh et al. 2015). In some cases, the immobilization of medicinal products (such as lactoperoxidase) into silver nanoparticles could enhance their medicinal efficacy (Sheikh et al. 2018).

Many marine-derived compounds are important to the nutraceutical industry. Nutraceuticals are bioactive chemicals having medical properties or additional health advantages, such as anticancer, anti-inflammatory, antioxidant, and antimicrobial activity, and many marine-based food components come under this category (Rasmussen and Morrissey 2007; Hamed et al. 2015; Kijjoa and Sawangwong 2004). Nutraceutical fortification of foods has become a popular means of offering nutritious food products to health-conscious customers. Consumer awareness of marine-based nutraceuticals has grown as a result of publications on their numerous health advantages, such as increased antioxidant activity and immunity (Ohr 2005). Currently marketed marine nutraceuticals include omega-3-rich fish and algal oils, chitin and chitosan, shark liver oil, marine enzymes and chondroitin from shark cartilage, sea cucumbers, and mussels (Rasmussen and Morrissey 2007). Chondroitin (a component of cartilage) has been shown to possess anti-inflammatory and anticancer properties, whereas omega-3 fatty acids are well-known for their wide range of health benefits, such as a reduced risk of cardiovascular disease and enhanced brain development in babies (Rasmussen and Morrissey 2007). Marine-based dietary components and nutraceuticals can be derived from a variety of sources, such as marine plants, microorganisms, and sponges, each of which has its own set of biomolecules that allow it to survive in its particular habitat (Rasmussen and Morrissey 2007) (Fig. 1.10).

As previously stated, several marine-derived compounds have been shown to have nutraceutical benefits. It is also commonly believed that marine resources provide the chance to uncover unique chemical diversity with exciting pharmacologically active molecules that could be utilized to treat bacterial, inflammation, cancer, parasitic infections, and several other ailments (Fajarningsih 2013).

Sponge extracts have bioactive compounds that are antiviral, anti-inflammatory, antibiotic, antifouling, antimalarial, anticancer, immunosuppressive or neurosuppressive (Sipkema et al. 2005). About 5000 therapeutically important compounds have been detected in sponges and about 15,000 or more marine organisms have been documented as possessing bioactive compounds (Sipkema et al. 2005). In addition to sponges, ascidians and gorgonian marine creatures have been also shown to exhibit antiviral and antiproliferative properties. In research, a

Fig. 1.10 Some marine invertebrates: (**a**) crown-of-thorns starfish, (**b**) bubble algae, (**c**) hard coral, (**d**) feather star

total of 65 marine species were collected off the coast of Mexico's Yucatan Peninsula, including 51 sponges from the phylum Porifera, 13 ascidians from the phylum Chordata, and one gorgonian from the phylum Cnidaria. They were chosen based on chemotaxonomical parameters. Each extract was tested in vitro for antiviral and antiproliferative activities against human adenovirus and five human tumor cell lines, including hepatocyte carcinoma, breast cancer adenocarcinoma, human lung carcinoma, pancreatic carcinoma, and human skin melanoma. In plaque tests, they were removed using organic solvents. Antiviral activity was found in 11 extracts from ten sponges, including *Ircinia felix*, *Ectyoplasia* sp., *Chondrilla* sp., *Myrmekioderma gyroderma*, *Agelas citrina*, *Monanchora arbuscula*, *Dysidea* sp., *Cinachyrella kuekenthali*, *Aaptos* sp., *and Spongia tubulifera*, *and Spongia tubulifera*, *Dysidea* sp., *Agelas citrina*, *Chondrilla* sp., and *Monanchora arbuscula* extracts demonstrated the strongest antiviral activity. The extract's IC_{50} values were lower than those reported for cidofovir (a drug used to treat human adenovirus infections) (Pech-Puch et al. 2020).

To date, antiproliferative activity has been demonstrated by four ascidians (*Trididemnum solidum*, *Polysyncraton* sp., *Clavelina* sp., and *Eudistoma amanitum*) and 21 sponges (*Tethya* sp., *Agelas citrina*, *Leucetta floridana*, *Forma hermatypica*, *Chondrilla caribensis*, *Dysidea* sp., *Myrmekioderma gyroderma*, *Clathria (Clathria) gomezae*, *Amphimedon compressa*, *Cinachyrella kuekenthali*, *Cliona varians*, *Monanchora arbuscula*, *Mycale laevis*, *Spongia tubulifera*, *Plakinastrella onkodes*, *Aaptos* sp., *Haliclona (Rhizoniera) curacaoensis*, *Aiolochroia crassa*, and *Scopalina ruetzleri*). The ascidian *Eudistoma amanitum* and the sponge *Haliclona (Rhizoniera) curacaoensis* had the most potent antiproliferative activity. In addition, greater than 50% of the extracts exhibited antiproliferative activity against the hepatocyte cancer cell line (Pech-Puch et al. 2020).

In addition to marine invertebrates and sponges, algae, sea cucumber, seagrasses, etc., also contribute as the source of drug production. Macroalgae including *Sargassum polycystum*, *Halymenia durvillaei*, *Caulerpa lentillifera*, *Caulerpa racemosa*, *Dictyota dichotoma*, *Kappaphycus alvarezii*, etc. have been shown to have anti-inflammatory, antioxidant, antibacterial, and anticancer properties. Several different nutraceutical compounds have been reported in these seaweeds. For instance, *S. polycystum* contains lutein, neophytadiene, and cis-vaccenic acid. *H. durvillaei* contains eucalyptol, oleic acid, and pentadecane. *C. lentillifera* contains canthaxanthin, oleic acid, and eicosane, *C. racemose* has monocaprin pseudoephedrine, and palmitic acid, *D. dichotoma* has squalene, saringosterol, and fucosterol, while *K. alvarezii* contains phthalic anhydride, 2-pentylthiophene, and furoic acid (Shah et al. 2022). The marine dinoflagellate *Gambierdiscus toxicus* has produced a series of new polyether antibiotics, gambieric acids, which are the most effective antifungal drugs yet to be discovered. For example, Gambieric acid A is 2000 times more active than amphotericin B, a therapeutically relevant antifungal drug with very mild toxicity in mice and cultured human cells (Nagai et al. 1992).

The sea cucumber is an abundant source of compounds with therapeutic properties, including amino acids, minerals, carotenoids, triterpene glycosides, chondroitin sulfates, bioactive peptides, vitamins, collagen, fatty acids, and gelatin. These compounds have exhibited therapeutic activities such as anticancer, antimicrobial, wound healing, anticoagulant, neuroprotective, and antioxidant. The most commonly known and used sea cucumber species include *Holothuria fuccogilva*, *Stichopus hermanni*, *Actinopyga mauritiana*, *Thelenota ananas*, and *Thelenota anax* (Pangestuti and Arifin 2018).

Seagrasses may contain bioactive compounds with industrial applications (Grignon-Dubois and Rezzonico 2013). Seagrasses, for example, have a high concentration of secondary metabolites (flavonoids, polyphenols, and fatty acids) that act as a defense mechanism against abiotic stresses (Custódio et al. 2016). Benito-González et al. (2019) report that *Halodule unnerves* and *Posidonia oceanica* extracts exhibit antifungal, antioxidant, and antiviral properties. In seagrass tissues, agents such as luteolin, chrysoeriol, and diosmetin are frequently found (Guan et al. 2017). Zosteric acid, which is found in the genus *Zostera* and has antifouling properties, is another agent (Vilas-Boas et al. 2017). However, a comprehensive evaluation of the biological roles of these metabolites is required due to the presence of hazardous chemicals. For instance, pyrrolizidine alkaloids are present in the grass subfamily *Pooideae (Poaceae)*, which has been used for a variety of biological purposes; nevertheless, it was recently discovered that this substance might cause hepatotoxicity in rats (Li et al. 2018).

Several aquatic vertebrates are also employed as model species in biomedical research. One such example is Zebrafish, which have been employed in more than 40,000 biomedical research investigations. Genetics, toxicology, drug development, pathobiology of human diseases, and cellular and developmental biology have all been transformed using transgenic fluorescent zebrafish lines. Due to the synthesis of fluorescent proteins in intracellular organelles, cells, and molecules of interest, these structures can be viewed and monitored instantaneously and in vivo. (Choe et al. 2021).

Despite the abundance of great biological activity and high potential of marine natural products, their advancement as medicinal agents has been slowed down as a result of several circumstances, such as the availability of low active molecules, a high level of chemical complexity and in certain situations, their high toxicity at therapeutic doses. To overcome some of the aforementioned challenges and advance some of the more promising compounds closer to the clinic and the market, many commercial companies have been established specifically to apply the ethos of biotechnology to the production of marine drugs. (Baerga-Ortiz 2009). Many potent new compounds derived from marine natural products are candidates for clinical trials. For example, aplidine is a cyclic depsipeptide which is isolated from the marine tunicate *Aplidium albicans*. The bioactive compound is in the trial against a patient with a solid tumor (Maroun et al. 2006). Bryostatins are produced by the marine invertebrate *Bugula neritina*, with several types being isolated from various populations of the same species, over 13 structurally related compounds have been obtained. The tumor-promoting phorbol esters are negatively impacted by bryostatin-1, a protein kinase C (PKC) activator. Additionally, bryostatin-1 modulates the immune system, causes myeloid and lymphoid cell lines to differentiate, produces platelet aggregation, and encourages hematopoiesis. It has shown considerable anticancer action in preclinical models against a variety of cell types and has also been proven to increase the antitumor effects of different chemotherapeutic drugs, including, vincristine, cytosine arabinoside, paclitaxel, etc. (Amador et al. 2003). The Phase II study of Bryostatin 1 in combination with the chemotherapy medication Vincristine in select patients for aggressive non-Hodgkin's lymphoma has been effective (Barr et al. 2009). Kahalalide F, cyclic depsipeptides, is isolated from a sacoglossan mollusc, *Elysia rufescens*. It is also obtained from a green alga, *Bryopsis* sp. Indeed, *E. rufescens* consumes *Bryopsis* sp. which shows that Kahalalide F is a part of the green alga it was consuming. Clinical studies for the potent anticancer agent have progressed to phase II for a variety of cancer types. The bioactive compound modifies the lysosomal and mitochondrial membranes and causes oncosis, which results in cell death (Piel 2010).

Additionally, several pharmaceutical firms, including Nereus Pharmaceuticals (San Diego, USA) and PharmaMar (Madrid, Spain) have developed strategies to locate cultivable marine microbes to maximize their growth, boost the production of bioactive compounds, and enhance chemical diversity through synthetic modification to produce marine drugs, Currently, Nereus Pharmaceuticals, a biotech firm focused on the development of marine drugs, is presently undertaking Phase I clinical studies for the proteasome inhibitor salinosporamide A (isolated directly from the fermentation of the marine actinomycete *Salinispora tropica*) for the treatment of solid tumors, lymphomas, and multiple myeloma (Baerga-Ortiz 2009; Fenical et al. 2009). PharmaMar, another biotech firm, manufactures and markets trabectedin (Yondelis®), a natural anticancer agent, which is developed through a semi-synthetic process that involves chemically modifying a naturally occurring fermentation product from *Pseudomonas fluorescens* to create the finished product (Baerga-Ortiz 2009; Cuevas and Francesch 2009). These advancements proved the crucial role of marine biotechnology in the production of drugs.

1.4 Bioenergy

Since the dawn of human civilization, carbon-based fuel has been the main source of energy. Due to expanding commercial businesses and the global population, there is a rising need for energy, which is squeezing fossil fuel resources and also adversely affecting our environment and posing dangerous consequences in the form of rising sea levels, harmful gases, rising temperatures, and declining biodiversity. Searching for and developing alternative energy sources that are also environmentally friendly is required to meet increasing energy demands.

One of the hot sectors for renewable energy is the manufacturing of biofuel from renewable biomass sources as a substitute strategy. Compared to fossil fuels, biofuel is affordable, environmentally beneficial, and holds the potential to substitute fossil fuels (Hossain and Jahan 2021). The development of cutting-edge technologies is expected to provide human civilization with renewable energy, particularly biofuels, on an affordable and sustainable scale (Hossain and Jahan 2021). Biofuels like biodiesel, bioalcohol, bio-oil, biogas, and syngas are produced from the biomass of living or dead organisms. Biomass contains carbon, which is used for biofuel production. Microalgae and macroalgae are marine resources that are used for biofuel production. Marine biomaterials are considered a good source of energy production. Marine biomaterials that are used for fuel production should be composed of a high level of lipids (ranging from simple polymers to complex polysaccharides) (Ali et al. 2020).

Fortunately, macro and microalgae have high lipid levels, which can produce a high amount of energy, and they have a fast growth rate that makes them suitable for biofuel production (Gosch et al. 2012; Ali et al. 2020). For example, water hyacinth can accumulate great biomass in a short time due to its fast growth rate, which makes it a potential renewable energy source that may substitute conventional fossil fuels. Besides, the research found that dried water hyacinth biomass can be manufactured into briquettes that can replace coal as the co-firing agent in power plants (Rezania et al. 2015). Marine resources, particularly algae, can be a potential and stable biomass source because the ocean contains a massive untapped algal resource that could reduce land costs and efficiently synthesize organic carbon through photosynthesis (Pogson et al. 2013; Hossain and Jahan 2021).

As well as being an energy source, algae can also contribute to the fixation of greenhouse gases (CO_2) by consuming them during the process of photosynthesis (Chen et al. 2015). The typical photosynthetic efficiency for algae is 6–8% which is significantly higher than the 1.8–2.2% of terrestrial plants (Chen et al. 2015; Aresta et al. 2005). Algal biomass can be converted into biofuels such as biogas, bioethanol, biodiesel, and bio-oils through anaerobic digestion, fermentation, transesterification, liquefaction, and pyrolysis (Chen et al. 2015). Microalgae appear to be the only biodiesel source with the ability to replace fossil fuels. Microalgae, unlike other oil crops, develop extraordinarily quickly and constitute rich oil content. Within 24 h, microalgae often quadruple their biomass. During exponential growth, biomass doubling durations can be as quick as 3.5 h. Microalgae can have an oil concentration of up to 80% by weight of dry biomass (Chisti 2007). In this context, marine algae might be a feasible and dependable source of biomass, as the ocean holds an untapped enormous algal resource that could reduce land expenses while simultaneously successfully synthesizing organic carbon via photosynthesis. (Chen et al. 2015). Biofuels from marine resources are cost-effective and reduce greenhouse gases, sulfur oxide, and hazardous matter emissions from the shipping industry (Tan et al. 2021).

1.5 Cultivation and Sustainable Collection Methods

The increasing number of species being introduced into in-vitro culture are directly related to the productive exploration of bioactive marine compounds. A well-regulated and controlled maintenance system for marine organisms is in high demand to ensure sustainable exploitation for industrial applications. Several photosynthetic marine organisms are heterotrophic. The bioactive compounds produced by marine organisms are markedly influenced by the type of growth nutrients and abundance of the organisms in the culture. Therefore, a sophisticated system is required to ensure the production of desired bioactive compounds in a cost-effective manner (Eriksen 2008).

To understand and harness the potential of marine organisms, it is crucial to develop and maintain pure cultures using preservation methods for biotechnological applications. To do so, it is vital to comprehend and replicate the naturally occurring environmental conditions for a particular organism (Joint et al. 2010; Khan et al. 2019; Ullah et al. 2017). After obtaining a pure culture, genetic screening and contaminant elimination are important to avoid biases in the result and the growth of a competitor organism in the same culture. For instance, the algal samples obtained from nature are often accompanied by zooplankton that could feed and eventually kill algae. One more thing that needs to be taken into consideration is timing. Some species die quickly and therefore, an adequate medium should be supplied so that the organism can multiply and lead to a sustainable pure culture.

One of the revolutionary techniques that have been recently developed is the Laser-Induced Forward Transfer Technique (LIFT). This system has allowed researchers to isolate single cells from a complex system, such as the ocean, to study the biodiversity of the environment and evaluate the physiology, genome, gene expression, and functions of an organism (Fig. 1.11). Interestingly, the system can be coupled with other microscopic approaches (e.g., fluorescent and Raman microscopy) to examine single microorganisms with particular functions, unmasking their activities in the natural reservoir. More details about the system have been explained elsewhere (Peng et al. 2022).

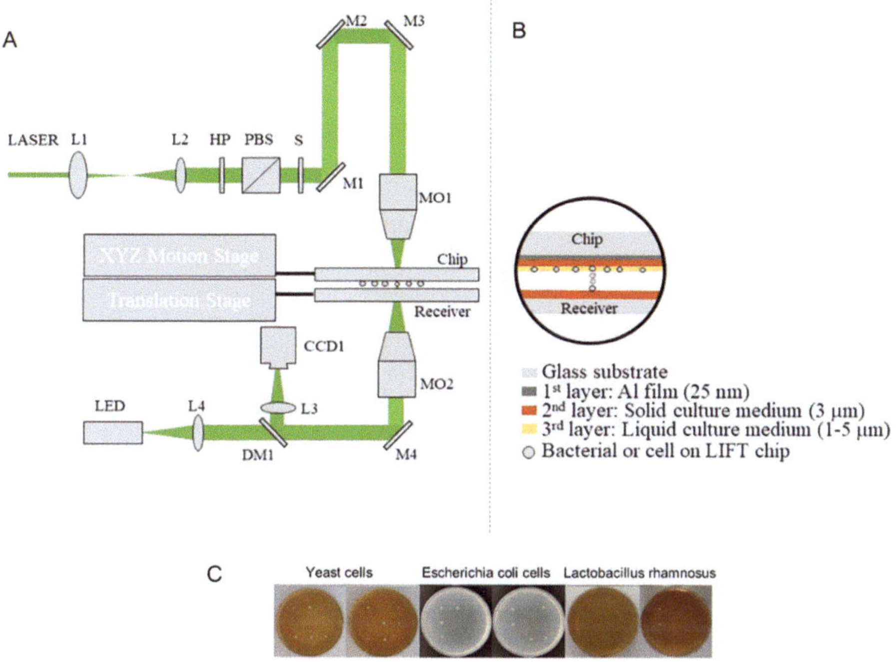

Fig. 1.11 Graphical illustration of the laser-induced forward transfer (LIFT) system. (**a**) This plot shows the mechanistic inner side of the system for sorting single cells out of the complex samples. (**b**) A three-layer structure is used for the isolation of a cell. (**c**) An example of the cells sorted and isolated by the LIFT system to the growth media

1.6 Recommendations

1. International partners are needed for joint expeditions to explore marine resources.
2. Novel culturing tools should be designed to bring uncultivable marine organisms into lab conditions.
3. Marine organisms' in-vitro cultivation for the mass production of industrially important products.
4. Since intraspecific changes cause variation in compounds and their concentrations in different marine environments, we recommended a wider exploration of marine organisms in different geological sites. This will help in cataloging the marine organisms at various sites and will help with their future characterization.
5. More investment and education should be brought to marine potential to promote and harvest the potential of marine biotechnology, which is a better tool for the holistic management of complex marine social-ecological systems.

6. Marine biotechnology should be integrated with other disciplines for a better understanding of the ocean system complexity, generate enough data for comprehension of the ocean capacity, and design pragmatic approaches that are solution-oriented, realistic, and practical.
7. Efficient biotechnological tools are required for bioactive compound identification, characterization, and isolation. Maintainable cultivation is required to bring promising organisms into the lab and manipulate their potential for selected compounds. In addition, sensitive biosensors are demanded to monitor the production of target compounds in the culture.

1.7 Conclusion

The marine environment is home to marine organisms that can be harvested and used in the medicinal, nutraceutical, and cosmeceutical industries due to their variety of primary and secondary metabolites due to their adaptability in harsh marine conditions. Current and future concerns, such as the exploitation of marine resources, climate change, and IUU fishing, among others, can be addressed while simultaneously improving humanity's quality of life by applying marine biotechnology to advance the discovery and characterization of novel marine pharmaceuticals. Doing so may reduce the pressure on the exploitation of marine resources while at the same time offering economic opportunities within the industries.

Funding This research work has been funded by the Skim Penyelidikan Lantikan Baharu (SLB2232) from the Universiti Malaysia Sabah to MDS.

References

Abdelnasser SM, Yahya SMM, Mohamed WF, Asker MMS, Abu Shady HM, Mahmoud MG, Gadallah MA (2017) Antitumor exopolysaccharides derived from novel marine *Bacillus*: isolation, characterization aspect and biological activity. Asian Pac J Cancer Prev 18(7): 1847–1854. https://doi.org/10.22034/APJCP.2017.18.7.1847

Ali MK, Mir SH, Hyder MKMZ, Yang W (2020) Harvesting of bioenergy and biomaterials from marine resources. In: Encyclopedia of marine biotechnology. Wiley, Hoboken, NJ, pp 711–736. https://doi.org/10.1002/9781119143802.ch27

Amador ML, Jimeno J, Paz-Ares L, Cortes-Funes H, Hidalgo M (2003) Progress in the development and acquisition of anticancer agents from marine sources. Ann Oncol 14(11):1607–1615. https://doi.org/10.1093/annonc/mdg443

Aresta M, Dibenedetto A, Barberio G (2005) Utilization of macro-algae for enhanced CO2 fixation and biofuels production: development of a computing software for an LCA study. Fuel Process Technol 86(14–15):1679–1693. https://doi.org/10.1016/j.fuproc.2005.01.016

Baerga-Ortiz A (2009) Biotechnology and biochemistry of marine natural products. P R Health Sci J 28(3):251–257

Bahrami Y, Zhang W, Franco C (2014) Discovery of novel saponins from the viscera of the sea cucumber *Holothuria lessoni*. Mar Drugs 12(5):2633–2667. https://doi.org/10.3390/md12052633

Barr PM, Lazarus HM, Cooper BW, Schluchter MD, Panneerselvam A, Jacobberger JW, Hsu JW, Janakiraman N, Simic A, Dowlati A, Remick SC (2009) Phase II study of bryostatin 1 and vincristine for aggressive non-Hodgkin lymphoma relapsing after an autologous stem cell transplant. Am J Hematol 84(8):484–487. https://doi.org/10.1002/ajh.21449

Benito-González I, López-Rubio A, Martínez-Abad A, Ballester A-R, Falcó I, González-Candelas L, Sánchez G, Lozano-Sánchez J, Borrás-Linares I, Segura-Carretero A (2019) In-depth characterization of bioactive extracts from *Posidonia oceanica* waste biomass. Mar Drugs 17(7):409

Chen H, Zhou D, Luo G, Zhang S, Chen J (2015) Macroalgae for biofuels production: progress and perspectives. Renew Sust Energ Rev 47:427–437. https://doi.org/10.1016/j.rser.2015.03.086

Ching CL, Senoo S (2008) Egg and larval development of a new hybrid grouper, tiger grouper *Epinephelus fuscoguttatus* giant grouper *E. lanceolatus*. Aquac Sci 56(4):505–512

Chisti Y (2007) Biodiesel from microalgae. Biotechnol Adv 25(3):294–306. https://doi.org/10.1016/j.biotechadv.2007.02.001

Choe CP, Choi S-Y, Kee Y, Kim MJ, Kim S-H, Lee Y, Park H-C, Ro H (2021) Transgenic fluorescent zebrafish lines that have revolutionized biomedical research. Lab Anim Res 37(1): 26. https://doi.org/10.1186/s42826-021-00103-2

Cragg GM, Newman DJ (2013) Natural products: a continuing source of novel drug leads. Biochim Biophys Acta 1830(6):3670–3695. https://doi.org/10.1016/j.bbagen.2013.02.008

Cuevas C, Francesch A (2009) Development of Yondelis® (trabectedin, ET-743). A semisynthetic process solves the supply problem. Nat Prod Rep 26(3):322. https://doi.org/10.1039/b808331m

Custódio L, Laukaityte S, Engelen AH, Rodrigues MJ, Pereira H, Vizetto-Duarte C, Barreira L, Rodríguez H, Albericio F, Varela J (2016) A comparative evaluation of biological activities and bioactive compounds of the seagrasses *Zostera marina* and *Zostera noltei* from southern Portugal. Nat Prod Res 30(6):724–728

Domínguez H (2013) Algae as a source of biologically active ingredients for the formulation of functional foods and nutraceuticals. In: Functional ingredients from algae for foods and nutraceuticals. Elsevier, Burlington, pp 1–19. https://doi.org/10.1533/9780857098689.1

Eriksen NT (2008) The technology of microalgal culturing. Biotechnol Lett 30(9):1525–1536. https://doi.org/10.1007/s10529-008-9740-3

Fajarningsih ND (2013) An emerging marine biotechnology: marine drug discovery. Squalen Bull Mar Fish Postharvest Biotechnol 7(2):89. https://doi.org/10.15578/squalen.v7i2.19

Fenical W, Jensen PR, Palladino MA, Lam KS, Lloyd GK, Potts BC (2009) Discovery and development of the anticancer agent salinosporamide A (NPI-0052). Bioorg Med Chem 17(6):2175–2180. https://doi.org/10.1016/j.bmc.2008.10.075

Freitas AC, Rodrigues D, Rocha-Santos TAP, Gomes AMP, Duarte AC (2012) Marine biotechnology advances towards applications in new functional foods. Biotechnol Adv 30(6):1506–1515. https://doi.org/10.1016/j.biotechadv.2012.03.006

Gosch BJ, Magnusson M, Paul NA, de Nys R (2012) Total lipid and fatty acid composition of seaweeds for the selection of species for oil-based biofuel and bioproducts. GCB Bioenergy 4(6):919–930. https://doi.org/10.1111/j.1757-1707.2012.01175.x

Grignon-Dubois M, Rezzonico B (2013) The economic potential of beach-cast seagrass–*Cymodocea nodosa*: a promising renewable source of chicoric acid. Bot Mar 56(4):303–311

Guan C, Parrot D, Wiese J, Sönnichsen FD, Saha M, Tasdemir D, Weinberger F (2017) Identification of rosmarinic acid and sulfated flavonoids as inhibitors of microfouling on the surface of eelgrass *Zostera marina*. Biofouling 33(10):867–880

Guimarães AM, Guertler C, do Vale Pereira G, da Rosa Coelho J, Costa Rezende P, Nóbrega RO, do Nascimento Vieira F (2021) *Nannochloropsis* spp. as feed additive for the pacific white shrimp: effect on midgut microbiology, thermal shock resistance and immunology. Animals 11(1):150. https://doi.org/10.3390/ani11010150

Gupta S, Abu-Ghannam N (2011) Recent developments in the application of seaweeds or seaweed extracts as a means for enhancing the safety and quality attributes of foods. Innovative Food Sci Emerg Technol 12(4):600–609

Hamed I, Özogul F, Özogul Y, Regenstein JM (2015) Marine bioactive compounds and their health benefits: a review. Compr Rev Food Sci Food Saf 14(4):446–465. https://doi.org/10.1111/1541-4337.12136

Harakeh SM, Khan I, Kumosani T, Barbour E, Almasaudi SB, Azhar EI, Rath M, Niedzweicki A (2015) The role of nutrients and phyto-compounds in the modulation of antimicrobial resistance. Front Cell Infect Microbiol 5(Jun). https://doi.org/10.3389/fcimb.2015.00053

Hossain J, Jahan R (2021) Biofuel: marine biotechnology securing alternative sources of renewable energy. Springer, Singapore, pp 161–194. https://doi.org/10.1007/978-981-15-8999-7_7

Hurst D, Børresen T, Almesjö L, De Raedemaecker F, Bergseth S (2016) Marine biotechnology strategic research and innovation roadmap: insights to the future direction of European marine biotechnology. Marine Biotechnology ERA-NET

IndustryARC (2020) Marine biotechnology market—forecast (2023–2028). Accessed on October 2022. https://www.industryarc.com/Report/16110/marine-biotechnology-market.html

Ismail H, Lemriss S, Ben Aoun Z, Mhadhebi L, Dellai A, Kacem Y, Boiron P, Bouraoui A (2008) Antifungal activity of aqueous and methanolic extracts from the Mediterranean sea cucumber, *Holothuria polii*. J Mycol Méd 18(1):23–26. https://doi.org/10.1016/j.mycmed.2008.01.002

Joint I, Mühling M, Querellou J (2010) Culturing marine bacteria—an essential prerequisite for biodiscovery. Microb Biotechnol 3(5):564–575. https://doi.org/10.1111/j.1751-7915.2010.00188.x

Khan I, Yasir M, Farman M, Kumosani T, Albasri SF, Bajouh OS, Azhar EI (2019) Evaluation of gut bacterial community composition and antimicrobial resistome in pregnant and non-pregnant women from Saudi population. Infect Drug Resist 12:1749–1761. https://doi.org/10.2147/IDR.S200213

Kijjoa A, Sawangwong P (2004) Drugs and cosmetics from the sea. Mar Drugs 2(2):73–82. https://doi.org/10.3390/md202073

Kong D-X, Jiang Y-Y, Zhang H-Y (2010) Marine natural products as sources of novel scaffolds: achievement and concern. Drug Discov Today 15(21–22):884–886. https://doi.org/10.1016/j.drudis.2010.09.002

Li YH, Tai WCS, Khan I, Lu C, Lu Y, Wong WY, Chan WY, Wendy Hsiao WL, Lin G (2018) Toxicoproteomic assessment of liver responses to acute pyrrolizidine alkaloid intoxication in rats. J Environ Sci Health C Environ Carcinog Ecotoxicol Rev 36(2):65–83. https://doi.org/10.1080/10590501.2018.1450186

Maroun JA, Belanger K, Seymour L, Matthews S, Roach J, Dionne J, Soulieres D, Stewart D, Goel R, Charpentier D, Goss G, Tomiak E, Yau J, Jimeno J, Chiritescu G (2006) Phase I study of Aplidine in a daily × 5 one-hour infusion every 3 weeks in patients with solid tumors refractory to standard therapy. A National Cancer Institute of Canada Clinical Trials Group study: NCIC CTG IND 115. Ann Oncol 17(9):1371–1378. https://doi.org/10.1093/annonc/mdl165

Mayer AMS, Glaser KB, Cuevas C, Jacobs RS, Kem W, Little RD, McIntosh JM, Newman DJ, Potts BC, Shuster DE (2010) The odyssey of marine pharmaceuticals: a current pipeline perspective. Trends Pharmacol Sci 31(6):255–265. https://doi.org/10.1016/j.tips.2010.02.005

Miyake H, Shibata H, Furushima Y (2011) Deep-sea litter study using deep-sea observation tools. In: Interdisciplinary studies on environmental chemistry, pp 261–269. Retrieved from https://www.researchgate.net/publication/266245893

Muir WM (2004) The threats and benefits of GM fish. EMBO Rep 5(7):654–659. https://doi.org/10.1038/sj.embor.7400197

Nagai H, Murata M, Torigoe K, Satake M, Yasumoto T (1992) Gambieric acids, new potent antifungal substances with unprecedented polyether structures from a marine dinoflagellate *Gambierdiscus toxicus*. J Org Chem 57(20):5448–5453. https://doi.org/10.1021/jo00046a029

Ohr LM (2005) Riding the nutraceuticals wave. Food Technol 59(8):95–96. Retrieved from https://agris.fao.org/agris-search/search.do?recordID=US201301029582

Ortega A, Geraldi NR, Alam I, Kamau AA, Acinas SG, Logares R, Gasol JM, Massana R, Krause-Jensen D, Duarte CM (2019) Important contribution of macroalgae to oceanic carbon sequestration. Nat Geosci 12(9):748–754

Othman AR, Kawamura G, Senoo S, Fui CF (2015) Effects of different salinities on growth, feeding performance and plasma cortisol level in hybrid TGGG (Tiger grouper, *Epinephelus fuscoguttatus* x giant grouper, *Epinephelus lanceolatus*) juveniles. Int Res J Biol Sci 4(3):15–20

Pangestuti R, Arifin Z (2018) Medicinal and health benefit effects of functional sea cucumbers. J Tradit Complement Med 8(3):341–351. https://doi.org/10.1016/j.jtcme.2017.06.007

Pears RJ, Choat JH, Mapstone BD, Begg GA (2006) Demography of a large grouper, *Epinephelus fuscoguttatus*, from Australia's Great Barrier Reef: implications for fishery management. Mar Ecol Prog Ser 307:259–272. https://doi.org/10.3354/meps307259

Pech-Puch D, Berastegui-Cabrera J, Pérez-Povedano M, Villegas-Hernández H, Guillén-Hernández S, Cautain B, Reyes F, Pachón J, Gómez P, Rodríguez J, Jiménez C, Sánchez-Céspedes J (2020) Antiviral and antiproliferative potential of marine organisms from the Yucatan Peninsula, Mexico. Front Mar Sci 7:607. https://doi.org/10.3389/fmars.2020.00607

Peng L, Bo L, Yun W, Kunxiang L, Yinping Z, Huang WE, Bei L, Haruyuki A (2022) Isolation and culture of single microbial cells by laser ejection sorting technology. Appl Environ Microbiol 88(3):e01165–e01121. https://doi.org/10.1128/aem.01165-21

Piel J (2010) The chemistry of symbiotic interactions. In: Comprehensive natural products II: chemistry and biology, vol 2. Elsevier, Oxford, pp 475–510. https://doi.org/10.1016/b978-008045382-8.00049-6

Pogson M, Hastings A, Smith P (2013) How does bioenergy compare with other land-based renewable energy sources globally? GCB Bioenergy 5(5):513–524. https://doi.org/10.1111/gcbb.12013

Poli A, Anzelmo G, Nicolaus B (2010) Bacterial exopolysaccharides from extreme marine habitats: production, characterization and biological activities. Mar Drugs 8(6):1779–1802. https://doi.org/10.3390/md8061779

Rasmussen RS, Morrissey MT (2007) Marine biotechnology for production of food ingredients. Adv Food Nutr Res 52:237–292. https://doi.org/10.1016/S1043-4526(06)52005-4

Rengasamy RRK, Radjassegarin A, Perumal A (2013) Seagrasses as potential source of medicinal food ingredients: Nutritional analysis and multivariate approach. Biomed Prev Nutr 3(4): 375–380.

Rezania S, Ponraj M, Din MFM, Songip AR, Sairan FM, Chelliapan S (2015) The diverse applications of water hyacinth with main focus on sustainable energy and production for new era: an overview. Renew Sustain Energy Rev 41:943–954. https://doi.org/10.1016/j.rser.2014.09.006

Rotter A, Barbier M, Bertoni F, Bones AM, Cancela ML, Carlsson J, Carvalho MF, Cegłowska M, Chirivella-Martorell J, Conk Dalay M, Cueto M, Dailianis T, Deniz I, Díaz-Marrero AR, Drakulovic D, Dubnika A, Edwards C, Einarsson H, Erdoğan A et al (2021) The essentials of marine biotechnology. Front Mar Sci 8:158. https://doi.org/10.3389/fmars.2021.629629

Ścieszka S, Klewicka E (2019) Algae in food: a general review. Crit Rev Food Sci Nutr 59(21): 3538–3547. https://doi.org/10.1080/10408398.2018.1496319

Shah MD, Venmathi Maran BA, Shaleh SRM, Zuldin WH, Gnanaraj C, Yong YS (2022) Therapeutic potential and nutraceutical profiling of North Bornean seaweeds: a review. Mar Drugs 20(2):101. https://doi.org/10.3390/md20020101

Shakouri B, Yazdi SK, Fashandi A (2010) Overfishing. In: 2010 2nd international conference on chemical, biological and environmental engineering, pp 229–234. https://doi.org/10.1109/ICBEE.2010.5649533

Shapawi R, Ching FF, Senoo S, Mustafa S (2019) Nutrition, growth and resilience of tiger grouper *(Epinephelus fuscoguttatus)* × giant grouper *(Epinephelus lanceolatus)* hybrid- a review. Rev Aquac 11(4):1285–1296

Sheikh IA, Yasir M, Khan I, Khan SB, Azum N, Jiffri EH, Kamal MA, Ashraf GM, Beg MA (2018) Lactoperoxidase immobilization on silver nanoparticles enhances its antimicrobial activity. J Dairy Res 85(4):460–464. https://doi.org/10.1017/S0022029918000730

Sipkema D, Franssen MCR, Osinga R, Tramper J, Wijffels RH (2005) Marine sponges as pharmacy. Mar Biotechnol (NY) 7(3):142–162. https://doi.org/10.1007/S10126-004-0405-5

Sørensen M, Gong Y, Bjarnason F, Vasanth GK, Dahle D, Huntley M, Kiron V (2017) Nannochloropsis oceania-derived defatted meal as an alternative to fishmeal in Atlantic salmon feeds. PLoS One 12(7):e0179907

Tan ECD, Hawkins TR, Lee U, Tao L, Meyer PA, Wang M, Thompson T (2021) Biofuel options for marine applications: technoeconomic and life-cycle analyses. Environ Sci Technol 55(11): 7561–7570. https://doi.org/10.1021/acs.est.0c06141

Tramper J, Battershill C, Brandenburg W, Burgess G, Hill R, Luiten E, Müller W, Osinga R, Rorrer G, Tredici M, Uriz M, Wright P, Wijffels R (2003) What to do in marine biotechnology? Biomol Eng 20(4–6):467–471. https://doi.org/10.1016/S1389-0344(03)00077-7

Uddin SA, Islam MM (2019) Blue biotechnology, renewable energy, unconventional resources and products as emerging frontiers at Sea. J Ocean Coast Econ 6(2):8. https://doi.org/10.15351/2373-8456.1100

Ullah R, Yasir M, Khan I, Bibi F, Sohrab SS, Al-Ansari A, Al-Abbasi F, Al-Sofyani AA, Daur I, Lee S-W, Azhar EI (2017) Comparative bacterial community analysis in relatively pristine and anthropogenically influenced mangrove ecosystems on the Red Sea. Can J Microbiol 63(8). https://doi.org/10.1139/cjm-2016-0587

United Nations (1992) Convention on biological diversity, concluded at Rio de Janeiro on 5 June 1992. In United Nations–Treaty Series No 30619

Varshney P, Mikulic P, Vonshak A, Beardall J, Wangikar PP (2015) Extremophilic micro-algae and their potential contribution in biotechnology. Bioresour Technol 184:363–372. https://doi.org/10.1016/j.biortech.2014.11.040

Vierros M, Suttle CA, Harden-Davies H, Burton G (2016) Who owns the ocean? Policy issues surrounding marine genetic resources. Limnol Oceanogr Bull 25(2):29–35

Vilas-Boas C, Sousa E, Pinto M, Correia-da-Silva M (2017) An antifouling model from the sea: a review of 25 years of zosteric acid studies. Biofouling 33(10):927–942

Zhang P, Xu Y, Liu Z, Xiang Y, Du S, Hew CL (1998) Gene transfer in red sea bream (*Pagrosomus Major*). In: New developments in marine biotechnology. Springer US, Boston, pp 15–18. https://doi.org/10.1007/978-1-4757-5983-9_4

Zhao Q, Xue Y, Wang JF, Li H, Long TT, Li Z, Wang YM, Dong P, Xue CH (2012) In vitro and in vivo anti-tumour activities of echinoside A and ds-echinoside A from *Pearsonothuria graeffei*. J Sci Food Agric 92(4):965–974. https://doi.org/10.1002/jsfa.4678

Biotechnological Utilization of the Marine Environment for Food, Drugs, and Energy

2

Riaz Ullah and Imran Khan

Abstract

Marine biotechnology is any technique that uses marine breathing creatures (or their parts) to make or modify things, or to engineer marine microbes for particular purposes. The marine environment, comprised of oceans and seas, covers more than two-thirds of the biosphere's exterior, which inhabit over 1,400,000 species and the most ancient forms of life. Marine microorganisms are of great significance because they have changed the global climate over time and control the atmosphere. Marine organisms adapt and survive in adverse environmental conditions, making them a huge reservoir for bioactive molecules with exceptional properties and high potential. Therefore, marine environments maintain an excess of a variety of bioactive molecules with distinctive characteristics and important capabilities for biotechnological purposes. Marine reservoirs are hotspots and provide a vital natural source of healthy food and functional food components with biological properties. The organisms' diversity in the marine environment is unknown and needs to be investigated and utilized. Modern marine biotechnology has focused on intensifying research on aquatic organisms and their secondary metabolites. Marine biotechnologists are interested in many marine organisms like crustaceans, macro and microalgae, fish and fish by-products, fungi and bacteria for healthy as well as functional food ingredients, marine drugs, and energy. Biotechnological approaches have a

R. Ullah
Department of Biological Sciences, National University of Medical Sciences, Rawalpindi, Pakistan

I. Khan (✉)
Department of Biotechnology, Abdul Wali Khan University Mardan, Khyber Pakhtunkhwa, Pakistan
e-mail: imrankhan31@awkum.edu.pk

M. D. Shah et al. (eds.), *Marine Biotechnology: Applications in Food, Drugs and Energy*, https://doi.org/10.1007/978-981-99-0624-6_2

crucial part in exploiting marine resources for food, functional food, biomedical purposes, and bioenergy.

2.1 Marine Biotechnology

Utilizing aquatic living things for the production of commodities and services is the goal of marine biotechnology. It is a multidisciplinary field that emerged from the combination of marine disciplines such as ocean science, marine biology, and marine engineering with modern research fields such as molecular and cellular biology, genomics, and proteomics (Freitas et al. 2012). Marine biotechnology also referred to as "blue biotechnology" includes the utilization of marine resources either as a source or as a target for biotechnological applications (Dixon et al. 2011). Generally, marine biotechnology is the application of science and technology to marine living organisms, and their parts, and products, for the development of goods and services. In addition, marine biotechnology also encompasses the development of biosensors from terrestrial organisms which are to be used in aquatic environments (Querellou et al. 2010).

2.2 Aquatic Environment

Oceans and seas are water bodies that cover more than 70% of the earth's surface and are the largest habitats dwelling on the most ancient forms of life. Marine microorganisms are of tremendous importance, have changed global climate over time, and control the atmosphere. The organisms' diversity in the oceans is untapped and needs to be unmasked and utilized. Using molecular approaches such as culture-independent techniques, meta-transcriptomics, proteomics, and metabolomics, these ecosystems are studied in detail and applied for food production, biomedical purposes, environmental benefits, and major scientific progress (Thakur and Thakur 2006).

The attention of the scientific community to marine biotechnology has been drawn because of the terrific opportunities offered by oceans and seas. As life originated in the sea, the diversity of higher taxons is much greater in the marine environment than on land. There are 14 unique maritime phyla compared with one endemic land-dwelling phylum. According to the recently completed marine life census, there are 0.24 million known marine species, while Bouchet (2006) predicted a total of 1.4–1.6 million marine species. Aquatic organisms inhabit the entire columns of water up to a maximum depth of 11 km and on seafloor up to 400 m in oceanic sediments. The aquatic environments and habitats are greatly variable in physical and chemical compositions, support and adapted life forms, and resulted in enormous genetic diversity on earth (Gamfeldt et al. 2015). Aquatic organisms produce a variety of enzymes, polymers, carbohydrates, and other vital molecules because of their adaptation to diverse environmental conditions.

Therefore, extra-terrestrial ecosystems represent a source of untapped natural products and enormous genetic richness, the blue biotechnology can harness to deliver untapped resources of relevance to food, pharmaceuticals, and energy.

The marine biotechnological techniques that have emerged since the 1980s have focused on intensifying the investigation of aquatic living creatures and their products of secondary metabolism (Rotter et al. 2020). The initial research was conducted on bioactive molecules extracted from typical taxon's that thrived in nonterrestrial environments such as immobile large organisms: animals of phylum Cnidaria, tunicates, sponges, and bryozoans, enlightening diverse biochemical compounds (De la Calle 2017). The pluricellular living things inhabiting various environments were already established to colonize intricate and specific microbes (De Oliveira et al. 2012; Simon et al. 2019). The microbial symbiotic relationship has not only played a vital role in the health and functions of the host, but mainly contributes to the host's secondary metabolites production that is used as weapons counter to hunters, infectious agents, and fouling creatures (Wilkins et al. 2019). Particularly, microorganisms colonizing the sponge constitute 40–60% of the sponge's body (Yarden 2014) and numerous biologically active compounds have been validated or anticipated to be derived from microbes (Gerwick and Fenner 2013). Additionally, microbes make up more than 85% of the living creatures of the marine ecosystem and are key players in its health and functioning, by carrying out biogeochemical cycles (Cristancho and López-Alvarez 2021; De la Calle 2017). Marine microorganisms are important in blue biotechnology for several reasons. Firstly, they produce an excess of biologically active metabolic substances with less moral and ecological issues for research and extensive production. Secondly, because of a wide range of genetic manipulation techniques, microorganisms have drawn attention to sustainable marine biotechnology applications and nearly 60% of bioactive secondary metabolites are currently derived from marine microbes (Faulkner 2001).

2.3 Applications in Food

The aquatic environment has been a source of human food since the beginning of human adaptation to the earth. This environment does not supply food for well-being, just like on the land under agricultural activities. Aquatic environments have been supplying food to humans since prehistoric periods, with fishing being an older practice than agriculture. These days fisheries and aquaculture deliver almost 50% of animal proteins (Béné et al. 2016). Whereas the contribution of the marine environment to the global food supply is far greater than the terrestrial environment. In addition, the seas and oceans offer a far richer variety of ingredients for food and possess higher potential than terrestrial environments.

Marine organisms adapt and survive in adverse environmental conditions, making them a huge reservoir for bioactive molecules with exceptional properties and high potential. Therefore, marine environments keep an excess of a variety of bioactive compounds with distinctive properties and important capabilities for biotechnological purposes. Some molecules possess valuable properties and are used as functional food ingredients, dietary supplements and prebiotics, while others are utilized in the food industry as preservatives, stabilizers, pigments and gelling agents, etc. (Boziaris 2014).

At the beginning of the twenty-first century, there was an increasing interest in functional food ingredients, probiotics, prebiotics, and dietary supplements (Shahidi 2009). Probiotics are living microorganisms that confer health benefits to the host. Functional foods can give additional benefits such as medical and physiological effects apart from nutritional effects. In addition, they contained constituents of bioactive molecules of specific amounts with medically established health values. Marine reservoirs are hotspots for phenomenal biodiversity and provide a vital natural source of healthy food and functional food components with biological properties. Marine scientists are interested in many organisms like crustaceans, macro and microalgae, fish and fish by-products, fungi and bacteria for healthy as well as functional food ingredients including, polysaccharides, proteins and peptides, phenolic compounds, chitins, vitamins, pigments, and lipids. Some of the marine organisms with biotechnological potential in food are described in detail.

2.3.1 Macroalgae

In recent years, marine macroalgae or seaweeds have gained immense interest because of their new applications in food additives, food components, and nutraceuticals, which are important in the development of many species of aquaculture (Stévant et al. 2017). The terminology "seaweed" comprises large and pluricellular maritime brown, green, and red algae, which have been known as a rich source of proteins, nondigestible polysaccharides, minerals, vitamins, and iodine. There has been an increasing need for sea polypeptides and fats in fish feed manufacturing for the past 20 years and weed extracts are becoming popular to be used as prophylaxis and therapeutics in fish and jellyfish aquaculture (Vatsos and Rebours 2015). In 2016, over 32,000,000 tons of macroalgae were produced, primarily through capture and aquaculture, with a 10% annual increase over the previous 4 years.

Macroalgae are known for their technological properties containing phycocolloids such as agar, alginates, and carrageenan, which are highly valued for gelling, thickening, and stabilizing and are used in cooking and baking. Algal metabolites, for example, proteins, vitamins, carbohydrates, and carotenoids are used as food and feed additives. The most biotechnologically important species such as *Chlorella vulgaris*, *Haematococcus pluvialis*, *Cyanobacteria*, *Dunaliella salina*, and *Spirulina maxima*, are successfully employed in the manufacturing of nutritional supplements and animal food additives.

2.3.2 Microalgae

Microalgae are known for their fast growth and short generation time. The fastest-growing species' biomass can be doubled more than once a day. The fascination with microalga biotechnological approaches is increasing during the past few decades because of their higher growth rate compared to land plants and might be thought of as reliable "cell factories" for the preparation of biologically active molecules. These bioactive compounds can be used for food, feed, and other biotechnological purposes (De Morais et al. 2015; De Vera et al. 2018; Skjånes et al. 2013). Several bioprocess technologies such as cultivation, harvesting, extraction, and isolation are combined to obtain active components from microalgae. Various culture parameters, for instance, the level of CO_2, O_2, light, temperature, pH, and nutrients are regulated to attain desired biomass activity and efficiency. The control of hunters is of immense importance in the large production of microalgae. The natural production capability of microalgae is engineered by adjusting factors or by putting on particular environmental stressors like changes in nitrate level, light intensity and spectrum, and salt concentration, to trigger the production of a desired bioactive substance (Chokshi et al. 2017; Markou and Nerantzis 2013; Smerilli et al. 2017; Vu et al. 2016; Yu et al. 2015). Microalgae, because of their content, are considered vital for the food and feed industries for both total biomass as well as nutrient constituents. Astaxanthin is a "super antioxidant" from *Haematococcus pluvialis* and its commercial-scale production has drawn global attention due to several applications such as nutraceuticals, beverages, and aquaculture feed additives (Shah et al. 2016). The molecular techniques are advancing very fastly and soon we will improve microalgal growth conditions, and various qualities and transfer whole metabolic pathways in interspecies and intraspecies (Noda et al. 2017; Nymark et al. 2019; Sharma et al. 2020; Slattery et al. 2018).

2.3.3 Bacteria, Archaea, and Fungi

Microorganisms inhabiting extreme environmental conditions are unique and novel in survival strategies, secondary metabolites production, and enzymatic catalysts of biotechnological importance. For instance, the biological catalysts produced by thermophiles and psychrophiles bacteria degrade polymers of monosaccharides, polymers of amino acids, and polymers of fatty acids used for several purposes in food large-scale production (Poli et al. 2017). Astaxanthin is among the carotenoids that have commercial value as food supplements for humans. A carotenoid biosynthetic gene cluster for astaxanthin has been identified in marine *Agrobacterium aurantiacum* (Misawa et al. 1995). *Paracoccus haeundaensis* is another astaxanthin-producing bacterium that is extracted and identified recently (Lee et al. 2004).

Cyanobacteria are important to produce secondary products which are capable of activities against microbes, protozoans, inflammations, coagulations, oxidations, cancer, and viruses. Hence, they are immense sources of biologically active substances for important food products (Demay et al. 2019). Marine fungi are also an important source of biosurfactants used in the food industry as stabilizers, texture and taste improvers, and shelf-life elongators (Pitocchi et al. 2020). In addition, marine fungal enzymes, apart from their diverse applications, are also used for food and beverages (Bonugli-Santos et al. 2015).

2.3.4 Thraustochytrids

Thraustochytrids comprise an important group of protists in eukaryotic microorganisms. The diversity of thraustochytrids is still uncovered, although a great number of species and strains have been discovered. The previous systematic approaches using morphological characteristics were insufficient. The development of molecular marker strategies such as 18S rRNA gene analysis has facilitated the characterization of thraustochytrids (Mo et al. 2002). The diversity of thraustochytrids at the species level to be fully tapped is still very far, owing to the limitations in 18S clone library preparation and the developing culturomic techniques. This group is an important source of bioactive substances including antibacterial, antifungal, antiviral, etc. For example, the extracts of Thraustochytrids, which were isolated from the mangrove ecosystem, showed antimicrobial activities *against Staphylococcus aureus*, *Klebsiella pneumonia*, and *Salmonella typhi*.

Thraustochytrids have been cultured from aquatic habitats such as coasts, open seas, and deep seas, rich in organic substances (Raghukumar 2002). Mangroves' environments dwell in this group of organisms, where they feed on decaying matter (Morabito et al. 2019), are primary consumers, efficient degraders of organic compounds, and importantly play an ecological role (Mo and Rinkevich 2001). Thraustochytrids are considered oleaginous microorganisms used for diverse biotechnological purposes. They are extensively utilized as a substantial nonterrestrial source of polyunsaturated fatty acids (PUFAs), mainly for the large-scale production of Omega-3 fatty acids DHA. Omega-3 fatty acids are nutrients provided in food and food supplements, they build and maintain a healthy body. The important characteristics of thraustochytrids make them an alternative to fish oil and are used as an environment-advantageous way out of overfishing. The US Food and Drug Administration (FDA) has approved *Schizochyrium* sp. oil for human and animal consumption.

2.3.5 Biotechnology's Role in Fish Farming

Fish farming (fisheries and aquaculture) is an important technique that gives rise to fisheries resources (fisheries products). It is an essential approach used for providing food, particularly in drought-prone regions. Aquaculture gained global importance when governments were struggling to ensure the availability of food in their communities (Kassam et al. 2011; Jamu and Brummett 2004). The world's rapidly increasing demand for marine foods is not being met sustainably by wild fish stocks

(Pauly et al. 2002). The world's population is increasing very rapidly, with a current population of 7.3 billion, which is projected to reach 8.5 billion by 2030, and 9.7 billion by 2050. To sustainably satisfy the needs of this expanded population, the world needs a substantial expansion in food production including marine food. Aquaculture growth not only makes more aquatic products for consumption available but also results in other economic, social, and environmental benefits for society. Together with fisheries, aquaculture, directly or indirectly, plays an essential role in the livelihoods of millions of people around the world.

Overall, employment in the fishing and aquaculture sector has grown faster than the world's population and employment in traditional agriculture. Today, aquaculture and capture fisheries directly employ over 45 million people, supporting the livelihoods of 8% of the world's population, and each sector provides about 50% of the world's aquatic food supply. It has been estimated that in 2008, aquaculture created about 11 million full-time jobs worldwide. Aquaculture has an important impact on the world economy and employment. For example, Indiana's aquaculture industry operates on a small scale and produces and distributes fish at the national and international levels. The industry makes $3,731,842 of equal value to labor income and $19,484,193 of added value. They generated an output wealth of $23,599,676 within the industry and a total of $37,892,895 with other supporting industries. Regarding the employment generated by fish farming, Indiana's aquaculture industry employed 280 people within the industry and 169 jobs in supporting industries, which are direct employees in the aquaculture industry. In the future, the dependence on aquaculture is rising, because the overall production is growing with the increasing global population.

The main challenges are:

1. To satisfy the demand for seafood for an expanded worldly population, while the natural habitats are decreasing.
2. Fish feed is expensive, world fish stock is declining, and causes environmental problems.
3. Aquatic animals are much more sensitive to their surroundings than terrestrial animals.
4. How to manage fish health in aquaculture.

Biotechnology is applied to tackle these challenges. Some of the applications of biotechnological tools are described below.

2.3.5.1 Biotechnology's Role in Fish Breeding

The gonadotropin-releasing hormone (GnRH) is the best available biotechnological application in the improvement of fish breeding. GnRH is an important regulator and initiator of the reproductive cascade in invertebrates (Bhattacharya et al. 2002). The GnRH was first isolated from pig and ship hypothalamic since then only one form of GnRH has been known from most placental mammals including humans. Conversely, 12 forms of GnRH have been structurally elucidated in nonmammals (except guinea pigs), among which seven or eight have been isolated from fish species (Carolsfeld et al. 2000; Mayekar et al. 2013). Relying on the variant's structure and biological activity, several analogs have been chemically synthesized, Salmon GnRH one of them, abundantly used in fish breeding and globally available under the trade name "Ovaprim". GnRH technology has successfully been used in induced fish breeding, and GnRH is the best accessible biotechnology tool.

2.3.5.2 Biotechnology's Role in Fish Feed

Marine biotechnology provides an important alternative for fish feed. Currently, the fish diet consists of proteins from fish processing. Along with a few other disadvantages, the elevated level of phosphorus in the fish diet causes environmental pollution. In response to this concern, researchers are using biotechnological plant-based protein sources to tackle phosphorus pollution (Adelizi et al. 1998).

2.3.5.3 Bioremediation

The diseases in aquacultural species are much more strongly connected to external environmental factors. Marine animals in forms are more susceptible to their immediate habitat than land animals. Regarding this intricate connection to avoid disturbances in this relationship, a biotechnological tool named bioremediation has been developed in aquaculture. It is the use of useful bacteria or probiotics to treat water or feeds the aquaculture. In other words, using natural processes to prevent the growth of unfriendly bacteria that are capable of causing diseases in farm animals.

2.3.5.4 Transgenesis

The biotechnological application of improving or modifying the traits of commercial animals for aquaculture. This is the introduction of an exogenous gene/DNA into the animal genome which leads to stable maintenance, transmission, and expression. The technique has already been successfully applied to numerous fish species. Using transgenesis, enhanced growth has been recorded especially in salmonoids (Diwan and Kandasamy 1997).

2.3.5.5 Biotechnology's Role in Fish Health Management

In aquaculture, fish's health is of prime importance. Biotechnology has played a vital role in fish health management. Several vaccines have developed against bacteria and viruses used in finfish aquaculture. In addition, new generation vaccines such as protein sub-unit and DNA vaccines are under development. For improving disease resistance in fish in shellfish species, biotechnological tools like diagnostic techniques, vaccines, and immunostimulants are gaining popularity worldwide. To prevent infections in farmed animals, one important approach is the early detection of the pathogen, biotechnological techniques such as gene probes and Polymerase Chain Reaction (PCR) have substantial potentials to be used.

A crucial issue in aquaculture is disease control. Bacterial pathogens such as *Vibrio* and *Aeromonas* are well-known infectious agents in many shrimp-producing countries. The new diagnostic tools such as monoclonal antibodies and DNA probes are biotechnological tools that are highly encouraging (Fig. 2.1).

Some issues of fishery management remain uncertain such as genetic variation, stock assessment, evolutionary processes of aquaculture, and wild species and

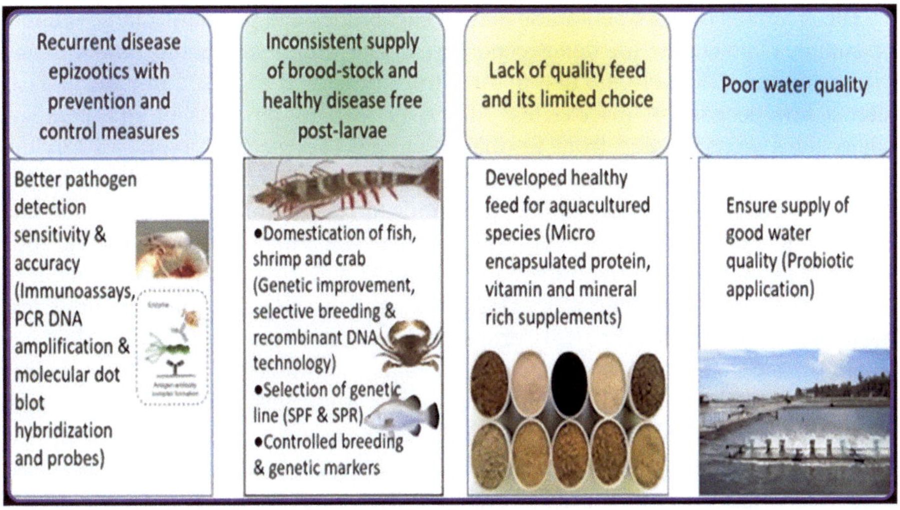

Fig. 2.1 Sustainable food supply through blue biotechnology

species recognizable proof. Molecular biology techniques; nuclear DNA, DNA chips (microarray), mitochondrial DNA, and microsatellite DNA markers are very promising to tackle these issues.

2.4 Applications in Drug

New diseases are emerging with the changing environment, and disease management patterns are changing continuously. The enormous increase in the world population has exhausted resources for the drug. Drug manufacturers are looking for new resources to develop new, safe, and effective drugs to meet the demand of the growing global population. More than two-thirds of the earth's surface is covered by water, and due to limited pharmacology research, the marine environment is almost unexplored for marine drugs. The marine environment is home to a variety of plants and animals, which produce many compounds that possess antibacterial, antifungal, immunomodulatory, anticancer, anti-inflammatory, antimicrobial, antimalarial, and analgesic properties. The bioactive compounds isolated from various marine resources and used as drugs are described below.

2.4.1 Metazoans

Marine environments serve as hot spots for natural products, and early research was concentrated on a marine group of metazoans like cnidarians, tunicates, sponges, and gastropods. These groups were the representatives of the explored aquatic ecosystem and sampling was relatively easy by underwater swimming (Molinski et al. 2009).

The creatures possess partial movement capabilities and generate a large number of complex molecules for defense purposes. The drugs vidarabine (ara-A®) and cytarabine (ara-C®) were the first marine drugs described and used in clinical trials after a few decades of discoveries. Moreover, two by-products of ribo-pentosyl nucleosides were isolated from Caribbean animal (Tectitethya crypta) of Phylum Porifera in 1950s (Bergmann and Feeney 1951). Another initial medicine of the marine environment was ziconotide, a manufactured form of omega-conotoxin and omega-3 acid ethyl esters (Lovaza). The prior one was isolated from the Pacific cone snail *Conus amagus* sold with the name "Prialt®" and used for the management of chronic pain, and the latter was extracted from fish and used for hyperlipidemia conditions. In addition, other early drugs were anticancer such as Eribulin, Mesylate (E7398), and macrolide commercially available as Halaven® and Monomethyl auristatin E from dolastatin peptides extracted from aquatic mollusc *Dolabella auricularia* marketed as Adcetris® (Jimenez et al. 2020). The medicine ecteinascidin-743 (ET-743), also called trabectedin and sold under the trade name YondelisR, was isolated from the Caribbean ascidian *Ecteinascidia turbinata*in in the1960s and was the first drug from marine environment that after 40 years was approved for cancer treatment (Aune et al. 2002). Various approaches, such as mariculture and total synthesis, were used to produce drugs on a commercial scale, but a semisynthetic process, beginning with cyanosafracin B, eventually solved the bottleneck. The antibiotic cyanosafracin B, was earlier produced using fermentation of *Pseudomonas fluorescens* (Cuevas and Francesch 2009; Cuevas et al. 2000). After the development of this technique, metazoan-derived marine compound discovery is increasing (Molinski et al. 2009; Rocha et al. 2011) and most of the natural compounds are extracted from sponges and cnidarians (Blunt et al. 2018; Carroll et al. 2022). In the last decade, there was an increase of approximately 200 new compounds reported each year, while the number of compounds from the widely explored other phyla such as molluscs, tunicates, and echinoderms range from 8 to 50 during this period (Carroll et al. 2022). Particularly, members from the phylum Porifera and Cnidaria can be cultivated widely to help medicine uncovering (Duckworth and Battershill 2003; Leal et al. 2013).

Sponges are an appealing source for bioproduction-focused growing owing to their plain body, high recreating ability, and high content of biologically active compounds. An effective approach to sponge farming has been planned (and frequently implemented) to address supply issues and make sure sustainable formation of molecules from sponge (Duckworth 2009), but there are still some challenges to collecting starting materials from a wild population at various stages of cultivation.

2.4.2 Microalgae

Microalgae grow very fast and their generation period is very fast. In the fastest growing species, their biomass becomes double more than once a day. For microalgal culture, the biomass yield has been reported at up to 20/kg/m^2/year

(Varshney et al. 2015). The precise growing frequency for microalgae is five to tenfold greater than land plants, and this has intensified researchers' curiosity in microalga biotechnology in the past 50 years. For the bioproduction of biologically active compounds used in various biotechnological applications, microalgal biomass is observed as genuine "cell factories" (De Vera et al. 2018). Microalgae are considered an important source of antimicrobials, and anticancer drugs. Marine dinoflagellates are a type of microalgae known for bioactive molecule production such as biological toxins and are among the largest, most highly complex and naturally influential compounds (Hinder et al. 2011). Biological toxins are of immense attention for their prospective practices in biomedical and pharmaceutical usages (De Vera et al. 2018).

2.4.3 Bacteria and Archaea

Various forms of microorganisms are thriving in marine habitats. An increasing number of microbes (bacteria and archaea) inhabiting diverse ranges of temperature and pH have been isolated. As these organisms are adapted to diverse environmental conditions and proliferate, consequently they have developed strategies involving cellular metabolic mechanisms.

Halophilic archaea are thriving in hypersaline habitats, such as salt deposits, salt lakes, and solar salterns. Carotenoids such as C40 and C50 are synthesized and deposited by haloarchaea. The carotenoid extracts of some halophilic archaea show bioactivity against oxidation, hemolysis, and cancer (Galasso et al. 2017; Giani et al. 2019).

Actinobacteria and cyanobacteria are types of bacteria known for their secondary metabolite production capacity. The number of natural compounds discovered from 1997 to 2008 were about 660, and more than one-third were originating from actinobacteria and cyanobacteria (Williams 2009). Marine habitats are of immense importance and a very promising source for biodiscovery. The number of antibacterial species identified from 2007 to 2017 were 177, and they were affiliated with 29 new genera and three new families (Subramani and Sipkema 2019). The members of phylum actinobacteria are an important source of biologically active molecules that possess activities against bacteria, cancer, inflammation, parasites, biofilm, and fouling (Bauermeister et al. 2018, 2019; Cartuche et al. 2019, 2020; Girão et al. 2019; Pereira et al. 2020; Prieto-Davó et al. 2016). Cyanobacteria are important producers of secondary metabolites and live in diverse environments. Bioactive metabolites from cyanobacteria possess anticancer, anticoagulant, antimicrobial, antiprotozoal, anti-inflammatory, antioxidant, and antiviral potential. Hence, cyanobacteria are immensely important sources of biologically active substances for clinical applications (Kini et al. 2020; Silva et al. 2018).

2.4.4 Fungi

Oceanic fungi are vastly distributed in oceans and seas, and inhabit various environmental niches; they are adapted to all tropic levels associated with organisms saprophytically, symbiotically, and parasitically (Poli et al. 2018; Raghukumar 2017; Wang et al. 2012). Aquatic fungi produce various bioactive compounds (Silber et al. 2016). The biotechnological potential of marine fungi is undeniable, and marine fungi account for 36% of all the 1277 natural bioactive compounds produced in 2016 (Faulkner 2001). The antibacterial and antifungal compounds isolated from fungi in 5 years (2010–2015) were 285 (Nicoletti and Andolfi 2018). Cephalosporin is the first bioactive compound group reported from oceanic fungus, which is a class of β-lactam antibacterial.

Most researchers working on fungi secondary metabolites have focused on genera such as *Cladosporium*, *Aspergillus*, *Penicillium*, and *Fusarium* (Imhoff 2016; Marchese et al. 2020). Oceanic fungi have been a promising source of pharmacologically important secondary bioactive substances (Imhoff 2016). They stand up for antibacterial, antiviral, anticancer, and antiplasmodial activities (Rajasekar et al. 2012).

2.4.5 Thraustochytrids

Thraustochytrids are a distinctive group of eukaryotes microbes belonging to protists, producing bioactive molecules including antimicrobials. The richness and evenness of thraustochytrids are considered to be unveiled and numerous species strains have been identified. Thraustochytrids remain a chief aquatic source of fatty acids containing multiple double bonds for various biotechnological applications. In addition, they are especially utilized for the industrial production of ω-3 fatty acids. Interestingly, ω-3 fatty acids are energy sources providing energy for the proper functions of the heart, lungs, blood vessels, and immune system. Some important long-chain lipids which possess multiple double bonds such as DHA (C22; 5 *n*3) and docosapentaenoic acid (DPA, C22:5 *n*6) are produced by thraustochytrids including *Schizochytrium* and *Aurantiochytrium* species (Heggeset et al. 2019). Recent studies have suggested that DHA can be effectively used against cancer in combination with other anticancer agents.

2.5 Applications in Energy

2.5.1 Microalgae

Regarding bioenergy, microalga has been generally considered for biodiesel production, because of its wide distribution and elemental arrangement (Nirbhay and Dolly 2011). Marine microalgae are important for producing biodiesel due to their high lipid content and doubling rate which exceeds most terrestrial plants. Various approaches are applied, particularly electroextraction, which is used without solvents and chemicals, and important biofuel compounds are obtained from microalgae (Goettel et al. 2013). The production of biodiesel with current technology is not financially feasible, although the technology is exceedingly popular and technically achievable. These cannot be densely cultivated because of their need for light—and this is one of the reasons that limit their application for large-scale manufacturing (Table 2.1). Additionally, the harvesting cost increases when biological mass is detached with a huge volume of H_2O. At present, the genomic manipulation of oceanic microalgae such as *Phaeodactylum tricornutum* and *Nannochloropsis* for biodiesel manufacture remains continuing to tackle the bottlenecks (Du et al. 2018; Nymark et al. 2019). Microalgae are biotechnologically important due to their high growth rate, and therefore, they are considered "cell factories" that produce necessary bioactive substances that are used in bioenergy. Because of their content, microalgae are well-thought-out as a favorable substrate for imperishable and inexhaustible energy (Moreno-Garcia et al. 2017; Suganya et al. 2016).

2.5.2 Thraustochytrids

Thraustochytrids are a protist group of great importance that thrive in the marine environment and produced saturated fatty acids that serve as a renewable source for biofuels and biodiesel. In addition, this group presents a likely fount for squalene and carotenoids, which are commercially important compounds (Aasen et al. 2016).

2.6 Formation of Secondary Metabolism Molecules Using Biotechnological Techniques

The frequency and efficiency of secondary metabolism substances formation by microbes mainly depend on the culturing conditions and upscaling abilities. Interestingly, the most important factor among the driving forces is the mimicking of the natural setting as well as the interactions with other microbes (Vallet et al. 2017). Under normal growth conditions, fungi and bacteria do not produce fascinating secondary metabolism substances (Reich and Labes 2017). New approaches are to be adopted to cultivate in-vitro whole organisms or cell types of particular marine environments to raise the cell biomass and reduce damage during sample collection (Barnay-Verdier et al. 2013; Ventura et al. 2018). In the case of sponges and corals, this is also of immense importance, harboring microorganisms. A suitable cultivation environment should make available optimum conditions for holobionts (Rinkevich 2011).

Cultivated microbes do not constantly remain to make the matching metabolites (Wijffels 2008). However, mimicking natural interactions such as co-culturing or mixed fermentation with other creatures has been established to elicit the formation of secondary metabolism products (Romano et al. 2018). In addition, the strategy of modifying quorum sensing has been applied in the antibiotic and cancer

Table 2.1 Enlisted are the major important oceanic taxonomic groups, their applications, and the difficulties of large-scale manufacture

Organisms	Application	Important phyla, genera/species	Challenges
Metazoans	Medicine	Tunicates—Chordata (*Ecteinascidia turbinata*), Mollusca (*Conus magus*), Sponges—Porifera (*Mycale hentscheli*), Cnidaria (*Sinularia* sp., *Clavularia* sp., *Pseudopterogorgia* sp.)	Sourcing and supply sustainability
Macroalgae and seagrasses	Food, feed, medicine, nutraceuticals, biofertilizers biomaterials, energy	Rhodophyta (*Euchema denticulatum*, *Porphyra*/ Pyropia spp., *Gelidium sesquipedale*, *Pterocladiella capillacea*, *Furcellaria lumbricalis*, *Palmaria* spp., *Gracilaria* spp.), Chlorophyta (*Ulva* spp.), Ochrophyta (*Laminaria hyperborea*, *Laminaria digitata*, *Undaria pinnatifida*, *Alaria* spp., Fucus spp.), seagrasses (*Zostera*, *Cymodocea*)	Sourcing and supply sustainability, yield optimization, large-scale processing and transport. Disease management
Microalgae	Sustainable energy, food, feed, biofertilizers, medicine	Chlorophyta (*Chlorella*, *Haematococcus*, *Tetraselmis*), Cryptophyta, Myzozoa, Ochrophyta (*Nannochloropsis*), Haptophyta (*Isochrysis*), Bacillariophyta (*Phaeodactylum*)	Bioprospecting and yield Optimization (1—increase in biomass/volume ratio, 2—increase yield of compound/extract production and 3—Improve solar-to-biomass energy conversion)
Bacteria and Archaea	Medicine, biomaterials, biofertilizers	Actinobacteria (*Salinispora tropica*), Firmicutes (*Bacillus*), Cyanobacteria (*Arthrospira*, *Spirulina*), Proteobacteria (*Pseudoalteromonas*, *Alteromonas*), Euryarchaeota (*Pyrococcus*)	Culturing for nonculturable species, yield optimization
Fungi	Medicine, food/ feed, biofertilizers	Ascomycota (*Penicillium*, *Aspergillus*, *Fusarium*, *Cladosporium*)	Limited in-depth understanding, yield optimization

(continued)

Table 2.1 (continued)

Organisms	Application	Important phyla, genera/ species	Challenges
Thraustochytrids	Food/feed, sustainable energy production	Bigyra (*Aurantiochytrium* sp.), Heterokonta (*Schizochytrium* sp.)	Limited in-depth understanding, yield optimization
Viruses	Medicine, biocontrol	Mycoviruses, bacteriophages	Limited in-depth understanding, yield optimization

therapeutical products industries (Bertrand et al. 2014). The communications can be altered to simulate the normal settings for biochemical environmental studies, such as the introduction of specific antagonistic organisms that stimulate the production of antibacterial (Bertrand et al. 2014; Bovio et al. 2019; Romano et al. 2018). Microalgae and bacteria including cyanobacterial interaction in co-culture as well as co-culture of microalgae and fungi have been demonstrated to favor the production of high content of lipids (Arora et al. 2019; Ferro et al. 2019; Gautam et al. 2019; Toyama et al. 2019). An improvement in microalgal growth in co-culture with another microalga (Ishika et al. 2019) along with fungi (Wang et al. 2019) and protozoa (Peng et al. 2016) has been documented.

2.7 A Genetic Engineering Approach to Marine Organisms

Currently, marine researchers are interested in fish biotechnology using methods including chromosomal manipulation and hormonal treatments, which can be used for the production of triploid, tetraploid, haploid gynogenetic, and androgenetic fish. As a result, individuals and lineages of sterile, mono sex or highly endogamic fish are produced. These approaches are used in fish culture for practical purposes to control the precocious sexual maturation in particular species, secondly to control reproductive processes for the larger production of specimens, and thirdly to obtain individuals of high commercial value belonging to a mono-sex line. The use of new technologies for gene transfer has ignited the interest of aquaculturists, which will lead to modified fish and will play an essential role in the specific program of fish production in near future.

To alter marine metabolite production in aquatic organisms, genetic engineering has been applied (Freitas et al. 2012). To modify microalgal metabolic pathways, for improved and high-standard products, genetic engineering has a documented application (Qin et al. 2012). In a study researchers used microalgae bioengineering for the development of sustainable energy. In another study, the strategy of digital microfluidic was used for efficient electroporation, resulting in high production and foreign gene expression without cell wall removal. The transformation of microalgal species such as *Haematococcus lacustris*, *Chlamydomonas reinhardtii*, *Chlorella* sp., *Dunaliella salina*, *Dunaliella bardawil*, *Nannochloropsis* sp.,

Symbiodinium sp., and *kessleri* was conducted utilizing *Agrobacterium tumefaciens* transformation method. However, the use of genetically modified, GM, microalgae for the production of important products like carotenoids and PUFA reflected a challenge (León-Bañares et al. 2004). Moreover, the use of GM microalgae is limited by concerns such as biosafety, competitiveness, and public acceptance (Freitas et al. 2012).

2.8 Marine Metagenomic Approach

Culturomic has revealed a fraction of marine microbial diversity and continues to offer new chemical structures with biological activities and provide a promising strategy to explore novel compounds. In addition, unveiling the marine genome, especially the unculturable organism, allow researchers to screen for new genes and to obtain new natural molecules exploiting the marine resources, therefore to fully exploit the marine environment, metagenomic approaches are applied. The flow chart of metagenomic approaches is presented in Fig. 2.2.

The genomes of more than 150 bacteria have been sequenced. Viruses are the most commonly found species among all biological species living in an aquatic environment. Using metagenomic approaches, a unique gene pool of marine viruses has been determined. The marine viral diversity investigated via metagenomics shows differences in the genetic makeup of hundreds of thousands of biological species, with genes different from other life forms.

2.9 Conclusion

Water bodies are the most important habitats for most ancient forms of life. Marine microorganisms are vital and have changed the global climate over time and control the atmosphere. Modern marine biotechnology that emerged in the 1980s has focused intensified research on aquatic organisms and their secondary metabolites. The initial studies were conducted on bioactive molecules extracted from representative taxa that thrived in marine environments like sessile large organisms: cnidarians, tunicates, sponges, and bryozoans, enlightening a unique chemical diversity of bioactive compounds. Blue biotechnology focuses on marine microorganisms for several reasons. Firstly, they produce an excess of bioactive metabolites with fewer moral and ecological issues for the study and industrial production. Second, microorganisms have attracted attention for sustainable marine biotechnology applications due to a wide range of genetic manipulation techniques; and third, nearly 60% of bioactive secondary metabolites are currently derived from marine microbial manipulation.

Marine resources are hotspots for remarkable biodiversity and provide a vital natural source of healthy food and functional food components with biological properties. Marine scientists are interested in many organisms like crustaceans, macro and microalgae, fish and fish by-products, fungi and bacteria for healthy as

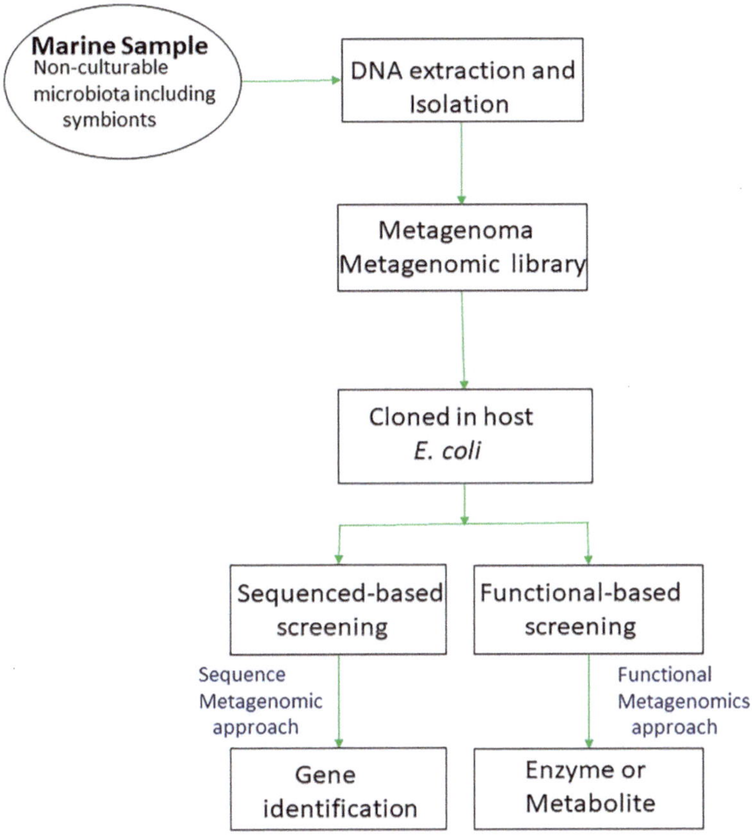

Fig. 2.2 Marine metagenomic approaches

well as functional food ingredients including, a polymer of monosaccharides, proteins and peptides, phenolic compounds, chitins, vitamins, pigments, and lipids. More than 70% of the earth's surface is covered by water, and due to limited pharmacology research, the marine environment is almost unexplored for marine drugs. The marine environment is home to a variety of plants and animals, which produce many compounds that possess antibacterial, antifungal, immunomodulatory, anticancer, anti-inflammatory, antimicrobial, antimalarial, and analgesic properties. Biodiesel production with current technology from microalga is not financially viable, although the technology is exceedingly popular and technically achievable. Because microalgae cannot be densely grown due to their need for light—and this is one of the reasons that lowers their application for large-scale manufacturing. Although the marine environment has been investigated since life existed on earth. There may still be more to be discovered in marine settings. Modern marine biotechnology techniques would be able to fully explore marine resources and would help to exploit them for the benefit of human beings.

References

Aasen IM, Ertesvåg H, Heggeset TMB, Liu B, Brautaset T, Vadstein O, Ellingsen TE (2016) Thraustochytrids as production organisms for docosahexaenoic acid (DHA), squalene, and carotenoids. Appl Microbiol Biotechnol 100(10):4309–4321. https://doi.org/10.1007/s00253-016-7498-4

Adelizi P, Rosati R, Warner K, Wu YV, Muench T, White M, Brown P (1998) Evaluation of fish-meal free diets for rainbow trout, Oncorhynchus mykiss. https://doi.org/10.1046/j.1365-2095.1998.00077.x

Arora N, Patel A, Mehtani J, Pruthi PA, Pruthi V, Poluri KM (2019) Co-culturing of oleaginous microalgae and yeast: paradigm shift towards enhanced lipid productivity. Environ Sci Pollut Res 26(17):16952–16973. https://doi.org/10.1007/s11356-019-05138-6

Aune GJ, Furuta T, Pommier Y (2002) Ecteinascidin 743: a novel anticancer drug with a unique mechanism of action. Anti-Cancer Drugs 13(6):545–555. https://doi.org/10.1097/00001813-200207000-00001

Barnay-Verdier S, Dall'Osso D, Joli N, Olivré J, Priouzeau F, Zamoum T, Merle P-L, Furla P (2013) Establishment of primary cell culture from the temperate symbiotic cnidarian, *Anemonia viridis*. Cytotechnology 65(5):697–704. https://doi.org/10.1007/s10616-013-9566-2

Bauermeister A, Velasco-Alzate K, Dias T, Macedo H, Ferreira EG, Jimenez PC, Lotufo T, Lopes NP, Gaudêncio SP, Costa-Lotufo LV (2018) Metabolomic fingerprinting of *Salinispora* from Atlantic oceanic islands. Front Microbiol 9:3021. https://doi.org/10.3389/fmicb.2018.03021

Bauermeister A, Pereira F, Grilo IR, Godinho CC, Paulino M, Almeida V, Gobbo-Neto L, Prieto-Davó A, Sobral RG, Lopes NP (2019) Intra-clade metabolomic profiling of MAR4 *Streptomyces* from the Macaronesia Atlantic region reveals a source of anti-biofilm metabolites. Environ Microbiol 21(3):1099–1112. https://doi.org/10.1111/1462-2920.14529

Béné C, Arthur R, Norbury H, Allison EH, Beveridge M, Bush S, Campling L, Leschen W, Little D, Squires D (2016) Contribution of fisheries and aquaculture to food security and poverty reduction: assessing the current evidence. World Dev 79:177–196. https://doi.org/10.1016/j.worlddev.2015.11.007

Bergmann W, Feeney RJ (1951) Contributions to the study of marine products. XXXII. The nucleosides of sponges. I. J Org Chem 16(6):981–987. https://doi.org/10.1021/jo01146a023

Bertrand S, Bohni N, Schnee S, Schumpp O, Gindro K, Wolfender J-L (2014) Metabolite induction via microorganism co-culture: a potential way to enhance chemical diversity for drug discovery. Biotechnol Adv 32(6):1180–1204. https://doi.org/10.1016/j.biotechadv.2014.03.001

Bhattacharya S, Dasgupta S, Datta M, Basu D (2002) Biotechnology input in fish breeding. Indian J Biotechnol 1:29–38. https://doi.org/10.1007/978-3-030-20944-5_4

Blunt JW, Carroll AR, Copp BR, Davis RA, Keyzers RA, Prinsep MR (2018) Marine natural products. Nat Prod Rep 35(1):8–53. https://doi.org/10.1039/c9np00069k

Bonugli-Santos RC, dos Santos Vasconcelos MR, Passarini MR, Vieira GA, Lopes VC, Mainardi PH, Dos Santos JA, de Azevedo DL, Otero IV, da Silva Yoshida AM (2015) Marine-derived fungi: diversity of enzymes and biotechnological applications. Front Microbiol 6:269. https://doi.org/10.3389/fmicb.2015.00269

Bouchet P (2006) The magnitude of marine biodiversity. In: The exploration of marine biodiversity: scientific and technological challenges. Fundación BBVA, Bilbao, pp 31–62. https://doi.org/10.1016/j.cub.2012.09.036

Bovio E, Garzoli L, Poli A, Luganini A, Villa P, Musumeci R, McCormack GP, Cocuzza CE, Gribaudo G, Mehiri M (2019) Marine fungi from the sponge *Grantia compressa*: biodiversity, chemodiversity, and biotechnological potential. Mar Drugs 17(4):220. https://doi.org/10.3390/md17040220

Boziaris IS (2014) Food ingredients from the marine environment. Marine biotechnology meets food science and technology. Front Mar Sci 1:66. https://doi.org/10.3389/fmars.2014.00066

Carolsfeld J, Powell JF, Park M, Fischer WH, Craig AG, Chang JP, Rivier JE, Sherwood NM (2000) Primary structure and function of three gonadotropin-releasing hormones, including a novel form, from an ancient teleost, herring. Endocrinology 141(2):505–512. https://doi.org/10.1210/endo.141.2.7300

Carroll AR, Copp BR, Davis RA, Keyzers RA, Prinsep MR (2022) Marine natural products. Nat Prod Rep 38:362–413. https://doi.org/10.1039/D0NP00089B

Cartuche L, Reyes-Batlle M, Sifaoui I, Arberas-Jiménez I, Piñero JE, Fernández JJ, Lorenzo-Morales J, Díaz-Marrero AR (2019) Antiamoebic activities of indolocarbazole metabolites isolated from *Streptomyces sanyensis* cultures. Mar Drugs 17(10):588. https://doi.org/10.3390/md17100588

Cartuche L, Sifaoui I, López-Arencibia A, Bethencourt-Estrella CJ, Nicolás-Hernández S, Lorenzo-Morales J, Piñero JE, Díaz-Marrero AR, Fernández JJ (2020) Antikinetoplastid activity of indolocarbazoles from *Streptomyces sanyensis*. Biomolecules 10(4):657. https://doi.org/10.3390/biom10040657

Chokshi K, Pancha I, Ghosh A, Mishra S (2017) Salinity induced oxidative stress alters the physiological responses and improves the biofuel potential of green microalgae *Acutodesmus dimorphus*. Bioresour Technol 244:1376–1383. https://doi.org/10.1016/j.biortech.2017.05.003

Cristancho M, López-Alvarez D (2021) Microbial diversity exploration of marine hosts at serrana bank, a coral atoll of the seaflower biosphere reserve. SeaFlower biosphere reserve: new findings and trends in the largest Caribbean marine protected area. https://doi.org/10.3389/fmars.2019.00338

Cuevas C, Francesch A (2009) Development of Yondelis®(trabectedin, ET-743). A semisynthetic process solves the supply problem. Nat Prod Rep 26(3):322–337. https://doi.org/10.1039/B808331M

Cuevas C, Pérez M, Martín MJ, Chicharro JL, Fernández-Rivas C, Flores M, Francesch A, Gallego P, Zarzuelo M, de la Calle F (2000) Synthesis of ecteinascidin ET-743 and phthalascidin Pt-650 from cyanosafracin B. Org Lett 2(16):2545–2548. https://doi.org/10.1021/ol0062502

De la Calle F (2017) Marine microbiome as source of natural products. Microb Biotechnol 10(6): 1293. https://doi.org/10.1111/1751-7915.12882

De Morais MG, Vaz BS, de Morais EG, Costa JAV (2015) Biologically active metabolites synthesized by microalgae. Biomed Res Int 2015. https://doi.org/10.1155/2015/835761

De Oliveira LS, Gregoracci GB, Silva GGZ, Salgado LT, Alves-Ferreira M, Pereira RC, Thompson FL (2012) Transcriptomic analysis of the red seaweed *Laurencia dendroidea* (Florideophyceae, Rhodophyta) and its microbiome. BMC Genomics 13(1):1–13. https://doi.org/10.1186/1471-2164-13-487

De Vera CR, Díaz Crespín G, Hernández Daranas A, Montalvão Looga S, Lillsunde K-E, Tammela P, Perälä M, Hongisto V, Virtanen J, Rischer H (2018) Marine microalgae: promising source for new bioactive compounds. Mar Drugs 16(9):317. https://doi.org/10.3390/md16090317

Demay J, Bernard C, Reinhardt A, Marie B (2019) Natural products from cyanobacteria: focus on beneficial activities. Mar Drugs 17(6):320. https://doi.org/10.3390/md17060320

Diwan AD, Kandasamy D (1997) Freezing of viable embryos and larvae of marine shrimp, Penaeus semisulcatus de Haan. Aquac Res 28:947–950. https://doi.org/10.1111/j.1365-2109.1997.tb01020.x

Dixon B, Jaspars M, McDonough N, Bart J (2011) A new dawn for marine biotechnology in Europe. Biotechnol Adv 29:453–456. https://doi.org/10.1016/j.biotechadv.2011.06.010

Du Z-Y, Alvaro J, Hyden B, Zienkiewicz K, Benning N, Zienkiewicz A, Bonito G, Benning C (2018) Enhancing oil production and harvest by combining the marine alga *Nannochloropsis oceanica* and the oleaginous fungus *Mortierella elongata*. Biotechnol Biofuels 11(1):1–16. https://doi.org/10.1186/s13068-018-1172-2

Duckworth A (2009) Farming sponges to supply bioactive metabolites and bath sponges: a review. Mar Biotechnol 11(6):669–679. https://doi.org/10.1007/s10126-009-9213-2

Duckworth A, Battershill C (2003) Sponge aquaculture for the production of biologically active metabolites: the influence of farming protocols and environment. Aquaculture 221(1-4): 311–329. https://doi.org/10.1016/S0044-8486(03)00070-X

Faulkner DJ (2001) Marine natural products. Nat Prod Rep 18(1):1R–49R. https://doi.org/10.1039/B006897G

Ferro L, Gojkovic Z, Muñoz R, Funk C (2019) Growth performance and nutrient removal of a *Chlorella vulgaris-Rhizobium* sp. co-culture during mixotrophic feed-batch cultivation in synthetic wastewater. Algal Res 44:101690. https://doi.org/10.1016/j.algal.2019.101690

Freitas AC, Rodrigues D, Rocha-Santos TA, Gomes AM, Duarte AC (2012) Marine biotechnology advances towards applications in new functional foods. Biotechnol Adv 30(6):1506–1515. https://doi.org/10.1016/j.biotechadv.2012.03.006

Galasso C, Corinaldesi C, Sansone C (2017) Carotenoids from marine organisms: biological functions and industrial applications. Antioxidants 6(4):96. https://doi.org/10.3390/antiox6040096

Gamfeldt L, Lefcheck JS, Byrnes JE, Cardinale BJ, Duffy JE, Griffin JN (2015) Marine biodiversity and ecosystem functioning: what's known and what's next? Oikos 124(3):252–265. https://doi.org/10.1111/oik.01549

Gautam K, Tripathi JK, Pareek A, Sharma DK (2019) Growth and secretome analysis of possible synergistic interaction between green algae and cyanobacteria. J Biosci Bioeng 127(2):213–221. https://doi.org/10.1016/j.jbiosc.2018.07.005

Gerwick WH, Fenner AM (2013) Drug discovery from marine microbes. Microb Ecol 65(4): 800–806. https://doi.org/10.1007/s00248-012-0169-9

Giani M, Garbayo I, Vílchez C, Martínez-Espinosa RM (2019) Haloarchaeal carotenoids: healthy novel compounds from extreme environments. Mar Drugs 17(9):524. https://doi.org/10.3390/md17090524

Girão M, Ribeiro I, Ribeiro T, Azevedo IC, Pereira F, Urbatzka R, Leão PN, Carvalho MF (2019) Actinobacteria isolated from *Laminaria ochroleuca*: a source of new bioactive compounds. Front Microbiol 10:683. https://doi.org/10.3389/fmicb.2019.00683

Goettel M, Eing C, Gusbeth C, Straessner R, Frey W (2013) Pulsed electric field assisted extraction of intracellular valuables from microalgae. Algal Res 2(4):401–408. https://doi.org/10.1016/j.algal.2013.07.004

Heggeset T, Ertesvåg H, Liu B, Ellingsen TE, Vadstein O, Aasen IM (2019) Lipid and DHA-production in Aurantiochytrium sp.–responses to nitrogen starvation and oxygen limitation revealed by analyses of production kinetics and global transcriptomes. Sci Rep 9(1):1–13. https://doi.org/10.1038/s41598-019-55902-4

Hinder SL, Hays GC, Brooks CJ, Davies AP, Edwards M, Walne AW, Gravenor MB (2011) Toxic marine microalgae and shellfish poisoning in the British Isles: history, review of epidemiology, and future implications. Environ Health 10(1):1–12. https://doi.org/10.1186/1476-069X-10-54

Imhoff JF (2016) Natural products from marine fungi—still an underrepresented resource. Mar Drugs 14(1):19. https://doi.org/10.3390/md14010019

Ishika T, Laird DW, Bahri PA, Moheimani NR (2019) Co-cultivation and stepwise cultivation of *Chaetoceros muelleri* and *Amphora* sp. for fucoxanthin production under gradual salinity increase. J Appl Phycol 31(3):1535–1544. https://doi.org/10.1007/s10811-018-1718-5

Jamu D, Brummett R (2004) Opportunities and challenges for African aquaculture. Use of genetically improved and alien species for aquaculture and conservation of aquatic biodiversity in Africa 68:1

Jimenez PC, Wilke DV, Branco PC, Bauermeister A, Rezende-Teixeira P, Gaudêncio SP, Costa-Lotufo LV (2020) Enriching cancer pharmacology with drugs of marine origin. Br J Pharmacol 177(1):3–27. https://doi.org/10.1111/bph.14876

Kassam L, Subasinghe R, Phillips M (2011) Aquaculture farmer organizations and cluster management: concepts and experiences, FAO Fisheries and aquaculture technical paper. no. 563. FAO, Rome, p 90

Kini S, Divyashree M, Mani MK, Mamatha BS (2020) Algae and cyanobacteria as a source of novel bioactive compounds for biomedical applications. In: Advances in cyanobacterial biology. Elsevier, San Diego, pp 173–194. https://doi.org/10.1016/B978-0-12-819311-2.00012-7

Leal MC, Calado R, Sheridan C, Alimonti A, Osinga R (2013) Coral aquaculture to support drug discovery. Trends Biotechnol 31(10):555–561. https://doi.org/10.1016/j.tibtech.2013.06.004

Lee JH, Kim YS, Choi TJ, Lee WJ, Kim YT (2004) Paracoccus haeundaensis sp. nov., a Gramnegative, halophilic, astaxanthin-producing bacterium. Int J Syst Evol Microbiol 54(5): 1699–1702. https://doi.org/10.1099/ijs.0.63146-0

León-Bañares R, González-Ballester D, Galván A, Fernández E (2004) Transgenic microalgae as green cell-factories. Trends Biotechnol 22(1):45–52. https://doi.org/10.1016/j.tibtech.2003.11.003

Marchese P, Garzoli L, Gnavi G, O'Connell E, Bouraoui A, Mehiri M, Murphy J, Varese GC (2020) Diversity and bioactivity of fungi associated with the marine sea cucumber *Holothuria poli*: disclosing the strains potential for biomedical applications. J Appl Microbiol 129(3): 612–625. https://doi.org/10.1111/jam.14659

Markou G, Nerantzis E (2013) Microalgae for high-value compounds and biofuels production: a review with focus on cultivation under stress conditions. Biotechnol Adv 31(8):1532–1542. https://doi.org/10.1016/j.biotechadv.2013.07.011

Mayekar TS, Salgaonkar AA, Koli J, Patil PR, Chaudhari A, Pawar N, Kamble S, Giri A, Phadke GG, Kapse P (2013) Biotechnology and its applications in aquaculture and fisheries. Aquafind-Aquat Fsh Database

Misawa N, Satomi Y, Kondo K, Yokoyama A, Kajiwara S, Saito T, Miki W (1995) Structure and functional analysis of a marine bacterial carotenoid biosynthesis gene cluster and astaxanthin biosynthetic pathway proposed at the gene level. J Bacteriol 177(22):6575–6584. https://doi.org/10.1128/jb.177.22.6575-6584.1995

Mo C, Rinkevich B (2001) A simple, reliable, and fast protocol for Thraustochytrid DNA extraction. Mar Biotechnol 3(2):100–102. https://doi.org/10.1007/s101260000069

Mo C, Douek J, Rinkevich B (2002) Development of a PCR strategy for thraustochytrid identification based on 18S rDNA sequence. Mar Biol 140(5):883–889. https://doi.org/10.1007/s00227-002-0778-9

Molinski TF, Dalisay DS, Lievens SL, Saludes JP (2009) Drug development from marine natural products. Nat Rev Drug Discov 8(1):69–85. https://doi.org/10.1038/nrd2487

Morabito C, Bournaud C, Maës C, Schuler M, Cigliano RA, Dellero Y, Maréchal E, Amato A, Rébeillé F (2019) The lipid metabolism in thraustochytrids. Prog Lipid Res 76:101007. https://doi.org/10.1016/j.plipres.2019.101007

Moreno-Garcia L, Adjallé K, Barnabé S, Raghavan G (2017) Microalgae biomass production for a biorefinery system: recent advances and the way towards sustainability. Renew Sust Energ Rev 76:493–506. https://doi.org/10.1016/j.rser.2017.03.024

Nicoletti R, Andolfi A (2018). The marine-derived filamentous fungi in biotechnology. Grand Chall Mar Biotechnol 157–189. https://doi.org/10.1007/978-3-319-69075-9_4

Nirbhay K, Dolly W (2011) Microalgae as second generation biofuel. A review. Agron Sustain Dev 31:605–629. https://doi.org/10.1007/s13593-011-0018-0

Noda J, Mühlroth A, Bučinská L, Dean J, Bones AM, Sobotka R (2017) Tools for biotechnological studies of the freshwater alga *Nannochloropsis limnetica*: antibiotic resistance and protoplast production. J Appl Phycol 29(2):853–863. https://doi.org/10.1007/s10811-016-1001-6

Nymark M, Volpe C, Hafskjold MCG, Kirst H, Serif M, Vadstein O, Bones AM, Melis A, Winge P (2019) Loss of ALBINO3b insertase results in truncated light-harvesting antenna in diatoms. Plant Physiol 181(3):1257–1276. https://doi.org/10.1104/pp.19.00868

Pauly D, Christensen V, Guénette S, Pitcher TJ, Sumaila UR, Walters CJ, Zeller D (2002) Towards sustainability in world fisheries. Nature 418(6898):689–695. https://doi.org/10.1038/nature01017

Peng L, Lan CQ, Zhang Z (2016) Cultivation of freshwater green alga *Neochloris oleoabundans* in non-sterile media co-inoculated with protozoa. Can J Chem Eng 94(3):439–445. https://doi.org/10.1002/cjce.22407

Pereira F, Almeida JR, Paulino M, Grilo IR, Macedo H, Cunha I, Sobral RG, Vasconcelos V, Gaudêncio SP (2020) *Antifouling napyradiomycins* from marine-derived actinomycetes *Streptomyces aculeolatus*. Mar Drugs 18(1):63

Pitocchi R, Cicatiello P, Birolo L, Piscitelli A, Bovio E, Varese GC, Giardina P (2020) Cerato-platanins from marine fungi as effective protein biosurfactants and bioemulsifiers. Int J Mol Sci 21(8):2913. https://doi.org/10.3390/ijms21082913

Poli A, Finore I, Romano I, Gioiello A, Lama L, Nicolaus B (2017) Microbial diversity in extreme marine habitats and their biomolecules. Microorganisms 5(2):25. https://doi.org/10.3390/microorganisms5020025

Poli A, Vizzini A, Prigione V, Varese GC (2018) Basidiomycota isolated from the Mediterranean Sea–Phylogeny and putative ecological roles. Fungal Ecol 36:51–62. https://doi.org/10.1016/j.funeco.2018.09.002

Prieto-Davó A, Dias T, Gomes SE, Rodrigues S, Parera-Valadez Y, Borralho PM, Pereira F, Rodrigues CM, Santos-Sanches I, Gaudêncio SP (2016) The Madeira Archipelago as a significant source of marine-derived actinomycete diversity with anticancer and antimicrobial potential. Front Microbiol 7:1594. https://doi.org/10.3389/fmicb.2016.01594

Qin S, Lin H, Jiang P (2012) Advances in genetic engineering of marine algae. Biotechnol Adv 30(6):1602–1613. https://doi.org/10.1016/j.biotechadv.2012.05.004

Querellou J, Børresen T, Boyen C, Dobson A, Höfle M, Ianora A, Jaspars M, Kijjoa A, Olafsen J, Rigos G (2010) Marine biotechnology: a new vision and strategy for Europe. Drukkerij De Windroos NV, Beernem

Raghukumar S (2002) Ecology of the marine protists, the *Labyrinthulomycetes* (Thraustochytrids and Labyrinthulids). Eur J Protistol 38(2):127–145. https://doi.org/10.1078/0932-4739-00832

Raghukumar S (2017) Fungi in coastal and oceanic marine ecosystems. Springer, Cham

Rajasekar T, Balaji S, Kumaran S, Deivasigamani B, Pugzhavendhan S (2012) Isolation and characterization of Marine fungal metabolites against clinical pathogens. Asian Pac J Trop Dis 2:S387–S392. https://doi.org/10.1016/S2222-1808(12)60187-X

Reich M, Labes A (2017) How to boost marine fungal research: a first step towards a multidisciplinary approach by combining molecular fungal ecology and natural products chemistry. Mar Genomics 36:57–75. https://doi.org/10.1016/j.margen.2017.09.007

Rinkevich B (2011) Cell cultures from marine invertebrates: new insights for capturing endless stemness. Mar Biotechnol 13(3):345–354. https://doi.org/10.1007/s10126-010-9354-3

Rocha J, Peixe L, Gomes NC, Calado R (2011) Cnidarians as a source of new marine bioactive compounds—an overview of the last decade and future steps for bioprospecting. Mar Drugs 9(10):1860–1886. https://doi.org/10.3390/md9101860

Romano S, Jackson SA, Patry S, Dobson AD (2018) Extending the "one strain many compounds" (OSMAC) principle to marine microorganisms. Mar Drugs 16(7):244. https://doi.org/10.3390/md16070244

Rotter A, Bacu A, Barbier M, Bertoni F, Bones AM, Cancela ML, Carlsson J, Carvalho MF, Cegłowska M, Dalay MC (2020) A new network for the advancement of marine biotechnology in Europe and beyond. Front Mar Sci 7:278. https://doi.org/10.3389/fmars.2020.00278

Shah M, Mahfuzur R, Liang Y, Cheng JJ, Daroch M (2016) Astaxanthin-producing green microalga *Haematococcus pluvialis*: from single cell to high value commercial products. Front Plant Sci 7:531. https://doi.org/10.3389/fpls.2016.00531

Shahidi F (2009) Nutraceuticals and functional foods: whole versus processed foods. Trends Food Sci Technol 20(9):376–387. https://doi.org/10.1016/j.tifs.2008.08.004

Sharma AK, Mühlroth A, Jouhet J, Maréchal E, Alipanah L, Kissen R, Brembu T, Bones A, Winge P (2020) The Myb-like transcription factor Phosphorus Starvation Response (PtPSR) controls conditional P acquisition and remodeling in marine microalgae. New Phytol 225:2380–2395. https://doi.org/10.1111/nph.16248

Silber J, Kramer A, Labes A, Tasdemir D (2016) From discovery to production: biotechnology of marine fungi for the production of new antibiotics. Marine drugs 14(7):137. https://doi.org/10.3390/md14070137

Silva T, Salomon PS, Hamerski L, Walter J, Menezes RB, Siqueira JE, Santos A, Santos JAM, Ferme N, Guimarães T (2018) Inhibitory effect of microalgae and cyanobacteria extracts on influenza virus replication and neuraminidase activity. PeerJ 6:e5716. https://doi.org/10.7717/peerj.5716

Simon J-C, Marchesi JR, Mougel C, Selosse M-A (2019) Host-microbiota interactions: from holobiont theory to analysis. Microbiome 7(1):1–5. https://doi.org/10.1186/s40168-019-0619-4

Skjånes K, Rebours C, Lindblad P (2013) Potential for green microalgae to produce hydrogen, pharmaceuticals and other high value products in a combined process. Crit Rev Biotechnol 33(2):172–215. https://doi.org/10.3109/07388551.2012.681625

Slattery SS, Diamond A, Wang H, Therrien JA, Lant JT, Jazey T, Lee K, Klassen Z, Desgagné-Penix I, Karas BJ (2018) An expanded plasmid-based genetic toolbox enables Cas9 genome editing and stable maintenance of synthetic pathways in *Phaeodactylum tricornutum*. ACS Synth Biol 7(2):328–338. https://doi.org/10.1021/acssynbio.7b00191

Smerilli A, Orefice I, Corato F, Gavalás Olea A, Ruban AV, Brunet C (2017) Photoprotective and antioxidant responses to light spectrum and intensity variations in the coastal diatom S keletonema marinoi. Environ Microbiol 19(2):611–627. https://doi.org/10.1111/1462-2920.13545

Stévant P, Rebours C, Chapman A (2017) Seaweed aquaculture in Norway: recent industrial developments and future perspectives. Aquac Int 25(4):1373–1390. https://doi.org/10.1007/s10499-017-0120-7

Subramani R, Sipkema D (2019) Marine rare actinomycetes: a promising source of structurally diverse and unique novel natural products. Mar Drugs 17(5):249. https://doi.org/10.3390/md17050249

Suganya T, Varman M, Masjuki H, Renganathan S (2016) Macroalgae and microalgae as a potential source for commercial applications along with biofuels production: a biorefinery approach. Renew Sust Energ Rev 55:909–941. https://doi.org/10.1016/j.rser.2015.11.026

Thakur NL, Thakur AN (2006) Marine biotechnology: an overview. Indian J Biotechnol 5:263–268

Toyama T, Hanaoka T, Yamada K, Suzuki K, Tanaka Y, Morikawa M, Mori K (2019) Enhanced production of biomass and lipids by *Euglena gracilis* via co-culturing with a microalga growth-promoting bacterium, *Emticicia* sp. EG3. Biotechnol Biofuels 12(1):1–12. https://doi.org/10.1186/s13068-019-1544-2

Vallet M, Vanbellingen QP, Fu T, Le Caer J-P, Della-Negra S, Touboul D, Duncan KR, Nay B, Brunelle A, Prado S (2017) An integrative approach to decipher the chemical antagonism between the competing endophytes *Paraconiothyrium variabile* and *Bacillus subtilis*. J Nat Prod 80(11):2863–2873. https://doi.org/10.1021/acs.jnatprod.6b01185

Varshney P, Mikulic P, Vonshak A, Beardall J, Wangikar PP (2015) Extremophilic micro-algae and their potential contribution in biotechnology. Bioresour Technol 184:363–372. https://doi.org/10.1016/j.biortech.2014.11.040

Vatsos IN, Rebours C (2015) Seaweed extracts as antimicrobial agents in aquaculture. J Appl Phycol 27(5):2017–2035. https://doi.org/10.1016/j.aquaculture.2016.08.001

Ventura P, Toullec G, Fricano C, Chapron L, Meunier V, Röttinger E, Furla P, Barnay-Verdier S (2018) Cnidarian primary cell culture as a tool to investigate the effect of thermal stress at cellular level. Mar Biotechnol 20(2):144–154. https://doi.org/10.1007/s10126-017-9791-3

Vu MTT, Douëtte C, Rayner TA, Thoisen C, Nielsen SL, Hansen BW (2016) Optimization of photosynthesis, growth, and biochemical composition of the microalga *Rhodomonas salina*—an established diet for live feed copepods in aquaculture. J Appl Phycol 28(3):1485–1500. https://doi.org/10.1007/s10811-015-0722-2

Wang G, Wang X, Liu X, Li Q (2012) Diversity and biogeochemical function of planktonic fungi in the ocean. Biol Mar Fungi 53:71–88. https://doi.org/10.1007/978-3-642-23342-5_4

Wang Q, Jin W, Zhou X, Guo S, Gao S-H, Chen C, Tu R, Han S-F, Jiang J, Feng X (2019) Growth enhancement of biodiesel-promising microalga *Chlorella pyrenoidosa* in municipal wastewater by polyphosphate-accumulating organisms. J Clean Prod 240:118148. https://doi.org/10.1016/j.jclepro.2019.118148
Wijffels RH (2008) Potential of sponges and microalgae for marine biotechnology. Trends Biotechnol 26(1):26–31. https://doi.org/10.1016/j.tibtech.2007.10.002
Wilkins LG, Leray M, O'Dea A, Yuen B, Peixoto RS, Pereira TJ, Bik HM, Coil DA, Duffy JE, Herre EA (2019) Host-associated microbiomes drive structure and function of marine ecosystems. PLoS Biol 17(11):e3000533. https://doi.org/10.1371/journal.pbio.3000533
Williams PG (2009) Panning for chemical gold: marine bacteria as a source of new therapeutics. Trends Biotechnol 27(1):45–52. https://doi.org/10.1016/j.tibtech.2008.10.005
Yarden O (2014) Fungal association with sessile marine invertebrates. Front Microbiol 5:228. https://doi.org/10.3389/fmicb.2014.00228
Yu X, Chen L, Zhang W (2015) Chemicals to enhance microalgal growth and accumulation of high-value bioproducts. Front Microbiol 6:56. https://doi.org/10.3389/fmicb.2015.00056

Grouper Hybridization: An Effective Biotechnological Tool for Food Security

3

Fui Fui Ching, Norfazreena Mohd Faudzi, and Shigeharu Senoo

Abstract

Fish is one of the most important animal protein sources for human consumption. However, it has been reported that the wild fish population is rapidly depleted. Therefore, the aquaculture industry getting high attention to producing a significant number of fish to replace wild fish stock. To produce more abundant, resilient, and healthy fish, application of biotechnology has been used in the aquaculture industry. Through biotechnology, cultured fish are improved for a variety of traits, including growth performance, feed conversion, disease resistance, tolerance to fluctuating water quality, excretion rate, flesh quality, fish quality, fecundity, and reproduction. Hybridization is an excellent example of the significant contribution of biotechnology in aquaculture, which offers the hope of producing aquatic organisms with valuable traits, with the progeny bearing the characteristics of hybrid vigour or positive heterosis. Mass production of hybrid grouper was first established in 2006 by Universiti Malaysia Sabah through hybridization of tiger grouper, *E. fuscoguttatus* ♀ × giant grouper, *E. lanceolatus* ♂. The method of grouper hybridization is inexpensive and more applicable to be used by fish farmers. The success of mass-produced hybrid grouper has reduced the cost of production, thus lowering the price and making them more accessible and affordable in achieving food security.

F. F. Ching (✉) · N. Mohd Faudzi
Borneo Marine Research Institute, Universiti Malaysia Sabah, Kota Kinabalu, Sabah, Malaysia
e-mail: cfuifui@ums.edu.my

S. Senoo
Borneo Marine Research Institute, Universiti Malaysia Sabah, Kota Kinabalu, Sabah, Malaysia

UMS-Kindai Aquaculture Development Centre, Kindai University, Shirahama, Wakayama, Japan

M. D. Shah et al. (eds.), *Marine Biotechnology: Applications in Food, Drugs and Energy*, https://doi.org/10.1007/978-981-99-0624-6_3

3.1 Introduction

More than 7.8 billion people in developed and developing countries will consume at least 17% of their daily average animal protein from fish in 2020 (FAO 2022). In addition to protein, fish is also recognized as an excellent source of essential fatty acids and micronutrients that are lacking in many diets (Toppe 2014). Fish is one of the most effective recyclers of feed into a high-quality food. It also produces less carbon dioxide than other types of animal production systems. Although the ocean appears to have no limits, it has long been noted that there are not enough fish to feed the ever-growing human population. In 2020, FAO issued a stark warning about the rapid decline of wild fish populations, which will have a significant impact on providing protein-rich food for the world's growing population, and highlighted the need to safeguard marine life as well (FAO 2020).

Due to the ongoing depletion of fish stocks, aquaculture—the fastest food production of various aquatic animals such as fish, crustaceans, molluscs, and aquatic plants—has become an important global sector that currently provides a significant amount of fish that are consumed to replace wild stocks (Tacon 2003). However, the global community has not made much headway in ensuring that everyone has access to healthy food throughout the year or in eliminating all forms of malnutrition (FAO 2021). Under these circumstances, aquaculture production must be enhanced several folds to meet the growing demand for affordable fish in larger quantities. Aquaculture has made a significant contribution to increasing global fish supply compared from the year 1986 (14.9 million tonnes, 14.6% of total fisheries and aquaculture production) to 2018 (82.1 million tonnes, 46.0% of total fisheries and aquaculture production) (Fig. 3.1) (FAO 2022). However, high prices of cultured fish remain the greatest potential risk to global food security due to high

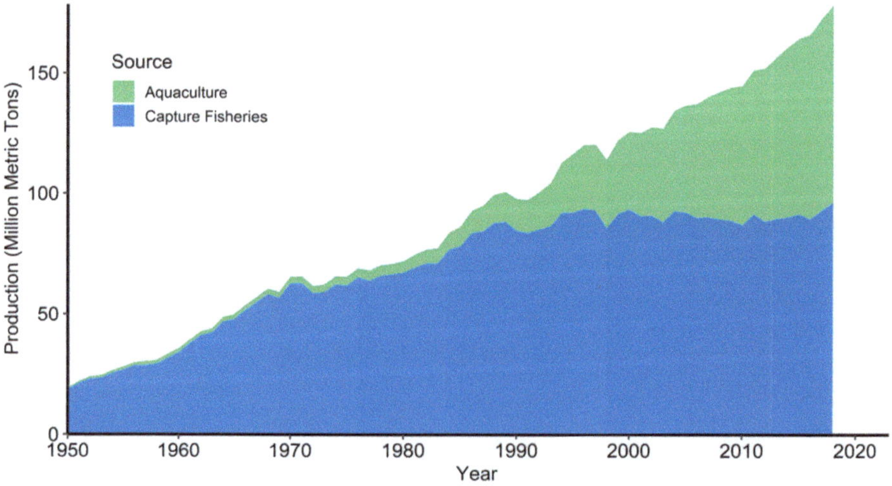

Fig. 3.1 World production of capture fisheries and aquaculture from 1950 to 2018. Source: FAO (2022)

production costs, low production output, disease outbreaks, poor water quality, high mortality, and slow fish growth (Boyd et al. 2022). Many food industries, including aquaculture, have increased the number of productivities by application of biotechnology, particularly through the application of genetic, cellular, and molecular technologies which allow fish farmers to produce more abundant, resilient, and healthier fish (Altunok et al. 2012).

3.2 Biotechnology in Aquaculture

Aquaculture is the breeding, raising, and harvesting of various aquatic animals including fish, crustaceans, shellfish, and aquatic plants. Fish breeding, in particular, can help improve food security by reducing pressure on wild fish stocks and producing a high-quality protein that would otherwise be difficult to obtain (Yue and Shen 2022). The most important and widely used biotechnology applications in aquaculture include the production of monosex, uniparental, and polyploid populations; molecular biology; transgenic fish; gene banking; and the development of natural products from aquatic organisms (Nasim 2010; Nwokwa 2012; Omole 2017). Through the applications of biotechnology, cultured fish are improved for a variety of traits, including growth performance, feed conversion, disease resistance, tolerance to fluctuating water quality, excretion rate, flesh quality, fish quality, fecundity, and reproduction (Dunham 2004). Among many successful uses of biotechnology in aquaculture is the production of all-male and all-female giant freshwater prawns (*Macrobrachium rosenbergii* De Man, 1879) (Levy et al. 2017; Aflalo et al. 2006) which has been shown to generate significantly greater growth performance, production yield, and profitability when compared to mix-sex production. This breakthrough is possible by combining the novel biotechnology knowledge on sex reversal with a mini-surgical technique for androgenic gland ablation as developed by Sagi et al. (1990). As a result, the all-male culture of giant freshwater prawn farming has been the most prolific and profitable system for decades, ensuring food security in many parts of the world where it has been introduced.

Another key contribution of biotechnology in aquaculture is the conservation of genetic resources. A decline in fish genetic diversity could result in extinction, and molecular markers, in particular, a powerful tool in biotechnology, have proven useful not just in breeding programmes, but also in conservation planning based on the genetic diversity of distinct fish species (Khatei et al. 2022). In most cases, wild fish are harvested as candidates for breeding in aquaculture owing to their high genetic variability and this helps to sustain aquaculture in a variety of ways (Li 2022). First, wild alleles can be reintroduced into farmed populations if inbreeding occurs because of as a result of bottleneck effects. Second, genetic resources from selected strains can be collected if unique features, such as disease resistance and fast growth, are desired. Therefore, maintaining and creating appropriate strains for sustainable aquaculture depend on the maintenance of biodiversity and genetic diversity of aquatic species. Biotechnology also plays a significant part in aquaculture fish health management, ensuring that protein from healthy fish is available for human consumption (Adams and Thompson 2006). Extremely low levels of aquatic pathogens are possible to be detected, identified, and quantified using molecular technologies such as the polymerase chain reaction (PCR), real-time PCR, and nucleic acid sequence-based amplification (NASBA), which have also given multiplex pathogen and host response screening a new dimension (Gil 2007; Hong et al. 2007; Chapela et al. 2018). The application of biotechnology in aquaculture also includes transgenesis, the introduction of exogenous genetic material (DNA) into a host genome, resulting in its stable maintenance, transmission, and expression as seen in both cold water (salmon, trout) and warm water (tilapia, carp) species (Maclean et al. 2002; Thorgaard et al. 2002; Williams et al. 2010; Fitzpatrick et al. 2011; Wakchaure et al. 2015; Kurdianto et al. 2016).

Aquaculture is the production of food fish, and grouper hybridization by biotechnology is the most successful aquaculture development, propelling the seafood industry to new heights in terms of supplying an appropriate supply of food fish for the world's rising population (Ching et al. 2018). Hybridization is simply one method for improving aquaculture output; it also requires an understanding of the broodstock's genetic structure, good broodstock management, and monitoring of the progeny's viability and fertility. Hybridization is a genetic alteration in which genes are transferred between species; the ramifications for biodiversity conservation and regulation are examined (Bartley et al. 2000).

3.3 Hybridization

Hybridization is an excellent example of the significant contribution of biotechnology in aquaculture, which offers the hope of producing aquatic organisms with valuable traits, with the progeny bearing the characteristics of hybrid vigour or positive heterosis (Ching et al. 2018). In general, preferred new progeny can result in a shorter growth and production cycle, higher survival, fast growth, better food conversion, improved flesh quality, high disease resistance, and the ability to tolerate a wider range of rearing environments, as has been reported in several marine and freshwater hybrid fish species (Bartley et al. 2000; Bunlipatanon and U-taynapun 2017). Hybridization is the process of crossing two or more closely related species. Crossing striped bass (*Morone saxatilis* Walbaum, 1792) × white bass (*M. chrysops* Rafinesque, 1820) (Zhang et al. 1994), channel catfish (*Ictalurus punctatus* Rafinesque, 1818) × blue catfish (*Ictalurus furcatus* Valenciennes, 1840) (Argue et al. 2014) or Nile tilapia (*Oreochromis niloticus*) × blue spotted tilapia (*O. leucostictus*) (Diedericks et al. 2021), tiger grouper (*Epinephelus fuscoguttatus*) × giant grouper (*E. lanceolatus*) (Ch'ng and Senoo 2008), orange-spotted grouper (*E. coioides*) × giant grouper (*E. lanceolatus*) (Koh et al. 2010) are just a few breakthroughs in aquaculture that have been shown to outperform parent fish. A hybrid fish with selected or favoured characteristics of each parent is one of the goals of aquaculture. Positive heterosis, or hybrid vigour, is the ultimate breeding goal when a hybrid exhibits traits that are better than either of its parents.

Dunham and Smitherman (1983) and Dunham (1996) reported significant improvement in the growth rates of 55% and 22%, respectively, in channel catfish (*I. punctatus*) and rainbow trout (*Oncorhynchus mykiss*) when hybridization techniques were used. Hybrid Coho salmon (*O. kisutch*) × Chinook salmon (*O. tshawytscha*) also attained significantly higher survival rates (50% above) compared to parental fish as reported by Blanc and Chevassus (1982). Several hybrid groupers have also been observed to have a lower food conversion ratio than parental fish, which helps aquaculture reduce overall production costs (Mohd Faudzi et al. 2018; Abang Zamhari and Yong 2021). In some situations, hybrids have been observed to exhibit greater tolerance to a wide range of water quality (Othman et al. 2015; Sutthinon et al. 2015), allowing for easier and more practical aquaculture production techniques. Other than in captivity, hybridization also occurs widely in fishes under natural conditions (Hubbs 1955) but is not well documented compared to those in captivity through aquaculture activity. External fertilization, the unequal abundance of the two parental species, competition for spawning grounds, and declining habitat complexity have all been identified as factors leading to the high occurrence of natural hybridization among closely related fish species (Campton 1987). It has been difficult to explore and hypothesize on the survivability of hybrid fish in nature because no scientific research on the subject has been published yet.

3.4 Grouper Hybridization

A hybrid grouper was first produced on a laboratory scale in 1999 between the brown marbled grouper (*E. fuscoguttatus*) and camouflage grouper (*E. polyphekadion*) followed by hybrid gold blotch grouper (*E. costae*), dusky grouper (*E. marginatus*) (Glamuzina et al. 2001) to combat the issue of high mortality of grouper larvae and complications of first feeding. The possibility of grouper hybridization was started by this breakthrough. However, no mass production has been reported since its first production on a laboratory scale.

Mass production of hybrid grouper was first established in 2006 through a crossbreed between tiger grouper (*E. fuscoguttatus*) and giant grouper (*E. lanceolatus*) (Ch'ng and Senoo 2008) (Fig. 3.2). Since the first production of hybrid grouper TGGG, hybrid groupers have taken the Asian aquaculture industry by storm due to their production success and superior organoleptic features, this unique hybrid grouper has quickly garnered appeal among aquaculturists and seafood customers, resulting in a high economic value (Ching et al. 2018). More than ten hybrid groupers have been produced to date, including hybrids between orange-spotted grouper (*E. coioides*), coral grouper (*E. corallicola*), mouse grouper (*Cromileptes altivelis*), camouflage grouper (*E. polyphekadion*), and red spotted grouper (*E. akaara*), among others (Liufu et al. 2007; Koh et al. 2010; Addin and Senoo 2011; Huang et al. 2016) (Table 3.1). Due to its fast growth, high survival, better feeding performances, and excellent tolerance to a wide range of rearing conditions, TGGG has been recognized as the most effective hybrid combination (Ch'ng and Senoo 2008; Othman et al. 2015; Shapawi et al. 2019).

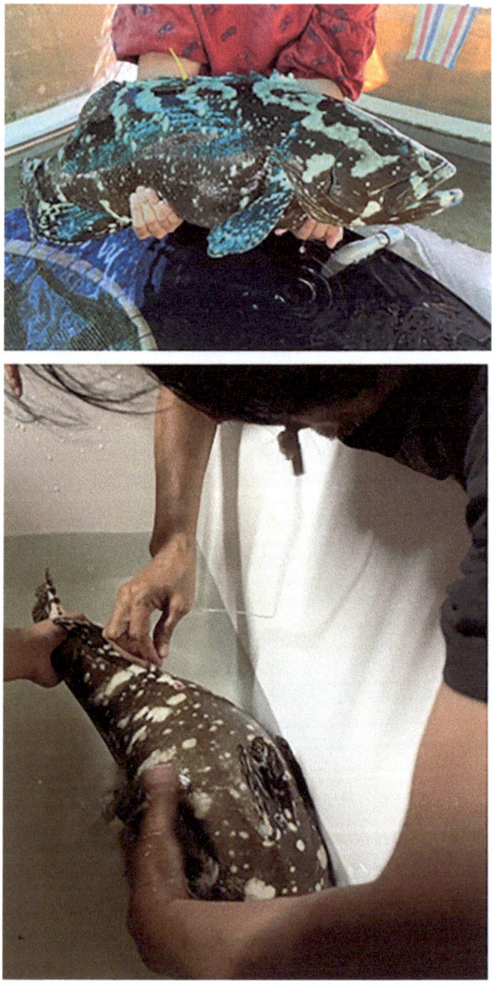

Fig. 3.2 A selected female grouper was cannulated to assess the maturity stage

The fast growth, high survival, and low FCR of hybrid TGGG particularly have been cited as the key reasons for its widespread acceptance among farmers (Tan 2021; Nankervis et al. 2021). Hybrid grouper has been mass-produced in Southeast Asia and China (Shapawi et al. 2019; Yang et al. 2022), and as a result, the price has been reduced and making them more accessible and affordable to low- and middle-income consumers. One of the most significant components in achieving food security is the availability of inexpensive food fish, which can be easily achieved owing to the established mass production of hybrid TGGG. Pure grouper species such as mouse grouper (*Cromileptes altivelis* Valenciennes, 1828), tiger grouper (*E. fuscoguttatus*), and red grouper (*E. akaara*) are typically considered luxury food items (Sutina et al. 2017) and commonly exported as live reef food fish (LRFF) (Sadovy 2003; Tupper and Sheriff 2008) and are clearly out of reach for many in

Table 3.1 List of hybrid groupers produced in the world

Species	Year	References
E. aeneus ♀ × *E. marginatus* ♂	1998	Glamuzina et al. (1999)
E. costae ♀ × *E. marginatus* ♂	2001	Glamuzina et al. (2001)
E. fuscoguttatus ♀ × *E. lanceolatus* ♂	2006	Ch'ng and Senoo (2008)
E. coioides ♀ × *E. fuscoguttatus* ♂	2007	Koh et al. (2008)
E. coioides ♀ × *E. lanceolatus* ♂	2007	Koh et al. (2010)
Cromileptes altivelis ♀ × *E. fuscoguttatus* ♂	2008	Addin and Senoo (2011)
Cromileptes altivelis ♀ × *E. lanceolatus* ♂	2008	Addin and Senoo (2011)
E. polyphekadion ♀ × *E. fuscoguttatus* ♂	2011	Addin and Senoo (2011)
E. corallicola ♀ × *E. fuscoguttatus* ♂	2011	Addin and Senoo (2011)
E. moara ♀ × *E. lanceolatus* ♂	2018	Chen et al. (2018)
E. bruneus ♀ × *E. akaara* ♂	2020	Kang et al. (2020)

areas where food security is a long way off. In comparison with hybrid TGGG, pure groupers have a greater mortality rate (Ching et al. 2016), slower growth (Pamungkas and Sari 2021), and a high deformation rate (Nagano et al. 2007), which have resulted in a scarcity of supply from the aquaculture industry and causes price hike.

The hybrid grouper production process has been well established and has undergone various improvements that made it possible for large-scale production in many Southeast Asian countries since its first production in 2006. The method established by Ch'ng and Senoo (2008) for hybrid grouper production has been praised for being simple to use, practical, repeatable, and affordable. Its rapid growth may also be seen as an example of how biotechnology advancements improve aquaculture production.

3.4.1 Method to Produce Grouper Hybridization

3.4.1.1 Selection

A matured female grouper was selected after being subjected to 2 months of intensive broodstock management. Potential female grouper broodstock was anaesthetized (25 ppm) with Transmore (alpha-methylquinoline) (Nika) before selection as prevention from stress during handling. A mature female typically has a larger abdomen as compared to an immature female and its sexual maturity was assessed through cannulation and checking on the abdominal area. The cannulation method is necessary because the grouper showed undifferentiated sexual maturity with immature females (Fig. 3.2). A biopsy analysis was used to confirm the presence of oocytes. The female grouper that was selected has a soft and inflated abdomen, a reddish urogenital papilla, and white unripe eggs that are oozing out of the oviduct by cannulation. Meanwhile, the potential male grouper broodstock is genially able to ooze out milt with gentle pressure near the genital pore. The cannulation is not essential to be performed on male grouper.

3.4.1.2 Hormone Injection

Hormones are chemical messengers that facilitate communication between various cell types (Hoga et al. 2018). These cells identify and carry out their specific functions through receptors, which are protein structures with molecular recognition expertise. Following the proximity and the hormone–receptor contact, several biochemical processes that result in certain biological responses take place. In aquaculture, hormones are used for artificial reproduction and sex reversal (Hoga et al. 2018). It is generally known that hormone injections are used to accelerate grouper females' ultimate maturation and ovulation (Marino et al. 2001; Park et al. 2002).

The selected female grouper was injected with commercially available human chorionic gonadotropin (HCG) (Profasi, Laboratories Serono, Switzerland) at a dosage of 500 IU/kg intraperitoneal injection at the basal part of the pectoral fin (Fig. 3.3). The fish was kept apart in separate net cages (1.5 × 1.5 × 1.5 m) or fibreglass tanks (1 tonne). During the isolation period, the water's temperature (°C), salinity (ppt), dissolved oxygen (DO) (mg/l), and pH were all between 28.0 and 29.5 °C, 29.0–30.0 ppt, 7.0–7.8 mg/l, and 6.0–7.5, respectively. The maturation stage of the oocytes was observed using the cannulation method (Fig. 3.4). When stripping begins, the female grouper's abdomen oozes out eggs when applied slight pressure (estimated at 44 h after hormone treatment).

3.4.1.3 Cryopreservation

The aquaculture sector has benefited from the cryopreservation of fish spermatozoa to facilitate artificial insemination in the production of fry. Sperm cryopreservation is

Fig. 3.3 A selected female grouper was injected at the basal part of the pectoral fin with the human chorionic gonadotropin (HCG) hormone

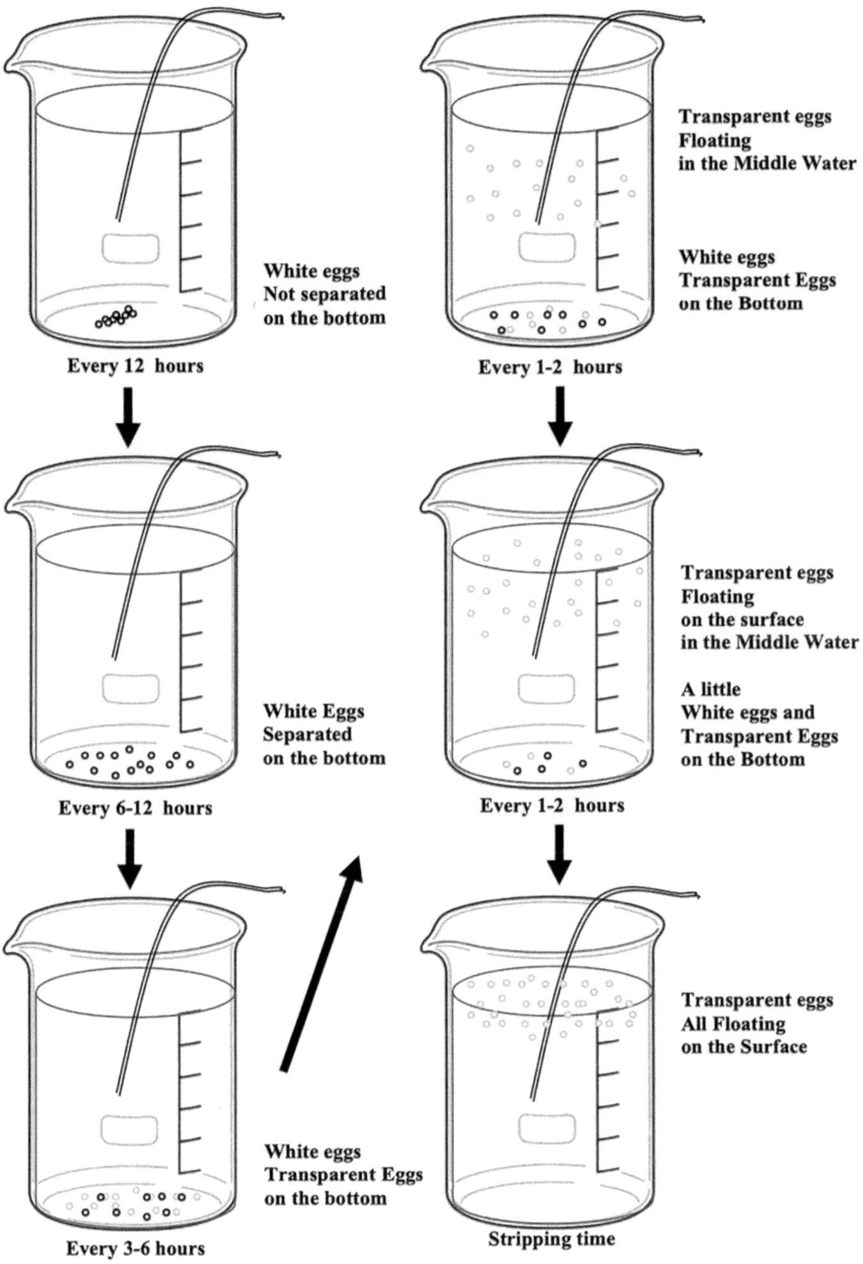

Fig. 3.4 Observation of oocytes through cannulation method

Fig. 3.5 Sperm of *E. lanceolatus* was collected for preservation

widely used to promote alternative methods of managing broodstock or to offer gametes year-round for fertilization outside of the typical reproductive seasons. As groupers are mostly protogynous hermaphrodite species, with individuals growing first as female and later as male, the collection of sperm (Fig. 3.5) is difficult due to the restricted availability of mature broodstock which hindered the development of seed production for this fish. To synchronize the availability of both sexes, cryopreservation enables long-term storage of sperm (Gwo 2011). It also permits the safe transportation of sperm samples over long distances without contamination (Koh et al. 2011). Additionally, this technique aids in the creation of hybrids (Kiriyakit et al. 2011; Koh et al. 2011).

Cryopreservation of fish sperm using liquid nitrogen has been widely studied (Agarwal 2011). The method of cryopreservation differs depending on the fish species. The method involves a series of interrelated processes; thus, errors in any particular process fail in the entire process. Sperm cryopreservation also requires the use of special types of equipment that are costly, challenging to procure and need expert handling by fish farmers. Thus, failure to scale up the cryopreservation of sperm has prevented the commercialization of cryopreserved sperm in the aquaculture industry (Benson et al. 2012).

Storage of refrigerated sperm is a simple and inexpensive method that can facilitate the management and reproduction programme in aquaculture. The short-term preservation method is more applicable to be used by the fish farmer to overcome the problem of shortage of sperm sources that can be obtained from other places or as one of the strategies to reduce workloads during production. Sperm preservation using an ice method was introduced which can keep the sperm sample within 1-month period and is easy to transport. To avoid water

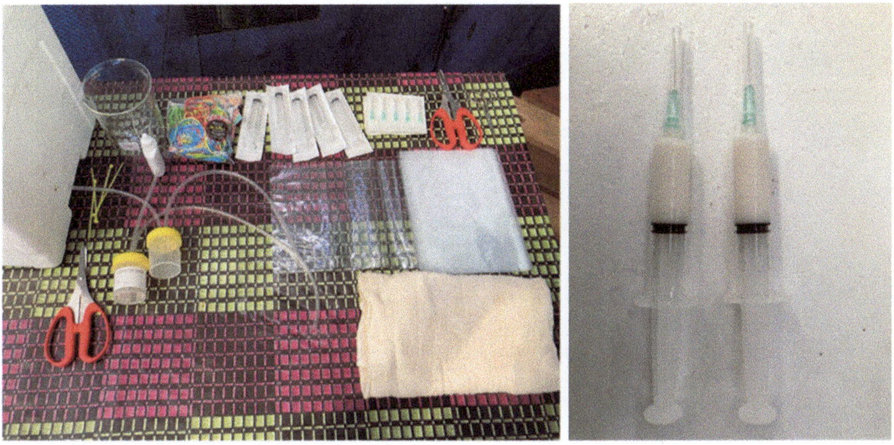

Fig. 3.6 Sperm preservation using the ice method

contamination, the fish sperm was preserved in dried syringes and packaged in plastic (Fig 3.6). The sperm was stored in a Styrofoam box with ice and kept refrigerated until used.

3.4.1.4 Egg and Larval Development of Hybrid Grouper

Most of the findings demonstrated that the eggs of hybrid grouper developed faster than non-hybrid grouper (Table 3.2 and Fig. 3.7). A hybrid of *E. fuscoguttatus* × *E. lanceolatus* hatched 18 h after fertilization (hAF) (Ch'ng and Senoo 2008). However, the parental species hatched at 24 hAF (*E. fuscoguttatus*) and 30 hAF (*E. lanceolatus*) (Ching et al. 2012; Garcia-Ortega et al. 2014). The hybrid of *E. polyphekadion* with *E. fuscoguttatus* and *E. corallicola* with *E. fuscoguttatus* also showed shorter periods to hatch compared to their parental fish (Addin and Senoo 2011). Other than that, a hybrid of *E. marginatus* × *E. aeneus* hatched in 28 hAF (Glamuzina et al. 1999), which is 5 h earlier than non-hybrid *E. marginatus* (Glamuzina et al. 1998). On the other hand, a hybrid of *E. coioides* × *E. fuscoguttatus* and *E. coioides* × *E. laceolatus* was reported to hatch at 18 hAF which means the hatching time was close to that of the parental fish, *E coioides* (20 hAF) (Kawahara et al. 1997; Koh et al. 2008, 2010).

According to previous studies, the size of hybrid eggs and larvae was reported to range between 0.75 to 0.90 mm and 1.50 to 2.00 mm (Glamuzina et al. 1998, 1999; Ch'ng and Senoo 2008; Koh et al. 2008, 2011). The size of eggs and larvae for hybrid fish was comparable to that observed in the parental species. High rates of larval production, fertilization, and hatching are strongly predicted by these factors. Hybridization of *E. fuscoguttatus* and *E. lanceolatus* showed greater fertilization and hatching rate that averaged 86.8% and 87.2%, respectively. This hybrid grouper is the most successful combination with higher survival and faster growth (Ch'ng and Senoo 2008), high tolerance towards disease, and a wide range of parameters (Othman et al. 2015; De et al. 2016).

Table 3.2 Summary of findings on parental fish and its hybrids

Species	Fertilization rate (%)	Egg diameter (mm)	Hatching hour	Hatching rate (%)	Larval size (mm)	References
E. fuscoguttatus	NA	NA	24	86.7	NA	Ching et al. (2012)
Cromileptes altivelis	90.5	0.80–0.86	24	73.8	NA	Senoo et al. (2002, 2004)
E. coioides	NA	NA	20	NA	NA	Kawahara et al. (1997)
E. marginatus	NA	0.85 ± 0.04	33	NA	1.52 ± 0.07	Glamuzina et al. (1998)
E. lanceolatus	NA	0.89 ± 0.01	30	NA	NA	Garcia-Ortega et al. (2014)
E. fuscoguttatus × *E. lanceolatus*	86.8	0.84 ± 0.03	18	87.2	2.00 ± 0.30	Ch'ng and Senoo (2008)
E. coioides × *E. lanceolatus*	91.0	0.81 ± 0.02	18	33.6	1.53 ± 0.01	Koh et al. (2010)
E. coioides × *E. fuscoguttatus*	93.9	0.83 ± 0.02	18	93.8	1.52 ± 0.01	Koh et al. (2008)
E. polyphekadion × *E. fuscoguttatus*	51.0	0.72 ± 0.02	20	23.3	1.62	Addin and Senoo (2011)
E. corallicola × *E. fuscoguttatus*	75.0	0.74 ± 0.02	20	50.3	1.83	Addin and Senoo (2011)
E. aeneus × *E. marginatus*	70	0.855	28	NA	1.76 ± 0.08	Glamuzina et al. (1999)

NA not available

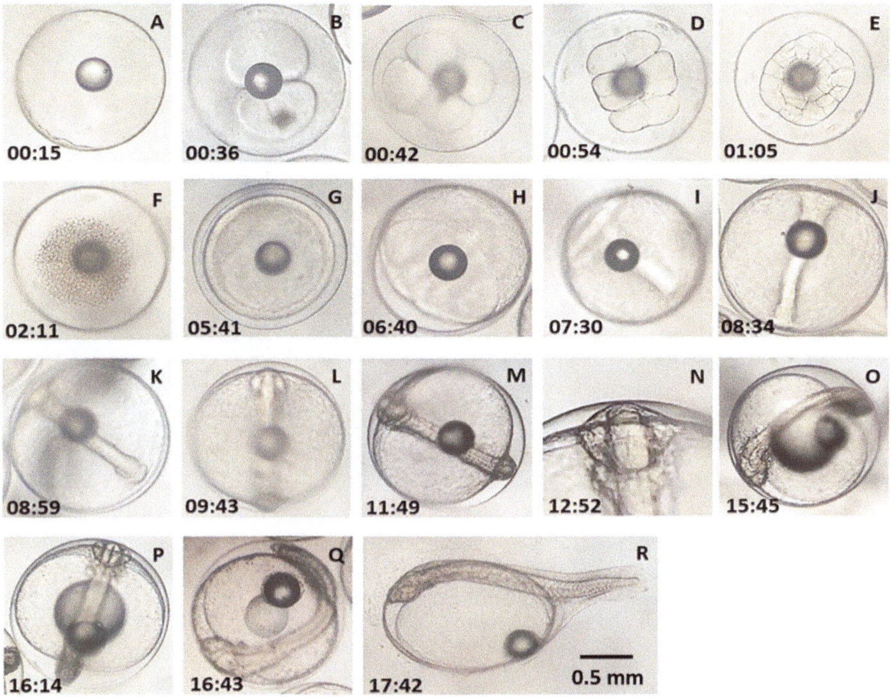

Fig. 3.7 Hybrid grouper of *E. fuscoguttatus* × *E. lanceolatus* embryo development; (**a**) fertilized egg; (**b**) 2-cell stage; (**c**) 4-cell stage; (**d**) 8-cell stage; (**e**) 32-cell stage; (**f**) morula stage; (**g**) gastrula stage; (**h**) embryo formation commenced; (**i**) blastopore nearly closed; (**j**) blastopore completely closed; (**k**) head and myomere formed; (**l**) optic vesicles appeared; (**m**) tail separated from yolk sac; (**n**) lens vesicle appeared; (**o**) embryo commenced moving; heart formed; (**p**) otocyst vesicle appeared; (**q**) hatching started; (**r**) hatched larvae

Numerous studies on hybridization have shown that most larval traits including the size of the yolk sac, the shape of the head and body, the presence of oil globules and pigmentation are almost identical to those of the *Epinephelus* species (Figs. 3.8 and 3.9). The best morphological traits for identification and rearing techniques were provided by the similarities of the grouper species. The findings suggest that the genetic characteristics of the parental species and environmental factors play a role in the success of hybridization for breeding programmes.

3.5 Biotechnology to Improve Quality of Grouper

3.5.1 Feed

The sustainability of hybrid grouper culture is limited due to the dependency on wild-caught prey fish as feed rather than commercially produced feed pellets. Additionally, Chor et al. (2020) highlighted that the usage of prey fish in sea cages

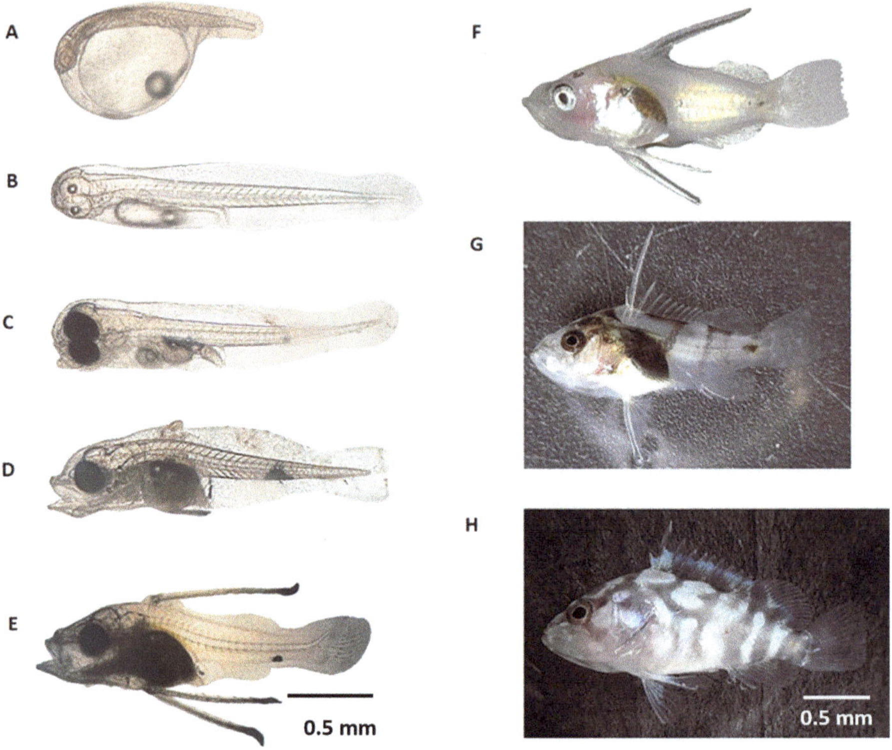

Fig. 3.8 Early stage of larval development on hybrid grouper, *E. fuscoguttatus* × *E. lanceolatus*

has led to environmental issues such as water pollution brought on by an increased nutrient load from uneaten feed and a high risk of disease transfer to the cultured fish. Based on its benefits for better growth and feed utilization of fish, pellet feed is therefore recommended as a sustainable replacement for the use of prey fish. It is obvious that improving the cost effectiveness of pelleted feeds for hybrid grouper farming will boost overall farm productivity and lessen overfishing pressure. Generally, fish grow effectively on natural feed in sufficient quantities to meet their metabolic needs. Amino acids, fatty acids, vitamins, minerals, and energy-providing macronutrients such as proteins, lipids, and carbohydrates are the important nutrients that support the growth and overall wellness of cultured fish (Hixson 2014). As hybrid grouper is a highly carnivorous species, a significant proportion of proteins and lipids are required for normal growth.

The fish use the amino acids from dietary proteins to synthesize new protein, maintain the protein that already exists in cells and tissues, and mobilize excess protein to energy. Since groupers have a restricted capacity for utilizing lipid-based energy, it is generally accepted that they require higher protein intakes than other fish. Meanwhile, dietary lipids provide energy and essential fatty acids that act as prostaglandin and steroid hormones. In most of the studies that have been conducted,

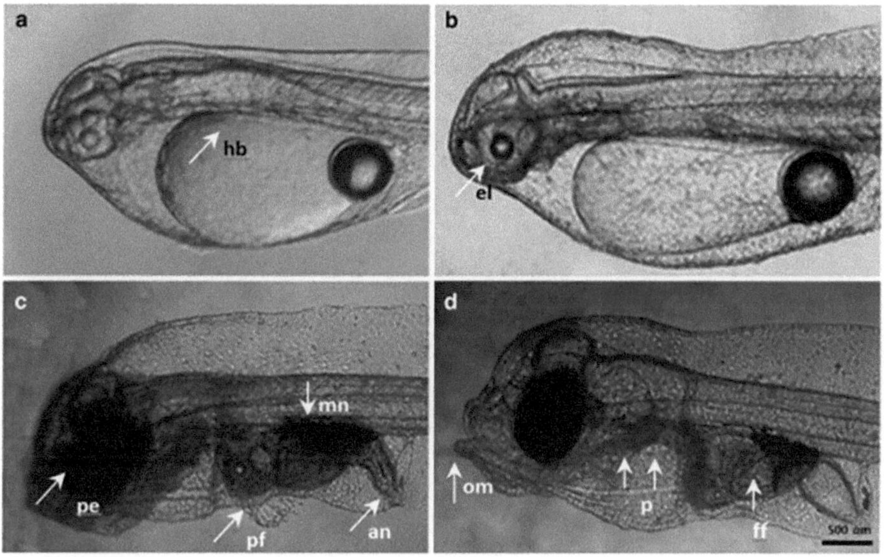

Fig. 3.9 Early stage of larval development on hybrid grouper; (**a**) visible heartbeat (hb) detected; (**b**) the onset of pigmentation around the eye lens (el); (**c**) well-pigmented eyes (pe), the first appearance of pectoral fins (pf), opened anus (an), and visible melanophores (mn); (**d**) functional and opened mouth (om), peristaltic movement detected in the oesophagus (p), first feeding detected (ff)

Table 3.3 Protein, lipid, and energy requirement of juvenile hybrid grouper in captivity

Body weight (g)					
Initial	Final	Protein (%)	Lipid (%)	Energy ($MJkg^{-1}$)	References
4.7	–	45–55	7	–	Jiang et al. (2015)
2.6	64	50	14	20.9	Rahimnejad et al. (2015)
3.6	34	53.5	11.1	33.2	Jiang et al. (2016)
6.6	111	50	16.5	20.5	Yong et al. (2019a)

it is suggested that hybrid grouper has a similar requirement for protein (40–50%) and lipid (8–16%) to other pure species (Jiang et al. 2015, 2016; Rahimnejad et al. 2015; Yong et al. 2019a) (Table 3.3). When the protein content in pellet feed is below the optimum level in captivity, the fish will consume more feed to obtain the necessary protein calories for development and metabolism. Optimizing feed composition promotes growth and aids in lowering feed amount which has economic implications.

According to the previous study, hybrid grouper has an improvement in weight gain, feed utilization, and protein retention efficiency when the dietary lipid increased from 8 to 16% (Yong et al. 2019a, b). Higher levels of lipid in the feed are commonly reported to reduce the feed intake of fish (Williams et al. 2004; Luo et al. 2005) but it was not observed in hybrid grouper (Yong et al. 2019a). The ability of the hybrid grouper to consume lipids efficiently and store lipids for later use was demonstrated by the higher level of lipid provided, which correlates to the higher lipid maintained in the fish body. Rahimnejad et al. (2015) reported a similar observation when the lipid in the feed for hybrid grouper increased from 7 to 14%. Both studies indicate that dietary lipid has a protein-sparing effect in the hybrid grouper.

In general, carnivorous fish have a low tolerance for dietary carbohydrates, and the hybrid grouper appears to be no exception (Nankervis et al. 2021). It has been demonstrated that increasing dietary carbohydrates above 7% leads to stunted growth and enlarging the liver (Li et al. 2019). Other than that, carbohydrates above 11.5% (9% of starch) have been reported to harm the antioxidant and non-specific immunological capabilities of the hybrid grouper (Li et al. 2020). Cassava has been used in both studies as their primary source of carbohydrates, which may point to a problem with cassava, particularly rather than with carbohydrates (Luo et al. 2016). A previous study also showed that the growth of hybrid grouper is not affected when corn starch is used as a carbohydrate source and the liver is enlarged due to the deposition of glycogen (Luo et al. 2016). However, the feed conversion ratio (FCR) was reported to be increased. The differences in results might be due to the complexity of the starch structure and the ability of fish to digest it (Lu et al. 2018). The demand for fish meal and fish oil, which are in short supply due to increased worldwide aquaculture feed production, has increased, pushing up costs while also putting pressure on wild small pelagic fisheries to fish meal and fish oil manufacturing. As a result, the aquaculture industry is looking into alternatives for fish meal and fish oil to produce low-cost feed but, at the same time, meet the nutritional requirements of the hybrid grouper. Since the hybrid grouper is a carnivorous fish with enzyme systems for protein-rich foods of animal origin, it may not be able to consume or digest plant-based feed effectively. However, it is possible to formulate a balanced nutrient of plant-based feed that can support the requirement of fish.

Several studies have documented the use of soybean protein concentrate (SPC), cottonseed protein concentrate, blood meal, mixed rendered animal protein, insect meal, and poultry by-product meal as an alternative to replace fish meal in the feed for hybrid grouper. Mohd Faudzi et al. (2018) demonstrated that increased SPC by more than 50% significantly reduced the growth, feed utilization, and digestibility of hybrid grouper. The study supplemented taurine, which is low in abundance in plant ingredients but important for growth and lipid metabolism (Koven et al. 2016; Lin and Lu 2020). The inclusion of cottonseed protein concentrate (CPC) at less than 15% level in the feed increased the growth of hybrid grouper (Ye et al. 2019). However, more than 15% of inclusion level resulted in a negative effect, especially on health, which was indicated by alteration in microbial population and inflammation in the intestine (Ye et al. 2019). The incorporation of black soldier fly (BSF) larvae meal at a 30% level in the feed has been reported to improve the growth of hybrid grouper (Mohamad-Zulkifli et al. 2019). Further investigation into the quality of protein replacement sources should be done to ensure the nutritional content, which may help in promoting better growth and protein utilization by the fish (Peisker 2001; Wang et al. 2004).

On the other hand, a growing interest in cheaper alternatives to replace fish oil has been documented. Palm oil is one of the potential vegetable oils to replace fish oil in marine fish species. It has been reported that the growth of hybrid grouper fed with refined, bleached, and deodorized palm olein (RBDPO) promoted better growth than crude palm oil (CPO), crude palm kernel oil (CPKO), corn oil (CO), coconut oil (COCO), and fish oil (FO) itself (Yong et al. 2019b). The use of COCO in the feed resulted in poorer growth of hybrid grouper (Fitriyani et al. 2015). As the grow-out period is the longest in fish farming, investigating the potential of vegetable oil to replace fish oil will be beneficial to the fish farmer in reducing costs. Gudid et al. (2020) revealed that CPO is an excellent source of vegetable oil to replace fish oil without significant changes in the fatty acid profiles of the marketable size of hybrid grouper.

3.5.2 Health and Disease

The sustainability of the aquaculture industry is seriously hindered by infectious diseases. The problem would reduce the amount of fish available in the market which would result in financial losses for the fish farmers. The use of antibiotics or drugs are prohibited due to consumer health risks associated with consuming fish that have antibiotic or drug residues as well as the environmental effects of effluent. Currently, there is still ongoing research focusing on the use of natural products to cure parasitic diseases for hybrid grouper (Shah et al. 2020; Venmathi Maran et al. 2021). It is challenging and expensive to prevent disease outbreaks in the culture system through water or food. Meanwhile, the use of vaccines against bacterial infection or pathogens is restricted to a few high-value species due to their cost. In grouper culture, innate disease resistance will be beneficial to enhance production. Therefore, hybridization is one of the methods that has proven to show greater resistance and lower mortality in grouper culture and it is the most advantageous trait in the aquaculture industry.

One of the common diseases that have received great attention from the grouper farming industry is vibriosis, which is caused by bacteria that belong to the genus of *Vibrio* (Bunlipatanon and U-taynapun 2017). Numerous marine fish species have been affected by these bacteria, which has also been affected by high mortality in the grouper culture industry (Sivaram et al. 2004; Albert and Ransangan 2013). Typically, *Vibrio* spp. can induce lethargy and skin ulcers in fish by skin or oral contact.

As the hybrid grouper showed better growth performance than the pure species, a study has been conducted to compare the resistance of hybrid grouper towards *Vibrio* with that of parental species. The hybrid grouper (body weight: 65 ± 6 g) has been challenged with *Vibrio vulnificus* and observation on immunological parameters, clearance time of the pathogen, and fish survival have been done. Bunlipatanon and U-taynapun (2017) revealed that the leucocyte number, lysozyme activity, and the ability to eliminate the bacteria were higher in the hybrid and giant grouper compared to tiger grouper. After 15 days of post-bacterial challenges, hybrid and giant groupers showed higher survival (>90%) than tiger grouper (57%).

Ebi et al. (2018) have noted that the hybrid grouper has high resistance to *Vibrio harveyi* VHJR7 at various concentrations. The cumulative mortality of hybrid grouper after 10 days of post-challenge ranged from 0 to 57%. According to the study, the hybrid grouper is vulnerable to *V. harveyi* at a concentration of 1.6×10^5 c.f.u/g body weight. The LD50 value of hybrid grouper is relatively higher than the parental species (Apines Amar et al. 2012; Xu et al. 2012; Ebi et al. 2018).

Resistance towards *Vibrio* spp. indicates the hybrid vigour characteristic of hybrid grouper. Implementation of hybridization in grouper production which is simple and relatively inexpensive may enhance not only the growth performance but also increase the tolerance of fish towards disease.

3.5.3 Water Quality

Water quality is one of the factors that determine the success or failure of an aquaculture operation. All the fish's physiological activities, including breathing, excretion of waste, feeding, and reproduction, occur in the water. Therefore, the health of the fish and production costs can be greatly impacted by the water quality in the culture system. Physical parameters, organic contaminants, biochemical risks, and biological contaminants are the factors that affect water quality and should be tracked.

However, to date, the ideal culture conditions in terms of temperature, dissolved oxygen, pH, and salinity have not been adequately examined for all the stages of development of the hybrid grouper. A range of values for certain water quality parameters for hybrid grouper that have been tested previously is stated in Table 3.4. The hybrid grouper can grow well in such conditions; temperature (26–30 °C), dissolved oxygen (5.00–7.50 mg/l), pH (7.5–7.8), and salinity (10–30 ppt).

Table 3.4 Water quality parameter of culturing hybrid grouper in captivity

Total length (cm)	Temperature (°C)	Dissolved oxygen (mg/l)	pH	Salinity (ppt)	References
0.20	27–29.5	5.5–7.2	7.5–8.6	31–32	Ch'ng and Senoo (2008)
73.9 ± 5.7	28–29.5	7.0–7.8	6–7.5	29–30	Luin et al. (2013)
5.5	28	6.5	7.5	30	Bunlipatanon and U-taynapun (2017)
20	22–34	–	–	30	De et al. (2016)
10–20	28–30	–	7.6–7.8	29–31	Leong-Seng et al. (2017)
–	26.5–29.5	5.2–6.8	6.0–7.0	28–31	Yong et al. (2019a, b)
–	28	–	–	10–20	Othman et al. (2015)

The temperature in culture conditions greatly affects fish appetite, feed utilization, and fish growth (Bendiksen et al. 2002; De et al. 2016). Finding the optimal temperature for fish will lower their energy requirements for survival and improve the efficiency of energy conversion from food to net energy (Van Ham et al. 2003). De et al. (2016) reported that the feed utilization of hybrid grouper decreased at lower temperatures. The finding is similar to the previous study on other grouper species such as orange-spotted grouper (*E. coioides*) (Lin et al. 2008) and humpback grouper (*C. altivelis*) (Sugama et al. 2004). In hybrid grouper, gastric emptying is shorter when the temperature is within the optimal range, which is associated with increased food consumption and growth (De et al. 2016). Meanwhile, a rise in temperature to higher than that lethal limit resulted in slow gastric emptying, low food consumption, and growth reduction in hybrid grouper.

Even though hybrid grouper were reported to have higher resistance to extreme water conditions, culturing the fish at low pH resulted in poor growth and body condition (Thalib et al. 2020). The reduction of growth in acidic water is also reported in other species such as yellowtail (*Seriola quinqueradiata*) (Lee et al. 2003) and Atlantic salmon (*Salmo salar* L.) (Fivelstad et al. 2015). Disturbance in osmoregulation and ionic concentration in the hybrid grouper has led to higher energy consumption, which contributes to the poor growth of the hybrid grouper (Thalib et al. 2020).

Unfavourable salinity in culturing the fish typically results in a modification to their physiological function, including metabolism, osmoregulation, and digestive enzymes (Othman et al. 2015; Sutthinon et al. 2015). It has been shown that fish can conserve energy which improves growth at the ideal salinity. On the other hand, changes in salinity over the ideal range will cause disruption and affect the performance of the fish. Based on salinity tests in hybrid grouper, a salinity range of 10–20 ppt is good for promoting better growth performance, feed utilization, and reducing the cortisol level due to its iso-osmotic condition (Othman et al. 2015).

As hybrid grouper have been widely cultured in sea cages, consideration of the accumulation of solid waste from uneaten feed or faeces should be taken. Increased levels of nitrogenous waste in the environment may harm fish and other aquatic animals. High levels of nutrients in the water (eutrophication) may cause the blooming of harmful algal blooms which in certain concentrations become toxic and drastically reduce dissolved oxygen, resulting in mortality. Adharini et al. (2021) suggested the use of *Ulva* sp. as a promising seaweed product to be used in the Integrated Multi Trophic Aquaculture (IMTA) system for environmentally friendly and sustainable grouper production.

3.6 Conclusion

Aquaculture produces food in a more environmentally friendly manner than commercial fishing and has been demonstrated to provide protein in many areas where it is limited owing to new technology and movement by Sustainable Development Goals (SDGs). The 2006 hybrid grouper aquaculture breakthrough not only benefited the aquaculture business but also helped the seafood industry grow significantly. The hybrid grouper is being produced in large quantities, which has reduced the scarcity of fish food supplies and hastened the growth of the aquaculture sector. Food security has been made possible in many parts of the world owing to the successful culture of hybrid grouper, not only by ensuring its supply but also by ensuring access to enough nutritious food.

References

Abang Zamhari DNJB, Yong ASK (2021) Dietary herbs supplementation improves growth, feed efficiency and apparent digestibility coefficient of hybrid grouper (*Epinephelus fuscoguttatus* × *Epinephelus lanceolatus*) juvenile. Aquac Res 52:5239–5246

Adams A, Thompson KD (2006) Biotechnology offers revolution to fish health management. Trends Biotechnol 24(5):201–205

Addin MA, Senoo S (2011) Production of hybrid groupers: spotted grouper, *Epinephelus polyphekadion* × tiger grouper, *E. fuscoguttatus* and coral grouper, *E. corallicola* × tiger grouper. In: 2011 International symposium on grouper culture—technological innovation and industrial development, Taiwan

Adharini RI, Murwantoko M, Probosunu N, Setiawan RY, Satriyo TB (2021) The effectiveness of seaweeds as biofilter for reducing wastewater nutrient and preventing water pollution from hybrid grouper culture. J Ilm Perikan Kelaut 13(2):133–143

Aflalo ED, Hoang TTT, Nguyen VH, Lam Q, Nguyen DM, Trinh QS, Raviv S, Sagi A (2006) A novel two-step procedure for mass production of all-male populations of the giant freshwater prawn *Macrobrachium rosenbergii*. Aquaculture 256:468–478

Agarwal NK (2011) Cryopreservation of fish semen. In: Himalayan aquatic biodiversity conservation and new tools in biotechnology. Transmedia Publication, Srinagar, pp 104–127

Albert V, Ransangan J (2013) Effect of water temperature on susceptibility of culture marine fish species to vibriosis. Int J Res Pure Appl Microbiol 3:48–52

Altunok M, Zerife P, Ramazan S, Adem TA, Volkan K (2012) Biotechnology and aquaculture in sustainable development. In: 3rd International Symposium on Sustainable Development (ISSD2012). Sarajevo, Bosnia and Herzegovina

Apines Amar MJS, Amar EC, Faisan JP Jr, Pakingking RV Jr, Satoh S (2012) Dietary onion and ginger enhance growth, hemato-immunological responses, and disease resistance in brown-marbled grouper, *Epinephelus fuscoguttatus*. AACL Bioflux 5(4):231–239

Argue BJ, Kuhlers DL, Liu Z, Dunham RA (2014) Growth of channel catfish (*Ictalurus punctatus*), blue catfish (*I. furcatus*), and their F1, F2, F3, and F1 reciprocal backcross hybrids in earthen ponds. J Anim Sci 92(10):4297–4305

Bartley DM, Rana K, Immink AJ (2000) The use of inter-specific hybrids in aquaculture and fisheries. Rev Fish Biol Fish 10:325–337

Bendiksen EÅ, Jobling M, Arnesen AM (2002) Feed intake of Atlantic salmon parr *Salmo salar* L. in relation to temperature and feed composition. Aquac Res 33:525–553

Benson JD, Woods EJ, Walters EM, Critser JK (2012) The cryobiology of spermatozoa. Theriogenology 78:1682–1699

Blanc JM, Chevassus B (1982) Interspecific hybridization of salmonid fish: II. Survival and growth up to the 4th month after hatching in F1 generation hybrids. Aquac Res 29(3–4):383–387

Boyd CE, McNevin AA, Davis RP (2022) The contribution of fisheries and aquaculture to the global protein supply. Food Secur 14:805–827

Bunlipatanon P, U-taynapun K (2017) Growth performance and disease resistance against *Vibrio vulnificus* infection of novel hybrid grouper (*Epinephelus lanceolatus* × *Epinephelus fuscoguttatus*). Aquac Res 48:1711–1723

Campton DE (1987) Natural hybridization and introgression in fishes: methods of detection and interpretation. In: Ryman N, Utter FM (eds) Population genetics and fishery management. University of Washington Press, Seattle, WA, pp 161–192

Ch'ng CL, Senoo S (2008) Egg and larval development of a new hybrid grouper, tiger grouper *Epinephelus fuscoguttatus* × giant grouper *E. lanceolatus*. Aquac Science 56(4):505–512

Chapela MJ, Ferreira M, Ruiz-Cruz A, Martin-Varela I, Fernández-Casal J, Garrido-Maestu A (2018) Application of real-time PCR for early diagnosis of diseases caused by *Aeromonas salmonicida*, *Vibrio anguillarum*, and *Tenacibaculum maritimum* in turbot: a field study. J Appl Aquac 30(1):76–89

Chen ZF, Tian YS, Wang PF, Tang J, Liu JC, Ma WH, Li WS, Wang XM, Zhai JM (2018) Embryonic and larval development of a hybrid between kelp grouper *Epinephelus moara* × giant grouper *E. lanceolatus* using cryopreserved sperm. Aquac Res 49:1407–1413

Ching FF, Nakagawa Y, Kato K, Murata O, Miyashita S (2012) Effects of delayed first feeding on the survival and growth of tiger grouper, *Epinephelus fuscoguttatus* (Forsskal, 1775) larvae. Aquac Res 43(2):303–310

Ching FF, Miura A, Nakagawa Y, Kato K, Sakamoto W, Takii K, Miyashita S, Senoo S (2016) Aeration rate adjustment at night to prevent sinking syndrome-related to death in the tiger grouper *Epinephelus fuscoguttatus* (Perciformes: Serranidae) larvae. Aquac Res 47(1):165–175

Ching FF, Othman N, Anuar A, Shapawi R, Senoo S (2018) Natural spawning, embryonic and larval development of F2 hybrid grouper, tiger grouper *Epinephelus fuscoguttatus* × giant grouper *E. lanceolatus*. Int Aquat Res 10:391–402

Chor WK, Andin VC, Gopalakrishnan CD, Amer A, Mohamed S, Matsumoto H, Leong-Seng L (2020) A comparative analysis of the effect of low-cost fish and commercially compounded feed on growth performance and organoleptic quality of hybrid grouper (*Epinephelus fuscoguttatus* × *Epinephelus lanceolatus*) in cage farming in Kuala Penyu, Sabah and nutritional costs. Borneo J Mar Sci Aquac 4:52–56

De M, Ghaffar MA, Bakar Y, Das SK (2016) Effect of temperature and diet on growth and gastric emptying time of the hybrid, *Epinephelus fuscoguttatus* × *E. lanceolatus*. Aquac Rep 14:118–124

Diedericks G, Martens H, Steenberge MV, Snoeks J (2021) Testing for hybridization between Nile tilapia (*Oreochromis niloticus*) and blue spotted tilapia (*Oreochromis leucostictus*) in the Lake Edward system. J Great Lakes Res 47:1446–1452

Dunham RA (1996) Contribution of genetically improved aquatic organisms to global food security. In: Proceedings of the International Conference on Sustainable Contribution of Fisheries to Food Security, (SCFFS'96), Government of Japan and FAO, Rome, pp 1–150

Dunham RA (2004) Aquaculture and fisheries biotechnology—genetic approaches. CABI Publishing, Wallingford, p 372

Dunham RA, Smitherman RO (1983) Crossbreeding channel catfish for improvement of body weight in earthen ponds. Growth 47:97–103

Ebi I, Muhammad Lal T, Ransangan J, Yong ASK, Shapawi R (2018) Susceptibility of hybrid grouper (*Epinephelus fuscoguttatus* ♀ × *Epinephelus lanceolatus* ♂) to *Vibrio harveyi* VHJR7. AACL Bioflux 11(1):31–42

FAO (2020) The state of world fisheries and aquaculture 2020. Sustainability in action. FAO, Rome

FAO (2021) The state of food security and nutrition in the world (SOFI). Transforming food systems for food security, improved nutrition and affordable healthy diets for all. FAO, Rome

FAO (2022) The State of Food and Agriculture 2022. Leveraging agricultural automation for transforming agrifood systems. FAO, Rome

Fitriyani, Kusdianto H, Sukarti K (2015) Effect of different dietary lipid sources on feed efficiency and feed conversion ratio of cantang grouper (*Epinephelus* sp.). Trop Fish Sci J 20:8–14

Fitzpatrick JL, Akbarashandiz H, Sakhrani D, Biagi CA, Pitcher TE, Devlin RH (2011) Cultured growth hormone transgenic salmon are reproductively out-competed by wild-reared salmon in semi-natural mating arenas. Aquaculture 312(1–4):185–191

Fivelstad S, Kvamme K, Handeland S, Fivelstad M, Olsen AB, Hosfeld CD (2015) Growth and physiological models for Atlantic salmon (*Salmo salar L.*) parr exposed to elevated carbon dioxide concentrations at high temperature. Aquaculture 436:90–94

Garcia-Ortega A, Daw A, Hopkins K (2014) Feeding hatchery-produced larvae of the giant grouper *Epinephelus lanceolatus*. In: Conference: hatchery technology for high quality juvenile production: proceedings of the 40th U.S.—Japan aquaculture panel symposium, Honolulu, HI

Gil LA (2007) PCR-based methods for fish and fishery products authentication. Trends Food Sci Technol 18(11):558–566

Glamuzina B, Skaramuca B, Glavic N, Kozul V, Dulcic J, Kraljevic M (1998) Egg and early larval development of laboratory reared dusky grouper, *Epinephelus marginatus* (Lowe, 1834) (Picies, Serranidae). Sci Mar 62(4):373–378

Glamuzina B, Glavic N, Kozul V, Tutman V, Skaramuca B (1999) Hybridization of Mediterranean groupers: *Epinephelus marginatus* × *E. aeneus* and early development. Aquac Res 30:625–628

Glamuzina B, Glavić N, Skaramuca B, Kozŭl V, Tutman P (2001) Early development of the hybrid *Epinephelus costae* ♀ × *E. marginatus* ♂. Aquac Res 198:55–61

Gudid SN, Leong-Seng L, Yong ASK, Yanuhar U, Mustafa S, Shapawi R (2020) Growth performance and organoleptic quality of hybrid grouper (*Epinephelus fuscoguttatus* × *Epinephelus lanceolatus*) fed palm-oil based diets at grow-out stage. Sains Malays 49(7): 1567–1576

Gwo JC (2011) Cryopreservation of sperm of some marine fishes. In: Tiersch TR, Mazik PM (eds) Cryopreservation in aquatic species, 2nd edn. World Aquaculture Society, Baton Rouge, LA, pp 459–481

Hixson SM (2014) Fish nutrition and current issues in aquaculture: the balance in providing safe and nutritious seafood, in an environmentally sustainable manner. J Aquat Res Dev 5:234

Hoga CA, Almeida FL, Reyes FGR (2018) A review on the use of hormones in fish farming: analytical methods to determine their residues. CyTA 16(1):679–691

Hong GE, Kim DG, Bae JY, Ahn SH, Bai SC, Kong IS (2007) Species-specific PCR detection of the fish pathogen, *Vibrio anguillarum*, using the amiB gene, which encodes N-acetylmuramoyl-L-alanine amidase. FEMS Microbiol Lett 269(2):201–206

Huang W, Liu Q, Xie J, Wang W, Xiao J, Li S, Lin H (2016) Characterization of triploid hybrid groupers from interspecies hybridization (*Epinephelus coioides* ♀× *Epinephelus lanceolatus* ♂). Aquac Res 47(7):2195–2204

Hubbs CL (1955) Hybridization between fish species in nature. Syst Zool 405:1–20

Jiang S, Wu X, Li W, Wu M, Luo Y, Lu S (2015) Effects of dietary protein and lipid levels on growth, feed utilization, body and plasma biochemical composition of hybrid grouper (*Epinephelus lanceolatus* ♂ × *E. fuscoguttatus* ♀) juveniles. Aquaculture 446:148–155

Jiang S, Wu X, Luo Y, Wu M, Jin Z, Yao W (2016) Optimal dietary protein level and protein-to-energy ratio for hybrid grouper (*Epinephelus fuscoguttatus* ♀ × *E. lanceolatus* ♂) juveniles. Aquaculture 465:28–36

Kang MJ, Noh CH, Choi HJ, Park DW, Hur SP (2020) Embryonic development and hatchability in the reciprocal hybrids of kelp grouper (*Epinephelus bruneus*) and red-spotted grouper (*Epinephelus akaara*). Ocean Polar Res 42(4):303–311

Kawahara S, Shams AJ, Al-Bosta AA, Mansor MH, Al-Baqqal AA (1997) Effects of incubation and spawning water temperature and salinity on egg development of the orange-spotted grouper (*Epinephelus coioides*, Serranidae). Asian Fish Soc 9(1997):239–250

Khatei A, Tripathy PS, Parhi J (2022) Molecular markers in aquaculture. In: Advances in fisheries biotechnology. Springer, Cham, pp 165–174

Kiriyakit A, Wenresti G, Gallardo G, Bart AN (2011) Successful hybridization of groupers (*Epinephelus coioides* × *Epinephelus lanceolatus*) using cryopreserved sperm. Aquaculture 320:106–112

Koh ICC, Shaleh SRM, Senoo S (2008) Egg and larval development of a new hybrid orange-spotted grouper *Epinephelus coioides* × tiger grouper *E. fuscoguttatus*. Aquac Sci 56(3): 441–451

Koh ICC, Sitti Raehanah MS, Akazawa N, Oota Y, Senoo S (2010) Egg and larval development of a new hybrid orange-spotted grouper *Epinephelus coioides* × giant grouper *E. lanceolatus*. Aquac Science 58(1):1–10

Koh ICC, Tanaka D, Itagane T, Tsuji M, Tsuchihashi Y, Ohta H (2011) Dry shipper cryopreservation of seven-band grouper, *Epinephelus septemfasciatus*. Aquac Res 44(1):59–66

Koven W, Peduel A, Gada M, Nixon O, Ucko M (2016) Taurine improves the performance of white grouper juveniles (*Epinephelus aeneus*) fed a reduced fish meal diet. Aquaculture 460:8–14

Kurdianto A, Faridah N, Yoshizaki G, Nuryati S, Setiawati M (2016) Growth, survival, and body composition of transgenic common carp *Cyprinus carpio* 3rd generation expressing tilapia growth hormone cDNA. HAYATI J Biosci 23(3):1–5

Lee S, Ishimatsu A, Sakaguchi H, Oda T (2003) Cardiac output during exposure to *Chattonella marina* and environmental hypoxia in yellowtail (*Seriola quinqueradiata*). J Mar Biol 142:391–397

Leong-Seng L, Firdaus RF, Shapawi R, Kawamura G (2017) Taste preference of hybrid grouper (*Epinephelus fuscoguttatus* × Epinephelus lanceolatus) for nucleoside and nucleotides. Borneo J Mar Sci Aquac 1:39–43

Levy T, Rosen O, Eilam B, Azulay D, Zohar I, Aflalo ED, Benet A, Naor A, Shechter A, Sagi A (2017) All-female monosex culture in the freshwater prawn *Macrobachium rosenbergii*—a comparative large-scale field study. Aquac Res 479:857–862

Li C (2022) Conservation of genetic resources for sustainable aquaculture. J World Aquacult Soc 53(1):4–7

Li S, Li Z, Zhang J, Sang C, Chen N (2019) The impacts of dietary carbohydrate levels on growth performance, feed utilization, glycogen accumulation and hepatic glucose metabolism in hybrid grouper (*Epinephelus fuscoguttatus* ♀ × *E. lanceolatus* ♂). Aquaculture 512:734351

Li S, Wang A, Li Z, Zhang J, Sang C, Chen N (2020) Antioxidant defenses and non-specific immunity at enzymatic and transcriptional levels in response to dietary carbohydrate in a typical carnivorous fish, hybrid grouper (*Epinephelus fuscoguttatus*♀ × *E. lanceolatus*♂). Fish Shellfish Immunol 100:109–116

Lin YH, Lu RM (2020) Dietary taurine supplementation enhances growth and nutrient digestibility in giant grouper *Epinephelus lanceolatus* fed a diet with soybean meal. Aquac Rep 18:100464

Lin X, Xie S, Su Y, Cui Y (2008) Optimum temperature for the growth performance of juvenile orange-spotted grouper (*Epinephelus coioides* H.). Chin J Oceanol Limnol 26(1):69–75

Liufu YZ, Zhao HZ, Liu XC, Lin HR, Huang GG, Zhang HF, Wang Y (2007) Preliminary study on the hybrid red-spotted grouper (*Epinephelus akaara*)♂× orange-spotted grouper (*Epinephelus coioides*)♀. Acta Sci Mat Univ Sunyatseni 46(3):72–75

Lu S, Wu X, Gao Y (2018) Effects of dietary carbohydrate sources on growth, digestive enzyme activity, gene expression of hepatic GLUTs and key enzymes involved in glycolysis-gluconeogenesis of giant grouper *Epinephelus lanceolatus* larvae. Aquaculture 484:343–350

Luin M, Ching FF, Senoo S (2013) Sexual maturation and gonad development in tiger grouper (*Epinephelus fuscoguttatus*) × giant grouper (*E. lanceolatus*) hybrid. J Aquat Res Dev 5:213

Luo Z, Liu YJ, Mai KS, Tian LX, Yang HJ, Tan XY, Liu DH (2005) Dietary L-methionine requirement of juvenile grouper at a constant dietary cystine level. Aquaculture 249:409–418

Luo Y, Wu X, Li W, Jiang S, Lu S, Wu M (2016) Effect of different corn starch levels on growth, protein input, and feed utilization of juvenile hybrid grouper (male *Epinephelus lanceolatus* × female *E. fuscoguttatus*). N Am J Aquac 78:168–173

Maclean N, Rahman MA, Sohm F, Hwang G, Iyerngar A, Ayad H, Smith A, Farahmand H (2002) Transgenic tilapia and the tilapia genome. Gene 295(2):265–277

Marino G, Azzuro E, Massari A, Finoia MG, Mandich A (2001) Reproduction of the dusky grouper from the southern Mediterranean. J Fish Biol 58:909–927

Mohamad-Zulkifli NFN, Yong ASK, Kawamura G, Lim LS, Senoo S, Devic E, Mustafa S, Shapawi R (2019) Apparent digestibility coefficient of black soldier fly (*Hermetia illucens*) larvae in formulated diets for hybrid grouper (*Epinephelus fuscoguttatus* ♀ × *Epinephelus lanceolatus* ♂). AACL Bioflux 12(2):513–522

Mohd Faudzi N, Yong ASK, Shapawi R, Senoo S, Biswas A, Taki K (2018) Soy protein concentrate as an alternative in replacement of fish meal in the feeds of hybrid grouper, brown-marbled grouper (*Epinephelus fuscoguttatus*) × giant grouper (*E. lanceolatus*) juvenile. Aquac Res 49:431–441

Nagano N, Hozawa A, Fujiki W, Yamada T, Miyaki K, Sakakura Y, Hagiwara A (2007) Skeletal development and deformities in cultured larval and juvenile seven-band grouper, *Epinephelus septemfasciatus* (Thunberg). Aquac Res 38(2):121–130

Nankervis L, Cobcroft JM, Nguyen NV, Rimmer MA (2021) Advances in practical feed formulation and adoption for hybrid grouper (*Epinephelus fuscoguttatus* ♀ × *E. lanceolatus* ♂) aquaculture. Rev Aquac 14(1):288–307

Nasim A (2010) Biotechnology: a powerful tool for human survival and sustainability. In: Gökçekus H, Türker U, LaMoreaux J (eds) Survival and sustainability. Environmental earth sciences. Springer, Berlin

Nwokwa MC (2012) The review of recent advances in fish genetics and biotechnology. Cont J Fish Aquat Sci 6(1):9–18

Omole IA (2017) Biotechnology as an important tool for improving fish productivity. Am J Biosci Bioeng 5(1):17–22

Othman AR, Kawamura G, Senoo S, Ching FF (2015) Effects of different salinities on growth, feeding performance and plasma cortisol level in hybrid TGGG (tiger grouper, *Epinephelus fuscoguttatus* × giant grouper, *Epinephelus lanceolatus*) juveniles. Int Res J Biol Sci 4(3):15–20

Pamungkas GY, Sari LA (2021) Enlargement technique of humpback grouper (*Cromileptes altivelis*) with floating nets cage. IOP Conf Ser Earth Environ Sci 679:012030

Park IS, Choi GC, Nam YK, Kim DS (2002) The effect of exogenous hormone treatment on spermiation in *Rhynchocypris oxycephalus* (Sauvage and Dabry). J World Aquac Soc 33(4): 494–500

Peisker M (2001) Manufacturing of soy protein concentrate for animal nutrition. In: Brufau J (ed) Feed manufacturing in the Mediterranean region improving safety from feed to fish, Zaragoza, pp 103–107

Rahimnejad S, Bang IC, Park JY, Sade A, Choi J, Lee SM (2015) Effects of dietary protein and lipid levels on growth performance, feed utilization and body composition of juvenile hybrid grouper *Epinephelus fuscoguttatus* × *E. lanceolatus*. Aquaculture 446:283–289

Sadovy Y (2003) While stocks last: the live reef food fish trade. Asian Development Bank, Manila, pp 1–143

Sagi A, Cohen D, Milner Y (1990) Effect of androgenic gland ablation on morphotypic differentiation and sexual characteristics of male freshwater prawns, *Macrobrachium rosenbergii*. Gen Comp Endocrinol 77(1):15–22

Senoo S, Shapawi R, Baidya A, Rahman RA (2002) Observation on eggs of mouse grouper, *Cromileptes altivelis* under rearing conditions. Aquac Sci 50(4):437–438

Senoo S, Shapawi R, Abdul Rahman R (2004) Induced spawning technique of mouse grouper, *Cromileptes altivelis*. Universiti Malaysia Sabah, Sabah, pp 1–23

Shah MD, Venmathi Maran BA, Haron FK, Ransangan J, Ching FF, Shaleh SRM, Shapawi R, Yong YS, Ohtsuka S (2020) Antiparasitic potential of *Nephrolepis biserrata* methanol extract against the parasitic leech *Zeylanicobdella arugamensis* (Hirudinea) and LC-QTOF analysis. Sci Rep 10(1):1–10

Shapawi R, Ching FF, Senoo S, Mustafa S (2019) Nutrition, growth and resilience of tiger grouper (*Epinephelus fuscoguttatus*) × giant grouper (*Epinephelus lanceolatus*) hybrid—a review. Rev Aquac 11(4):1285–1296

Sivaram V, Babu MM, Immanuel G, Murugadass S, Citarasu T, Marian MP (2004) Growth and immune response of juvenile greasy groupers (*Epinephelus tauvina*) fed with herbal antibacterial active principle supplemented diets against *Vibrio harveyi* infections. Aquaculture 237:9–20

Sugama K, Trijoko I, Maha Setiawati S (2004) Effect of water temperature on growth, survival and feeding rate of humpback grouper (*Cromileptes altivelis*) larvae. In: Rimmer MA, McBride S,

Williams KC (eds) Advances in grouper aquaculture, ACIAR monograph 110. Australian Centre for International Agricultural Research, Canberra, pp 61–66

Sutina, Yunus R, Seniwati (2017) Market structure of grouper fish in Indonesia. Sci Int (Lahore) 29(2):219–222

Sutthinon P, Thongprajukaew K, Saekhow S, Ketmanee R (2015) Juvenile hybrid grouper (*Epinephelus coioides* × *E. lanceolatus*) are euryhaline and can grow in a wide range of salinities. Aquac Int 23:671–682

Tacon AGJ (2003) Fish farming. In: Encyclopedia of food sciences and nutrition, 2nd edn. Academic Press, Amsterdam, pp 2479–2486. isbn:9780122270550.

Tan J (2021) Backcross breeding between TGGG hybrid grouper (*Epinephelus fuscoguttatus* × *E. lanceolatus*) and giant grouper (*E. lanceolatus*). J Surv Fish Sci 7(2):49–52

Thalib YA, Razali RS, Mohamad S, Zainuddin RA, Rahmah S, Abdul Ghaffar M, Nhan HT, Liew HJ (2020) Environmental changes affecting physiological responses and growth of hybrid grouper—the interactive impact of low pH and temperature. Environ Pollut 271:116375

Thorgaard GH, Bailey GS, Williams D, Buhler DR, Kaattari SL, Ristow SS, Hansen JD, Winton JR, Bartholomew JL, Nagler JJ, Walsh PJ, Vijayan MM, Devlin RH, Hardy RW, Overturf KE, Young WP, Robison BD, Rexroad C, Palti Y (2002) Status and opportunities for genomics research with rainbow trout. Comp Biochem Physiol B Biochem Mol Biol 133(4):609–646

Toppe J (2014) Fish trade and human nutrition. The role of fish in nutrition and food security. Food and Agriculture Organization of the United Nations. C0FI;FT/XIV/2014/4

Tupper M, Sheriff N (2008) Capture-based aquaculture of groupers. In: Lovatelli A, Holthus PF (eds) Capture-based aquaculture. Global overview. FAO Fisheries technical paper. No. 508. FAO, Rome, pp 217–253

Van Ham EH, Berntssen MHG, Imsland AK, Parpoura AC, Wendelaar Bonga SE, Stefansson SO (2003) The influence of temperature and ration on growth, feed conversion, body composition and nutrient retention of juvenile turbot (*Scophthalmus maximus*). Aquaculture 217:547–558

Venmathi Maran BA, Josmeh D, Tan JK, Yong YS, Shah MD (2021) Efficacy of the aqueous extract of *Azadirachta indica* against the marine parasitic leech and its phytochemical profiling. Molecules 26(7):1908

Wakchaure R, Ganguly S, Qadri K, Praveen PK, Mahajan T (2015) Importance of transgenic fish to global aquaculture: a review. Fish Aquac J 6:4

Wang H, Johnson LA, Wang T (2004) Preparation of soy protein concentrate and isolate from extruded-expelled soybean meals. JAOCS 81(7):713–717

Williams KC, Irvin S, Barclay M (2004) Polka dot grouper, *Cromileptes altivelis* fingerlings require high protein and moderate lipid diets for optimal growth and nutrient retention. Aquac Nutr 10: 125–134

Williams LJ, Hartman N, Cavazotte F (2010) Method variance and marker variables: a review and comprehensive CFA marker technique. Organ Res Methods 13(3):477–514

Xu XD, Xie ZY, Ouiang JLO, Shi-Feng W, Yong-Can Z (2012) Isolation and identification of bacterial pathogens from *Epinephelus fuscoguttatus* with scale-peeled syndrome disease. Mar Sci, Qingdao—Chinese Edition 36:74–80. [in Chinese]

Yang X, Zhi X, Song Z, Wang G, Zhao X, Chi S, Tan B (2022) Flesh quality of hybrid grouper (*Epinephelus fuscoguttatus* ♀ × *Epinephelus lanceolatus* ♂) fed with hydrolyzed porcine mucosa-supplemented low fishmeal diet. Anim Nutr 8:114–124

Ye H, Xu M, Chen L, Tan X, Chen S, Zou C, Sun Z, Liu Q, Ye C, Wang A (2019) Effects of dietary plant protein sources influencing hepatic lipid metabolism and hepatocyte apoptosis in hybrid grouper (*Epinephelus lanceolatus* ♂ × *Epinephelus fuscoguttatus* ♀). Aquaculture 506:437–444

Yong ASK, Mohd Faudzi N, Senoo S, Shapawi R (2019a) Optimum level of dietary protein and lipid for hybrid grouper, brown-marbled grouper (*Epinephelus fuscoguttatus*) × giant grouper (*Epinephelus lanceolatus*). J Sustain Sci Manag 14:1–15

Yong ASK, Syed Mubarak NS, Shapawi R (2019b) Effects of partial replacement of fish oil with different vegetable oils on growth, feed utilisation and fatty acid profile of hybrid grouper juvenile (*Epinephelus fuscoguttatus* × *Epinephelus lanceolatus*). J Oil Palm Res 31(1):110–121

Yue K, Shen Y (2022) An overview of disruptive technologies for aquaculture. Aquac Fish 7(2): 111–120

Zhang Q, Reign RC, Wolters WR (1994) Growth and body composition of pond raised hybrid striped basses, *Morone saxatilis* × *M. chrysops* and *M. saxatilis* × *M. mississippiensis*, fed low and moderate levels of dietary lipid. Aquaculture 125(1–2):119–129

The Application of Molecular Markers in Fish Breeding and Aquaculture

4

Kenneth Francis Rodrigues, Helena Biun, Wilson Thau Lym Yong, Grace Joy Wei Lie Chin, Fui Fui Ching, and Rafidah Othman

Abstract

Over the past two decades, the global dependence of aquaculture is evident from the data indicating global aquaculture production more than tripled in live-weight volume from 34 million tons in 1997 to 179 million tons in 2018. Concurrently, there has been increasing pressure to improve the production efficiency of conventional production systems. The long-term economic sustainability of the aquaculture industry is founded on the selection, development, maintenance, and management of high-quality germplasm. The traditional approaches for selecting broodstock from wild populations solely on the basis of phenotypic traits have been replaced by the use of molecular markers offer a cost-effective solution for identifying superior individuals from the wild, which can be recruited into breeding programs. Maintenance of broodstock via the process of selective breeding can be managed effectively by the application of marker-assisted selection and the linkage of quantitative traits to genomic loci. The emergence of low-cost whole-genome sequencing platforms and the availability of genomic data at publicly available databases have contributed to a significant reduction in the cost of marker development which improves precision and accuracy. The following review will emphasize the different types of markers that have been developed for commercially important fishes, the methods

K. F. Rodrigues (✉) · W. T. L. Yong · G. J. W. L. Chin
Biotechnology Research Institute, Universiti Malaysia Sabah, Kota Kinabalu, Sabah, Malaysia
e-mail: kennethr@ums.edu.my

H. Biun
Likas Fisheries Complex, Department of Fisheries Sabah, Kota Kinabalu, Sabah, Malaysia

F. F. Ching · R. Othman
Borneo Marine Research Institute, Universiti Malaysia Sabah, Kota Kinabalu, Sabah, Malaysia

M. D. Shah et al. (eds.), *Marine Biotechnology: Applications in Food, Drugs and Energy*, https://doi.org/10.1007/978-981-99-0624-6_4

employed for marker development and application, as well as the limitations of the currently available technologies.

4.1 Introduction to Molecular Markers

The term "molecular markers" can be broadly applied to refer to a diverse range of molecules which can be leveraged to establish a statistical linkage between a specific phenotype and its molecular progenitor. This can extend to the genome (Allal and Nguyen 2022), transcriptome (Bao et al. 2022), proteome (Tian et al. 2020), or metabolome (Koubová et al. 2022). Within the context of a breeding program, the term "marker" generally refers to a unique genomic locus which has been linked to a specific trait. The primary reason for this qualification of the term "marker" is linked to the fact that DNA markers exhibit characteristics of stability, traceability, and reproducibility as compared to RNA, protein, or metabolic markers. The laboratory methods associated with DNA markers are far less complex as compared to their counterparts in terms of laboratory setup and technical skills.

The historical process of DNA marker development commenced with the discovery of the enzyme DNA polymerase and its successful application to the amplification of DNA templates using short single-stranded DNA oligonucleotides via the polymerase chain reaction (PCR). This method superseded the laborious method of restriction fragment length polymorphism, which was one of the first methods used for DNA fingerprinting using unamplified DNA. PCR technology spawned a diverse range of DNA markers, which included random amplified polymorphic DNA (RAPD) (Mahboob et al. 2019), amplified fragment length polymorphism (AFLPs) (Cheng et al. 2010), microsatellites (Tesfaye et al. 2021), and restriction fragment length polymorphisms (RFLPs) (Algammal et al. 2020). These early markers, although not as reliable as the current ones, served as a driving force for the development of more accurate technologies with an emphasis on reliability, reproducibility, and accuracy. The invention of Sanger sequencing technology enhanced the reproducibility of DNA markers as short sequences of DNA could now be read, archived, and distributed across public databases for curation across geographic boundaries. Currently, the emergence of high-throughput next-generation sequencing (NGS) platforms offers breeders an even higher degree of reliability in terms of establishing a linkage between genotype and phenotype. NGS data can be processed rapidly using bioinformatics platforms in order to zero in on specific polymorphisms in both coding and non-coding regions of the genome and to design and develop panels of DNA markers that can be adopted by commercial fish breeding companies in their field laboratories. The synergistic approach between research institutions and commercial breeding companies, whereby each element of the marker discovery, design, and deployment process is critical to the success of the aquaculture industry.

The acceptance of molecular markers by the breeding industry is evident from the large number of genome sequencing projects (Table 4.1) that focus on economically

Table 4.1 The ten species that contribute to the aquaculture industry and their genomic data

No.	Fish species	Genome availability	References
1	Grass carp (*Ctenopharyngodon idellus*)	Complete genome	Wu et al. (2022)
2	Silver carp (*Hypophthalmichthys molitrix*)	Complete genomes	Zhou et al. (2021)
3	Tilapia (*Oreochromis niloticus*)	Complete genome	Cádiz et al. (2020)
4	Common carp (*Cyprinus carpio*)	Complete genome	Xu et al. (2014)
5	Bighead Carp (*Hypophthalmichthys nobilis*)	Complete genome	Fu et al. (2021)
6	Indian Carp (*Labeo catla*)	Draft genome	Sahoo et al. (2020)
7	Atlantic salmon (*Salmo salar*)	Draft genome	Lien et al. (2016)
8	Striped catfish (*Pangasianodon hypophthalmus*)	Complete genome	Gao et al. (2021)
9	Pacific white shrimp (*Litopenaeus vannamei*)	Complete genome	Zhang et al. (2019)
10	Pacific oyster (*Crassostrea gigas*)	Chromosomal assembly	Peñaloza et al. (2021)

important fish species, both in terms of profitability and in terms of food security. Genomes of hundreds of fish species can now be deduced using genome survey sequences (GSS) using data that is currently available at public databases such as the NCBI GenBank and the Ensemble genome browser for vertebrates.

4.2 Attributes of Genomic Molecular Markers for Application in Aquaculture and Breeding

The primary criteria which must be considered prior to the selection of a genomic molecular marker in any breeding program are as follows:

(a) The availability of prior genetic information pertinent to the species which may be in the form of whole-genome sequence data, mitochondrial DNA data, or transcriptomic data.
(b) The relative ease of application in field laboratories where a large number of samples have to be processed, especially in the case of linkage studies.
(c) The reliability and reproducibility of the data in terms of statistical linkages between markers and quantitative traits across populations of the same species.

Genomic data are essential for the design of locus-specific markers which can be tested and validated prior to large-scale deployment. Loci can be categorized into non-coding and coding loci. The non-coding loci, which are typically designated as type II molecular markers, encompass simple sequence repeats (SSRs) (Huang et al. 2022a, b), microsatellites (MS) (Weng et al. 2021), inter-simple sequence repeat (ISSR) (Gu et al. 2019), transposable genetic elements (Kon et al. 2020), single nucleotide polymorphisms (SNP) (Perez-Enriquez et al. 2018; Salem et al. 2018), and introns (Song et al. 2021). Coding loci, commonly referred to as type I molecular markers, represent specific genes, which are likely to be associated with specific traits such as disease resistance (Vela-Avitúa et al. 2022), fecundity (Li et al. 2021), higher growth rates (Valenza-Troubat et al. 2022), and gender (Zhu et al. 2022). The choice of molecular marker is determined by multiple factors, which include the specific application, availability of prior genetic information, cost effectiveness, and the availability of expertise. Microsatellites are very useful for the characterization of population structure in both wild and hatchery-reared populations, they are cost-effective and require a low-to-moderate level of technical expertise, and the laboratory processes, which include DNA extraction, PCR, and fragment analysis, can be automated. SNPs, on the other hand, are highly reliable, but come with an associated cost as microarrays have to be customized for each species. Mitochondrial DNA loci are effective in determining parentage as they are maternally inherited, they can be applied to evaluate the range of haplotypes within a population; however, one of the limitations is mitochondrial heteroplasmy, which may be observed in populations and contribute to errors in determining the real structure of a natural population. Recently, NGS-based approaches have become cost-effective as the cost of sequencing and data analysis has dropped significantly (Table 4.2).

4.3 Quantitative Traits of Relevance to Aquaculture

Quantitative trait locus (QTL) analysis is a statistical technique that combines phenotypic data (quantifiable traits) and genotypic data (DNA sequence data) to explain the genetic basis of variance in complex traits. QTL analysis enables researchers in fish breeding to link complicated characteristics to specific groups of genetic loci. The intention of this research is to determine the action, interaction, diversity, and location of these loci and map them into a linkage map for application to breeding programs. Two things are required for fish molecular geneticists to undertake a QTL study. Firstly, they require two or more fish varieties that differ genetically in terms of the attributive or traits of interest. For instance, they may choose lines fixed for certain alleles controlling growth rates (one fast and one slow). Second, scientists require genetic markers that differentiate between these parental lineages. For genotyping, molecular markers are preferred since they are unlikely to be influenced by the trait of interest. Diverse classes of markers, such as single nucleotide polymorphisms (SNPs), simple sequence repeats (SSRs, or microsatellites), restriction fragment length polymorphisms (RFLPs), and transposons, can be leveraged for this purpose. Then, in order to conduct the QTL analysis, the parental strains are mated to obtain heterozygous (F1) individuals, which are then bred using a variety of different methods. The phenotypes and genotypes of the F2 population are subsequently scored. In contrast, unlinked markers will not demonstrate a significant correlation with phenotype. The primary objective of QTL analysis has been to determine whether phenotypic variations are largely caused by a small number of loci with relatively large effects or by a large number of loci with relatively low effects. It appears that a significant part of the phenotypic variation in many quantitative traits can be explained by a few loci with

Table 4.2 Diverse molecular markers and their applications in fish breeding and genetic selection

No.	Molecular marker	Utility and methods of application	References
1	Simple sequence repeats (SSR)	SSR can be identified by analyzing whole-genome sequences. Once identified, locus-specific primers can be designed to flank the SSR and subsequent analysis can be performed via PCR and fragment analysis	Huang et al. (2022a, b), Gandomkar et al. (2021), and Tian et al. (2021)
2	Microsatellites	Microsatellites are effective in determining the structure of natural populations, breeding populations, and monitoring stock restoration programs. The general method involves PCR and fragment analysis	Saillant et al. (2022), Guo et al. (2022), Divya et al. (2022), Fazzi-Gomes et al. (2021), Klütsch et al. (2021), White et al. (2021), Weng et al. (2021), Wang et al. (2021), Yagishita and Kume (2021), and Cossu et al. (2021)
3	Single nucleotide polymorphism	SNP analysis can be done using whole-genome sequence data or microarrays designed for a specific fish	Casanova et al. (2022), Abecia et al. (2022), Deeg et al. (2022), Ciezarek et al. (2022), Sandoval-Castillo et al. (2022), and Sinclair-Waters et al. (2022)
4	Double digest RADseq	DNA is digested with restriction enzymes followed by high-throughput sequencing. The method is cost-effective and permits the rapid identification of QTLs	Miller-Crews et al. (2021)
5	Mitochondrial DNA markers	Mitochondrial DNA analysis involves PCR of a specific locus followed by DNA sequencing and analysis Typical loci include the cytochrome oxidase (*Cox I*) or the control region (D-loop)	Yüncü et al. (2021), Nam et al. (2022), Ayadi et al. (2022), Mar-Silva et al. (2022), Wallace et al. (2022), Mzingirwa et al. (2019), Aziz et al. (2016), Barasa et al. (2016), Zhu et al. (2016), Kuo et al. (2014), and An et al. (2013)
6	RNA-based markers	RNA markers can be developed based on expressed sequence tags or cDNA, which are obtained using transcriptomic approaches. Prior information on genes associated with specific traits facilitates the development of RNA markers	Chu et al. (2021), Hsu et al. (2021), and Feng et al. (2018)

significant effects, with the remainder being ascribed to a large number of loci with minor effects. Once QTL has been detected, molecular approaches can be used to limit them to candidate genes.

The basis for the selection of coding loci for deployment as molecular markers is based on prior genetic association as reported in the target species or in a closely

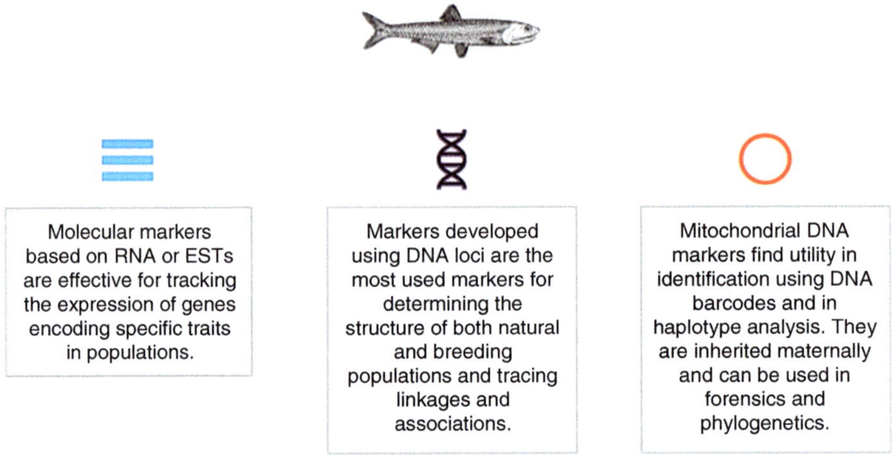

Fig. 4.1 A graphical representation of the three most commonly used molecular markers in aquaculture. DNA markers based on genomic and mitochondrial DNA loci are the most widely used on account of their stability and amenability to technical analysis. Analysis with RNA-based markers requires a higher degree of technical expertise, but they are extremely useful in validating the actual expression of genes associated with traits

related species. The markers, in this case, are designed so as to flank the coding locus of interest. The advantage of markers developed using this approach is that the presence of the gene can be traced over successive generations. The disadvantage of this approach is that any mutation at the primer binding site will result in a false-negative result during the process of PCR amplification. Markers based on non-coding loci are generally neutral as they are not under selection pressure. However, there is a possibility that they may be lost as a result of genetic recombination during meiosis. The large-scale genetic testing in field laboratories associated with fish breeding facilities must be streamlined in order to ensure high throughput and ease of transferability of data. In the case of traits which are regulated by multiple genes, the parental lines need not be different for the phenotype in question; rather, they must include diverse alleles, which are subsequently reassorted by recombination to generate a range of phenotypic values in the derived population. Consider a trait that is controlled by four genes, with the uppercase alleles increasing the trait's value and the lower-case alleles decreasing the trait's value. In this case, individuals with the XXYYppqq and xxyyPPQQ genotypes may have a similar phenotype if the effects of the alleles of the four genes are comparable. Members of the F1 generation (XxYyPpQq) would have an intermediate phenotype and be invariant. The F2 generation, or the offspring produced by a backcross between an F1 individual and either parent, would be variable. The F2 offspring would have somewhere between zero and eight uppercase alleles, while the backcross offspring would have between four and eight uppercase alleles.

The graphical representation of a QTL is a linkage group that indicates the molecular markers that are associated with a specific trait (Fig. 4.1). The molecular

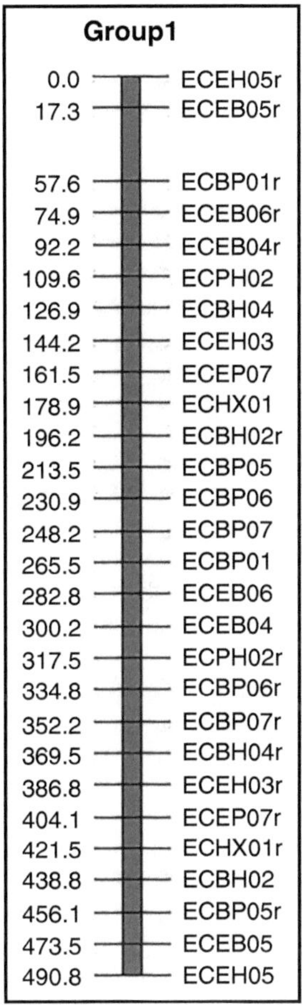

Fig. 4.2 A linkage map depicting the QTLs for gender in *Epinephelus coioides* (female). The numbers on the left side indicate the proximity in centi-Morgans (cM) and the labels on the right side indicate the molecular markers that are linked to the trait. Linkage maps are developed by testing for the presence or absence of a specific locus in a population and its association with a specific trait

markers are tested in the parental genotypes as well as in the sib or half-sib populations in order to determine the degree of linkage between individual loci. Genetic recombination can lead to loss of linkage between loci as they recombine during meiosis. The greater the distance between loci (cM), the more likely they are to disassociate during meiosis. The linkage map is not an indication of the location on a chromosome, but rather a hypothetical construct based on the probability of linkage (Fig. 4.2).

As with most methodologies, QTL analysis has limitations. For instance, QTL investigations need extremely high sample sizes and can only map the changes between the founding parental lines. Some loci will go undiscovered because these genotypes are unlikely to carry segregating alleles of considerable influence at every locus, contributing to heterogeneity in natural populations. In addition, the individual alleles that segregate, primarily in inbred lines, may not be relevant to wild relatives. However, other alleles at these regions are likely to be of relevance. Thus, the objective of multiple studies is to identify loci rather than specific alleles. Exclusions include genes associated with specific phenotypes in fishes.

Feed conversion efficiency (FCE) is one of the most relevant traits as far as the commercial aquaculture industry is concerned. The conversion of biomass within an organism is governed by multiple genes which are associated with the digestibility of formulated feed, and manufacturers can apply these data in order to modify formulations based on species-specific requirements. The transition from fishmeal-based fish feed to vegetable-based formulated fish feed has been proposed as a sustainable solution for the aquaculture industry and there have been several comparative studies which have supported this assertion. For instance, in the case of hybrid groupers, soy protein concentrate was tested as an alternative to fish meal and a formulated feed with 30% soy protein was determined to be ideal based on the data obtained from a transcriptomic study of genes associated with multiple cellular processes (Chen et al. 2020). The approach involving analysis of differentially expressed genes has provided the developers of fish feed with high-resolution data pertaining to the precise physiological responses to the diverse components in formulated feed, as reported in Yellowtail Kingfish (Dam et al. 2020). A study conducted on Atlantic salmon reported on specific metabolic changes in response to feeding with soybean meal, which has been reported to reduce bile acid levels and hypercholesterolemia (Kortner et al. 2013). Investigations into the inclusion of microbial probiotics have also been conducted in white shrimp (Chien et al. 2020), Atlantic salmon, and tilapia (Cano-Lozano et al. 2022; Gopi et al. 2022; Zabidi et al. 2021). The genetic basis for FCE is based on the association of specific genomic features with the trait in question. Genome-Wide Association Studies (GWAS) form the basis for the establishment of associations between the genotype and phenotype and have been conducted extensively in the model organism *Danio rerio* (Wallis et al. 2022). The methods applied in these studies have been extended to a large number of endangered and commercially developed fish species. The genetic basis for sexual dimorphism has been validated in Yellow Catfish by identifying a specific gene (*statb5*), which contributed to increased growth in transgenic catfish and zebrafish that had been engineered to express the protein stat5b (Huang et al. 2022a, b). The genomes of three species of carp have been analyzed extensively. A high-resolution linkage map of *Cyprinus carpio* identified 28,426 single nucleotide polymorphism (SNP) markers, among which 17 QTLs and nine candidate genes which are involved in carbohydrate metabolism, fat accumulation, digestion, growth regulation, cell proliferation and differentiation, lipid metabolism, and energy deposition based on the gene ontology (Zhang et al. 2021). Bulked segregant analysis is an effective method for the characterization of QTLs (Taslima et al. 2020) and the extension of this method to transgenic Yellow River Carp led to the discovery of 28 genes linked to growth, of which two genes encoding BR serine/threonine-protein kinase 2 (BRSK2) and eukaryotic translation-initiation factor 2-alpha kinase 3 (*Eif2ak3*) had a strong correlation with growth and final body weight (Luo et al. 2018). Linkage maps have proven to be highly effective in mapping quantitative traits to loci, as in the case of the crucian carp, in which seven candidate genes with linkages to digestion, signal transduction, energy metabolism, and biosynthesis were identified using a high-resolution linkage map (Pang et al. 2017). High-throughput methods based on microarrays have proven to be effective for the detection of QTLs in the Nile tilapia in which candidate genes previously reported to be associated with feed-efficiency traits were located in these QTL regions, including neurotrophic tyrosine kinase, receptor, type 3a, growth hormone-releasing hormone, and eukaryotic translation-initiation factor 4E family member 3 (Barría et al. 2021).

The emergence of novel pathogens in aquaculture systems (Ziarati et al. 2022) and the development of reservoirs due to intensive culture (Johnstone et al. 2022) coupled with the indiscriminate application of antibiotics (Abdalla et al. 2022; Raharjo et al. 2022) pose a significant challenge to the aquaculture industry, and the selection of broodstock with an innate resistance to microbial pathogens is one of the key goals of a breeding program. GWAS has proven to be effective in the identification of genes related to immunity and resistance in fishes (Lund et al. 2022) and crustaceans (Lǚ et al. 2022). Investigations into the molecular mechanisms of resistance in aquatic species have been directed toward the identification of genes which are differentially expressed in challenge tests involving pathogens, and the elucidation of the pathways that direct the immune response. Microsatellite markers linked to protein coding genes for Hepcidin-2 (*hamp2*), Hepcidin antimicrobial peptide 1 (*hamp1*), Progranulin (pgrn2), Progranulin receptor (*pgrn1*), and piscidin 4 (*TP4*) genes were established for tilapia strains farmed in Taiwan after challenge with *Streptococcus iniae* with a total of 55 genotypes that were linked to resistance to the pathogen (Chen et al. 2021). Genomic prediction models based on the best linear unbiased prediction (BLUP), which rely on data obtained from SNP and challenge tests, have proven to be effective in identifying Tilapia genotypes which are resistant to *Streptococcosis agalactiae* in Thailand and both the BLUP and best linear unbiased estimation (BLUE) models have proven to be useful in predicting random and fixed effects in pedigreed populations (Sukhavachana et al. 2021). Recently, a GWAS study conducted across natural populations of the red-spotted grouper identified five genes: ephrin type-A receptor 7 (*EPHA7*), oxysterol-binding protein-related protein 2 (*OSBPL2*), glypican 5 (*GPC5*), cadherin 4 (*CDH4*), and POU domain protein (*POU3F1*) all of which are involved in cellular processes such as receptor binding, cell division, cell division, and the neuroendocrine system (Yang et al. 2021). Challenge tests in European sea bass with *Vibrio anguillarum* identified QTLs related to stress response and body weight but not to disease resistance (Chatziplis et al. 2020), and a similar study conducted in Norway, involving Atlantic salmon challenged with the most infectious variant of salmonid alphavirus (SAV), led to the identification of QTL linked to the innate and adaptive immune response (Hillestad et al. 2020). The linkage between microsatellite markers and tolerance to white spot syndrome virus (WSSV) has revealed a linkage between disease tolerance and genetic diversity among populations of *Penaeus* monodon (Mondal et al. 2019). A high-resolution map of Atlantic salmon that had been challenged with piscine myocarditis virus

(PMCV) which is the causative agent of cardiomyopathy syndrome in farmed fish identified several QTLs linked to resistance and concluded that genetic factors contributed to this trait. Identification of a specific region of the chromosome or specific genes can be leveraged for the development of molecular markers for rapid genotyping (Hillestad and Moghadam 2019). Interspecific hybrids of channel catfish and blue catfish have been studied as a model for the discovery of QTLs linked to Enteric septicemia of catfish (ESC), a devastating disease which is caused by *Edwardsiella ictaluri*. Blue catfish are highly resistant as compared to channel catfish, which are susceptible. The development of a breeding population by interspecific backcrossing followed by GWAS led to the discovery of two genomic regions that were significantly associated with disease resistance, both of which originated from blue catfish (Tan et al. 2018). The identification of a single gene linked to a specific trait greatly improves the traceability and reliability of a molecular marker as the trait can be ascertained in the parental genotypes and the progeny. Genes involved in innate and adaptive immunity are ideal for marker development as polymorphisms, both synonymous and non-synonymous, can provide the basis for linkage studies. The sequencing and analysis of the *Ctenopharyngodon idella* Toll-like receptor 7 (*citlr7*) gene, revealed 11 SNPs which could be correlated to resistance to grass carp reovirus (GCRV) (Su et al. 2018). Challenge tests in Pacific oysters led to the identification of markers linked to resistance to Ostreid herpesvirus (OsHV), which can be leveraged for the design and development of selective breeding programs (Gutierrez et al. 2018).

Sex-linked markers are of particular interest to breeders as they facilitate the process of selection of monosex populations for grow out as well as founders for the establishment of breeding programs. Sex-determining regions have been extensively studied in Mozambique tilapia (Tao et al. 2021), Nile tilapia (Conte et al. 2017; Yu et al. 2014; Harvey et al. 2003), Cichlids (Curzon et al. 2021a, b), Pacific halibut (Jasonowicz et al. 2022), Southern catfish (Zheng et al. 2022), Russian sturgeon (Curzon et al. 2022), Rock bream (Gong et al. 2022), and Atlantic salmon (Gabián et al. 2019; Lubieniecki et al. 2015a). The identification of sex-linked markers in fish that undergo protogyny and protandry (Pla et al. 2022) is complex as several factors, such as environmental cues (Soyano et al. 2022; Delbes et al. 2022; García-Cruz et al. 2020), climate change (Geffroy and Wedekind 2020; Edmands 2021), developmental stage (Boddington et al. 2021), epigenetics (He et al. 2020), or a combination of multiple factors (Rajendiran et al. 2021) can contribute to gender reversal. The effectiveness of the MAS-based approach has been reported to be successful in some species such as tilapia (Curzon et al. 2021a, b) and salmon (Anglès d'Auriac et al. 2014). However, the inherent genetic instability (Podlesnykh et al. 2017; Lubieniecki et al. 2015a, b) and the complex interaction of multiple factors make the process of development of molecular markers for sexual determination a long and tedious one.

4.4 Recruitment of Broodstock from Wild Populations Based on Molecular Markers

The assessment of the diversity of natural fish populations is critical to their effective management. Recruitment involves the selection of wild germplasm with desirable traits with the intention of developing a pedigreed population for large-scale commercial production. Individuals from an admixed population can be selected randomly for captive breeding, following which their phenotypic characteristics such as growth, resistance to disease, and FCR can be characterized under controlled conditions. Individuals selected from wild populations must be identified based on their morphological features, with the assistance of an expert taxonomist, or alternatively, using DNA barcoding (Munguia-Vega et al. 2022; Zainal Abidin et al. 2021; Collins et al. 2021). Aquatic species which are recruited for aquaculture are derived from natural populations which can be geographically distinct and genetically diverse. The general assumption made by fish breeders is that an admixed population can lead to progeny with a higher level of fitness, as has been reported in species ranging from salmon (Besnier et al. 2022), Nile tilapia (Geraerts et al. 2022) to striped bass (LeBlanc et al. 2020), although this assumption must be treated with caution as it can also contribute to an overall loss of fitness in natural populations, especially in the case of restocking programs (Castellani et al. 2018; Carlon et al. 2021; Blackwell et al. 2021). The genetic diversity of an admixed population poses challenges to marker development as a result of the high allele frequency (Leitwein et al. 2018), genetic divergence (Zhao et al. 2018), and cryptic diversity (Lee and Munroe 2021; Shao et al. 2021).

The practical approach to recruitment of individuals from wild populations must be preceded by an estimation of genetic diversity and the contextualization of this information within management units (MU) and ecologically significant units (ESU) (Cardeñosa et al. 2014; Lyons et al. 2019). Fish conservation biologists routinely assess and analyze the structure of extant populations using a variety of molecular tools which include both coding and non-coding regions of the nuclear and mitochondrial genomes in order to develop management strategies for the conservation of native populations which may be threatened with loss of diversity or decrease in numbers (Barreto et al. 2016). The data obtained from these investigations can be extended to breeding programs which can be incorporated into the overall management strategy for conservation in conjunction with commercial exploitation. Natural populations are likely to exhibit signatures of cryptic speciation (Hoekzema and Sidlauskas 2014; Weiss et al. 2013) if separated by geographic or reproductive barriers as is the case with allopatric, peripatric (Gotoh et al. 2009), parapatric (Berner et al. 2017), or sympatric (Salisbury and Ruzzante 2022) speciation. The phenomenon of speciation has implications for aquaculture and breeding, as in most cases, the evolution of reproductive barriers, either physiological or genetic, can limit the range of breeding combinations to a specific population.

Mitogenomes offer a reliable preliminary insight into the structure of populations based on analysis of haplotypes. The availability of a large number of mitogenomes at publicly available databases (Pathak et al. 2019; Nagpure et al. 2015) and advances in high-throughput DNA sequencing of mitochondrial genomes (Hirase et al. 2016) and innovative strategies for metabarcoding (Miya et al. 2015) have facilitated the development of high-resolution mapping of population structure. The cost involved in mitogenome analysis is far lower than whole-genome sequencing and can provide breeders with sufficient information for incorporation into planning and design of breeding programs.

Higher-resolution mapping of population structure can be carried out using microsatellites and the data from these studies can be used to assess if populations are actively migrating out of their range and bear evidence of outbreeding. These populations are likely to exhibit a higher level of genetic fitness as compared to inbreeding populations and can be recruited for development of breeding programs. Assessment of recruits on the basis of specific genomic features such as resistance to disease is another parameter which can be relied upon to select individuals from the wild. This must be done on the basis of trait specific markers which can be applied to screen the entire population.

4.5 Factors to Consider When Developing Molecular Markers

The development of genomic molecular markers for application to aquaculture depends on several factors that determine the complexity of the process. The following questions must be addressed prior to designing the experimental setup:

(a) Existing genomic information in public databases: the availability of genome sequence data, either in the form of whole genomes, transcriptomes or cDNA as well as experimental data from linkage studies is a major contributor to marked development. In addition to the target species, molecular data from closely related species can also provide additional information.
(b) Genomic complexity: polyploid genomes, genomes containing a large number of features such as repeat units or regions of complexity can present a challenge to market development. In these cases, it is advisable to focus on genomic regions which are less likely to present difficulties during high-throughput analysis.
(c) Life cycle of the organism: a majority of commercially cultivated fishes have a short life cycle which can be completed in captivity; however, some species may have complex life cycles which are based on migration from saline to freshwater environments such as the anadromous Hilsa Shad Tenualosa (Asaduzzaman et al. 2019; Giri et al. 2022) and vice versa as in the case of the tropical eels (Arai 2020; Sudo and Yada 2022) and may require that certain developmental stages be completed in different ecological niches. Some fishes such as groupers (Khasanah et al. 2019; Oh et al. 2013) undergo physiological changes during sex reversal and these must be taken into account when breeding protogynous species. Protogyny is regulated by subtle environmental cues that may include social interaction and population density (Chen et al. 2019), and management of captive spawning in these cases may necessitate the application of specific

hormones (Wang et al. 2017a, b; Lee et al. 2014) or the regulation of environmental cues such as salinity, photoperiod, and diet.

(d) The number of populations which are available for sampling: Breeders may not have access to all wild populations in a given geographical area due to statutory limitations from the local wildlife authorities. This can be overcome by designing breeding programs in collaboration with local authorities in order to ensure that restocking and maintenance of genetic purity ensures a tradeoff between conservation and commercial application.

(e) Amenability to captive breeding in a controlled environment: not all fishes may be compatible with the confined environment of a captive breeding facility. Facility design should take into consideration the unique environmental habitats of each fish species. These may include the incorporation of design elements which ensure optimal water quality parameters, lighting, simulation of tidal flow rates, and low stocking densities which are all representative of the natural milieu from which fish has been relocated.

4.6 Development of Molecular Markers Based on Genomic Data

The general approach to marker development commences with the establishment of the breeding population, following which the downstream processes such as DNA extraction, library preparation, genome sequencing, QTL mapping or GWAS and design of analytical tools follow. One possible approach involves the random selection of mature breeding individuals from different populations (Fig. 4.3), followed by relocation to a breeding facility or hatchery. An admixed population is generated by random mating, followed by selection of individuals from the F1 population based on a specific challenge, which may be resistance to a specific pathogen. The surviving individuals are selected and subjected to genome sequencing and the sequences are translated into markers for application to subsequent generations. This method is based on an a priori assumption that outbreeding results in greater fitness and the aim is to select outbred individuals purely on the basis of a specific trait. The advantage is that this approach can be adopted quickly. The disadvantage being that outbreeding may not necessarily lead to a genetically superior population.

Another approach is based on the development of F1 populations based on sibs and half-sibs (Fig. 4.4). In this case, the common factor can be either the dam or the sire. The process involves whole-genome sequencing of each of the three parental genotypes, followed by comparative analysis and the identification of unique markers for each individual. Following this, the F1 hybrid population is developed by artificial spawning, and both sibs and half-sibs are screened for the markers derived from the parental genotypes. This method is time-consuming, but the major advantage is that linkage groups can be established and the frequency of recombination at each locus can be ascertained.

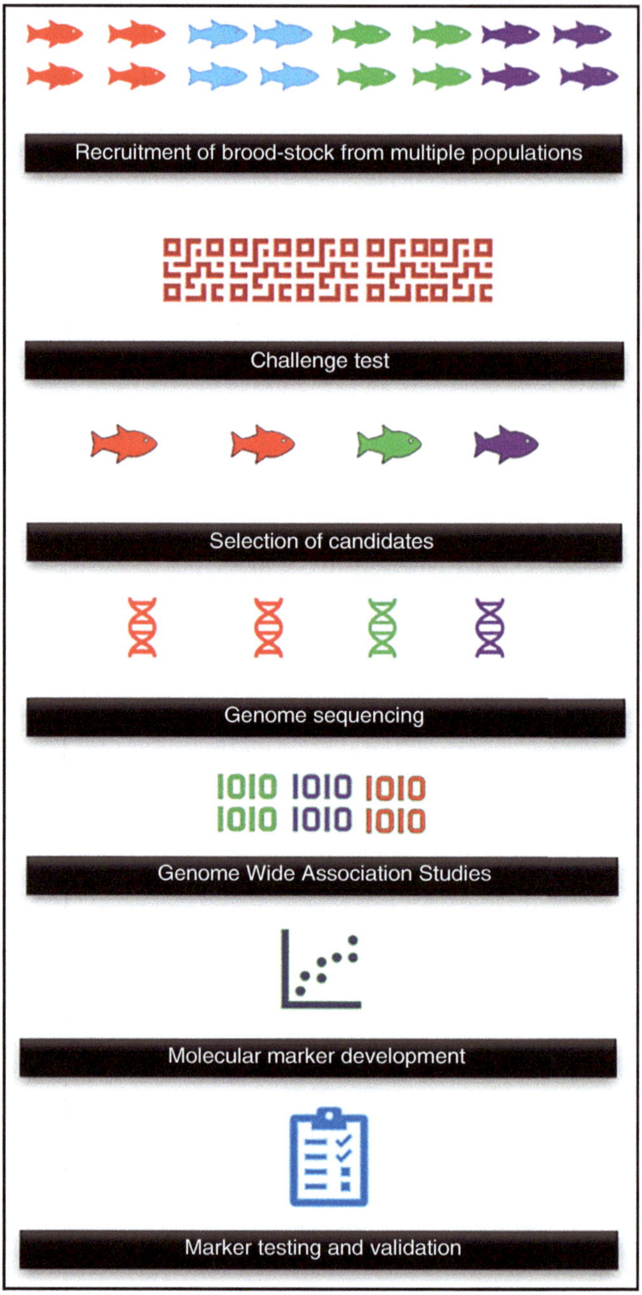

Fig. 4.3 Overview of one of the possible approaches to marker development. Four populations of a single species are established at a breeding facility. Random breeding is conducted to create an admixed population, following which the F1 population is subjected to a single challenge test. The individuals who overcome the challenge are selected for genome sequencing, GWAS, QTL mapping, and marker development

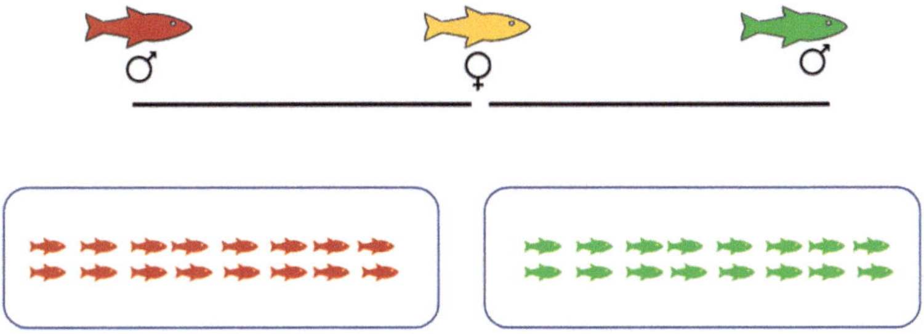

Fig. 4.4 Another possible approach can involve the breeding of half-sib population. This facilitates the development of QTLs based on linkage groups in sib and half-sibs

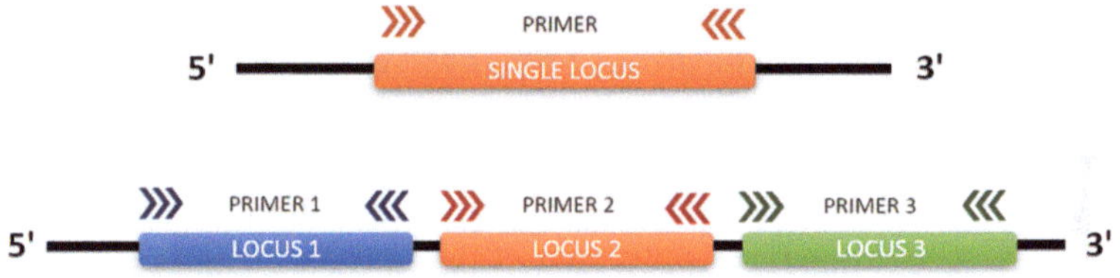

Fig. 4.5 Design of locus-specific PCR primers based on the linkage of these loci to a specific trait

There may be practical limitations when it comes to an actual breeding program which may include issues related to fecundity as well as the diversity of approaches which can be adopted to overcome the limitations, these include promiscuous breeding in the event that admixtures of natural populations breed readily, polygyny, or polyandry in populations where either the sire or dam has a lower level of fecundity, or there is a deficit in the number of mature individuals with the capability to breed.

Genomic data, or specifically genomic loci, can be mapped by developing locus-specific primers pairs for PCR (Fig. 4.5). In cases, where more than one locus contributes to the manifestation of a trait, individual markers can be designed for each of these loci and tested in the parental genotypes and F1 and subsequent generations. The inheritance of these markers can be used as the basis for the selection of progeny based solely on the marker-derived data or in combination with other phenotypic markers. This method is highly effective for the identification of discrete traits but has limited applicability to continuous traits which are governed by multiple loci or a combination of genetic and environmental factors.

4.7 Molecular Markers in Breeding Programs

Molecular markers can be applied to trace the inheritance of specific alleles during the process of pyramiding for hybrid development. For example, if individuals from six different populations have been selected on the basis of phenotypic traits in order to create a pedigreed population containing a combination of all the alleles.

The experimental design for achieving this objective involves the selection of the six individuals which exhibit the desired phenotype (Fig. 4.6). This is followed by whole-genome sequencing of the individuals. Comparison of the six complete genome datasets will enable the identification of loci which are linked to the trait. Once this has been successfully established, the next step is to develop markers, either PCR or microarray based which can be used to screen the first generation of F1 hybrids. This generation can then be used for selection of the desired traits followed by development of the F2 and finally the F3 generation. The similar approach can be used to mapping of traits in the case of backcrosses of fish hybrids.

A similar approach can be applied to the development of hybrids based on backcrossing of wild recruits with a recurrent parent from an elite commercial variety (Fig. 4.7). The practical application of this approach relies on backcrossing

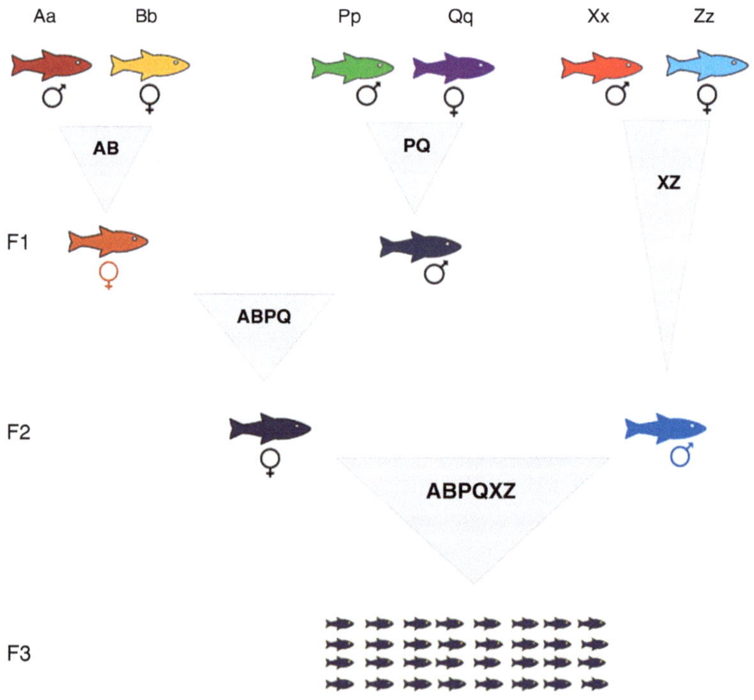

Fig. 4.6 Molecular markers can be applied for the development of pedigreed lines. In this case, the desired traits are linked to the alleles *A*, *B*, *P*, *Q*, *X*, and *Z*. PCR primers developed to amplify these specific alleles are applied at every stage in order to select individuals who have inherited the respective alleles

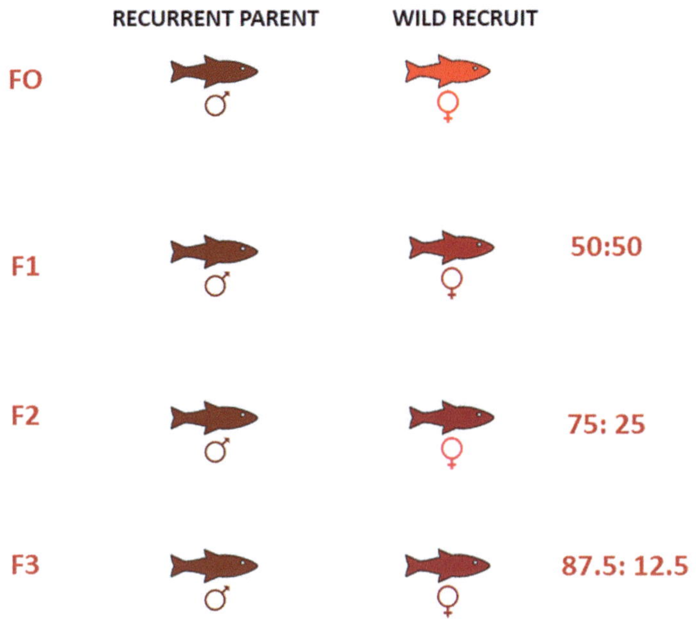

Fig. 4.7 Molecular markers can be used to the levels of integration over subsequent generations in a backcrossing program

each generation with the original parent, assuming that the recurrent parent remains fertile and has the ability to produce viable sperm for each breeding cycle. Molecular markers can be applied at each of the generations in order to determine the degree of introgression. Assuming that there is a steady reduction of wild-type genes over subsequent generations, there will be a point at which the genome of the hybrid will be almost identical to the recurrent parent. This approach is adopted in cases where wild-type genes such as those for resistance to novel pathogens have to be recruited into an elite or pedigreed population as a component of a genetic enhancement strategy.

4.8 Software Used for Analysis of Data from Breeding Experiments

The data obtained from the application of molecular markers are of two general types. The first type comprises binary data, which is obtained via PCR and is recorded as either the presence or absence of a specific locus. The second type of data is in the form of DNA sequences, which must be analyzed to determine variations in the form of mutations, insertions and deletions (indels), transitions, and transversions. Both of these data types can be analyzed using different approaches. One of the most widely used approaches is linkage analysis. The linkage disequilibrium (LD) between phenotypes and genetic markers spanning the entire genome is the foundation of Genome-Wide Association Studies (GWAS). Aside from the genetic relationship between the genetic markers and the causal mutations, a number of additional factors, such as selection and nonrandom mating formatting population structure, also contribute to the LD. Numerous techniques have been created along with the related software, such as the multiple loci mixed model (MLMM). Software packages use a variety of techniques to lower the learning curve. The Genomic Association and Prediction Integrated Tool (GAPIT) implements eight models, including the general linear model (GLM), the mixed linear model (MLM), the compressed MLM, the MLMM, SUPER (Settlement of Mixed Linear Models Under Progressively Exclusive Relationship), FarmCPU (Fixed and Random Model Circulating Probability Unification), and BLINK (Bayesian-information and Linkage-disequilibrium Iteratively Nested Keyway). (Wang et al. 2022; Lup et al. 2022). Random mutagenesis is a popular method for genetically dissecting biological processes, followed by screening for desirable phenotypes. However, in the past, finding the causal gene required time-consuming mapping methods based on iterative linkage analysis. Using next-generation sequencing (NGS) data from a mapping population that segregates for the mutant phenotype, mapping-by-sequencing expedites this procedure by effectively connecting the mutant phenotype to a specific candidate chromosomal region. The Easy Map software can be applied to perform mapping by sequencing (Lup et al. 2022). Alternative computational approaches include statistical analysis using the effective-median-based Mendelian randomization (MR) framework, for correlation of casual genes with complex phenotypes (Jiang et al. 2022). Higher levels of efficiency can be achieved by using microarray-based methods (Balagué-Dobón et al. 2022) that incorporate customized software that leads to higher-resolution mapping. Data from the highly cited method RADseq can be analyzed using customized pipelines based on the open-source platform R (Seki 2021). The rapid advances in bioinformatics and the development of user interfaces which can be utilized by technical staff who may not be familiar with programming languages have provided opportunities for aquaculture companies to streamline the deployment of molecular methods on a large scale.

4.9 Factors Affecting the Application of Molecular Markers

Genetic recombination and the loss of linkage during meiosis, as well as epigenetic factors, contribute to the reduction in the accuracy of DNA-based molecular markers. Current evidence indicates that traits in general are governed by multiple factors which include translational and transcriptional regulatory elements (Mehta et al. 2021), signaling pathways (Dvergedal et al. 2020), variability in gene expression in response to environmental cues (Moustakas-Verho et al. 2020; Verta et al. 2020), and although recent developments in predictive transcriptomics have provided some degree of clarity (Mulugeta et al. 2019), the larger picture still remains a complex one which involves the interplay of a wide range of molecular interactions and determines phenotype and genotypes associations. Although methods for epigenetic analysis are available, they are not as reliable as the currently utilized markers based on DNA. Over the next few decades, the development of more efficient DNA sequencing methods in combination with metabolomics and proteomics will shape our understanding of the factors which contribute to genome evolution and the reliability of molecular markers.

4.10 Future Direction

Precision aquaculture will become increasingly relevant within the context of a growing global population and the associated impact on natural resources. Molecular markers will play an increasingly important role in the characterization of wild fishery resources, selection of broodstock, and the identification of QTLs, which in turn will facilitate the effective management of breeding programs and increase the efficiency of aquaculture operations. The rapid decline in the cost of DNA sequencing coupled with the simplicity of data analysis and interpretation using artificial intelligence will usher in a new era in aquaculture which will be based on efficiency and scalability to meet increasing market demands.

References

Abdalla T, Al-Rumaithi H, Osaili TM, Hasan F, Obaid RS, Abushelaibi A, Ayyash MM (2022) Prevalence, antibiotic-resistance, and growth profile of *Vibrio* spp. isolated from fish and shellfish in Subtropical-Arid Area. Front Microbiol 13:861547. https://doi.org/10.3389/fmicb.2022.861547

Abecia JE, King AJ, Luiz OJ, Crook DA, Wedd D, Banks SC (2022) Diverse parentage relationships in paternal mouthbrooding fishes. Biol Lett 18(5):20210576. https://doi.org/10.1098/rsbl.2021.0576

Algammal AM, Mohamed MF, Tawfiek BA, Hozzein WN, El Kazzaz WM, Mabrok M (2020) Molecular typing, antibiogram and PCR-RFLP based detection of *Aeromonas hydrophila* complex isolated from *Oreochromis niloticus*. Pathogens 9(3):238. https://doi.org/10.3390/pathogens9030238

Allal F, Nguyen NH (2022) Genomic selection in aquaculture species. Methods Mol Biol 2467:469–491. https://doi.org/10.1007/978-1-0716-2205-6_17

An HS, Lee JW, Park JY, Myeong JI, An CM (2013) Mitochondrial and microsatellite DNA analyses showed comparative genetic diversity between parent and offspring populations of Korean black rockfish in a hatchery facility. Genet Mol Res 12(4):6389–6401. https://doi.org/10.4238/2013.December.9.2

Anglès d'Auriac MB, Urke HA, Kristensen T (2014) A rapid qPCR method for genetic sex identification of *Salmo salar* and *Salmo trutta* including simultaneous elucidation of interspecies hybrid paternity by high-resolution melt analysis. J Fish Biol 84(6):1971–1977. https://doi.org/10.1111/jfb.12401

Arai T (2020) Ecology and evolution of migration in the freshwater eels of the genus *Anguilla* Schrank, 1798. Heliyon 6(10):e05176. https://doi.org/10.1016/j.heliyon.2020.e05176

Asaduzzaman M, Igarashi Y, Wahab MA, Nahiduzzaman M, Rahman MJ, Phillips MJ et al (2019) Population genomics of an Anadromous Hilsa Shad *Tenualosa ilisha* species across its diverse migratory habitats: discrimination by fine-scale local adaptation. Genes 11(1):46. https://doi.org/10.3390/genes11010046

Ayadi ZEM, Tazerouti F, Gastineau R, Justine J-L (2022) Redescription, complete mitochondrial genome and phylogenetic relationships of *Hexostoma thynni* (Delaroche, 1811) Rafinesque, 1815 (Monogenea, Hexostomatidae). Parasite 29:29. https://doi.org/10.1051/parasite/2022030

Aziz NMA, Esa Y, Arshad A (2016) DNA barcoding and phylogenetic analysis of Malaysian groupers (Subfamily: Epinephelinae) using mitochondrial Cytochrome c oxidase I (COI) gene. J Environ Biol 37(4 Spec No):725–733

Balagué-Dobón L, Cáceres A, González JR (2022) Fully exploiting SNP arrays: a systematic review on the tools to extract underlying genomic structure. Brief Bioinform 23(2):bbac043. https://doi.org/10.1093/bib/bbac043

Bao X, Wang W, Yuan T, Li Y, Chen X, Liu X et al (2022) Transcriptome profiling based on larvae at different time points after hatching provides a core set of gene resource for understanding the immune response mechanisms of the egg-protecting behavior against *Vibrio anguillarum* infection in *Amphioctopus fangsiao*. Fish Shellfish Immunol 124:430–441. https://doi.org/10.1016/j.fsi.2022.04.030

Barasa JE, Abila R, Grobler JP, Agaba M, Chemoiwa EJ, Kaunda-Arara B (2016) High genetic diversity and population differentiation in *Clarias gariepinus* of Yala Swamp: evidence from mitochondrial DNA sequences. J Fish Biol 89(6):2557–2570. https://doi.org/10.1111/jfb.13150

Barreto SB, Nunes LA, da Silva AT, Jucá-Chagas R, Diniz D, Sampaio I et al (2016) Is *Nematocharax* (Actinopterygii, Characiformes) a monotypic fish genus? Genome 59(10): 851–865. https://doi.org/10.1139/gen-2015-0166

Barría A, Benzie JAH, Houston RD, De Koning D-J, de Verdal H (2021) Genomic selection and genome-wide association study for feed-efficiency traits in a farmed Nile tilapia (*Oreochromis niloticus*) population. Front Genet 12:737906. https://doi.org/10.3389/fgene.2021.737906

Berner D, Ammann M, Spencer E, Rüegg A, Lüscher D, Moser D (2017) Sexual isolation promotes divergence between parapatric lake and stream stickleback. J Evol Biol 30(2):401–411. https://doi.org/10.1111/jeb.13016

Besnier F, Ayllon F, Skaala Ø, Solberg MF, Fjeldheim PT, Anderson K et al (2022) Introgression of domesticated salmon changes life history and phenology of a wild salmon population. Evol Appl 15(5):853–864. https://doi.org/10.1111/eva.13375

Blackwell T, Ford AGP, Ciezarek AG, Bradbeer SJ, Gracida Juarez CA, Smith AM et al (2021) Newly discovered cichlid fish biodiversity threatened by hybridization with non-native species. Mol Ecol 30(4):895–911. https://doi.org/10.1111/mec.15638

Boddington DK, Wakefield CB, Fisher EA, Fairclough DV, Harvey ES, Newman SJ (2021) Age, growth and reproductive life-history characteristics infer a high population productivity for the sustainably fished protogynous hermaphroditic yellowspotted rockcod (*Epinephelus areolatus*) in north-western Australia. J Fish Biol 99(6):1869–1886. https://doi.org/10.1111/jfb.14889

Cádiz MI, López ME, Díaz-Domínguez D, Cáceres G, Yoshida GM, Gomez-Uchida D, Yáñez JM (2020) Whole genome re-sequencing reveals recent signatures of selection in three strains of farmed Nile tilapia (*Oreochromis niloticus*). Sci Rep 10(1):11514. https://doi.org/10.1038/s41598-020-68064-5

Cano-Lozano JA, Villamil Diaz LM, Melo Bolivar JF, Hume ME, Ruiz Pardo RY (2022) Probiotics in tilapia (*Oreochromis niloticus*) culture: potential probiotic *Lactococcus lactis* culture conditions. J Biosci Bioeng 133(3):187–194. https://doi.org/10.1016/j.jbiosc.2021.11.004

Cardeñosa D, Hyde J, Caballero S (2014) Genetic diversity and population structure of the pelagic thresher shark (*Alopias pelagicus*) in the Pacific Ocean: evidence for two evolutionarily significant units. PLos One 9(10):e110193. https://doi.org/10.1371/journal.pone.0110193

Carlon DB, Robertson DR, Barron RL, Choat JH, Anderson DJ, Schwartz SA, Sánchez-Ortiz CA (2021) The origin of the parrotfish species *Scarus compressus* in the Tropical Eastern Pacific: region-wide hybridization between ancient species pairs. BMC Ecol Evol 21(1):7. https://doi.org/10.1186/s12862-020-01731-3

Casanova A, Heras S, Abras A, Roldán MI, Bouza C, Vera M et al (2022) Genomic hatchery introgression in brown trout (*Salmo trutta* L.): development of a diagnostic SNP panel for monitoring the impacted Mediterranean rivers. Genes 13(2):255. https://doi.org/10.3390/genes13020255

Castellani M, Heino M, Gilbey J, Araki H, Svåsand T, Glover KA (2018) Modeling fitness changes in wild Atlantic salmon populations faced by spawning intrusion of domesticated escapees. Evol Appl 11(6):1010–1025. https://doi.org/10.1111/eva.12615

Chatziplis D, Oikonomou S, Loukovitis D, Tsiokos D, Samaras A, Dimitroglou A et al (2020) QTL for stress and disease resistance in European sea bass, *Dicentrarhus labrax* L. Animals 10(9): 1668. https://doi.org/10.3390/ani10091668

Chen J, Xiao L, Peng C, Ye Z, Wang D, Yang Y et al (2019) Socially controlled male-to-female sex reversal in the protogynous orange-spotted grouper, *Epinephelus coioides*. J Fish Biol 94(3): 414–421. https://doi.org/10.1111/jfb.13911

Chen Y, Liu W, Ma J, Wang Y, Huang H (2020) Comprehensive physiological and transcriptomic analysis revealing the responses of hybrid grouper (*Epinephelus fuscoguttatus*♀ × *E. lanceolatus*♂) to the replacement of fish meal with soy protein concentrate. Fish Physiol Biochem 46(6):2037–2053. https://doi.org/10.1007/s10695-020-00851-3

Chen C-C, Huang C-W, Lin C-Y, Ho C-H, Pham HN, Hsu T-H et al (2021) Development of disease-resistance-associated microsatellite DNA markers for selective breeding of tilapia (*Oreochromis* spp.) farmed in Taiwan. Genes 13(1):99. https://doi.org/10.3390/genes13010099

Cheng L, Liu L, Yu X, Wang D, Tong J (2010) A linkage map of common carp (*Cyprinus carpio*) based on AFLP and microsatellite markers. Anim Genet 41(2):191–198. https://doi.org/10.1111/j.1365-2052.2009.01985.x

Chien C-C, Lin T-Y, Chi C-C, Liu C-H (2020) Probiotic, Bacillus subtilis E20 alters the immunity of white shrimp, *Litopenaeus vannamei* via glutamine metabolism and hexosamine biosynthetic pathway. Fish Shellfish Immunol 98:176–185. https://doi.org/10.1016/j.fsi.2020.01.014

Chu P-Y, Li J-X, Hsu T-H, Gong H-Y, Lin C-Y, Wang J-H, Huang C-W (2021) Identification of genes related to cold tolerance and novel genetic markers for molecular breeding in Taiwan tilapia (*Oreochromis* spp.) via transcriptome analysis. Animals 11(12):3538. https://doi.org/10.3390/ani11123538

Ciezarek A, Ford AGP, Etherington GJ, Kasozi N, Malinsky M, Mehta TK et al (2022) Whole genome resequencing data enables a targeted SNP panel for conservation and aquaculture of *Oreochromis cichlid* fishes. Aquaculture 548:737637. https://doi.org/10.1016/j.aquaculture.2021.737637

Collins RA, Trauzzi G, Maltby KM, Gibson TI, Ratcliffe FC, Hallam J et al (2021) Meta-Fish-Lib: a generalised, dynamic DNA reference library pipeline for metabarcoding of fishes. J Fish Biol 99(4):1446–1454. https://doi.org/10.1111/jfb.14852

Conte MA, Gammerdinger WJ, Bartie KL, Penman DJ, Kocher TD (2017) A high quality assembly of the Nile Tilapia (*Oreochromis niloticus*) genome reveals the structure of two sex determination regions. BMC Genomics 18(1):341. https://doi.org/10.1186/s12864-017-3723-5

Cossu P, Mura L, Scarpa F, Lai T, Sanna D, Azzena I et al (2021) Genetic patterns in *Mugil cephalus* and implications for fisheries and aquaculture management. Sci Rep 11(1):2887. https://doi.org/10.1038/s41598-021-82515-7

Curzon AY, Shirak A, Benet-Perlberg A, Naor A, Low-Tanne SI, Sharkawi H et al (2021a) Gene variant of barrier to autointegration factor 2 (Banf2w) is concordant with female determination in Cichlids. Int J Mol Sci 22(13):7073. https://doi.org/10.3390/ijms22137073

Curzon AY, Shirak A, Zak T, Dor L, Benet-Perlberg A, Naor A et al (2021b) All-male production by marker-assisted selection for sex determining loci of admixed *Oreochromis niloticus* and *Oreochromis aureus* stocks. Anim Genet 52(3):361–364. https://doi.org/10.1111/age.13057

Curzon AY, Shirak A, Meerson A, Degani G, Hurvitz A, Ben-Naim N et al (2022) Cross-species conservation of a transposase-linked element enables genetic sexing of commercial populations of Russian sturgeon (*Acipenser gueldenstaedtii*). Anim Genet 53(3):441–446. https://doi.org/10.1111/age.13188

Dam CTM, Ventura T, Booth M, Pirozzi I, Salini M, Smullen R, Elizur A (2020) Intestinal transcriptome analysis highlights key differentially expressed genes involved in nutrient metabolism and digestion in yellowtail kingfish (*Seriola lalandi*) fed terrestrial animal and plant proteins. Genes 11(6):621. https://doi.org/10.3390/genes11060621

Deeg CM, Sutherland BJG, Ming TJ, Wallace C, Jonsen K, Flynn KL et al (2022) In-field genetic stock identification of overwintering coho salmon in the Gulf of Alaska: evaluation of nanopore sequencing for remote real-time deployment. Mol Ecol Resour 22(5):1824–1835. https://doi.org/10.1111/1755-0998.13595

Delbes G, Blázquez M, Fernandino JI, Grigorova P, Hales BF, Metcalfe C et al (2022) Effects of endocrine disrupting chemicals on gonad development: mechanistic insights from fish and mammals. Environ Res 204(Pt B):112040. https://doi.org/10.1016/j.envres.2021.112040

Divya PR, Joy L, Poulose S, Jency PME, Basheer VS, Kathirvelpandian A et al (2022) Genetic stock identification in *Perna viridis* (Linnaeus 1758) from the Indian Peninsula by using microsatellite markers. Mol Biol Rep 49(4):3357–3363. https://doi.org/10.1007/s11033-022-07178-7

Dvergedal H, Harvey TN, Jin Y, Ødegård J, Grønvold L, Sandve SR et al (2020) Genomic regions and signaling pathways associated with indicator traits for feed efficiency in juvenile Atlantic salmon (*Salmo salar*). Genet Sel Evol 52(1):66. https://doi.org/10.1186/s12711-020-00587-x

Edmands S (2021) Sex ratios in a warming world: thermal effects on sex-biased survival, sex determination, and sex reversal. J Hered 112(2):155–164. https://doi.org/10.1093/jhered/esab006

Fazzi-Gomes PF, da Paz Aguiar J, Marques D, Fonseca Cabral G, Moreira FC, Rodrigues MDN et al (2021) Novel microsatellite markers used for determining genetic diversity and tracing of wild and farmed populations of the Amazonian giant fish *Arapaima gigas*. Genes 12(9):1324. https://doi.org/10.3390/genes12091324

Feng B, Yi SV, Zhang M, Zhou X (2018) Development of novel EST-SSR markers for ploidy identification based on de novo transcriptome assembly for *Misgurnus anguillicaudatus*. PLoS One 13(4):e0195829. https://doi.org/10.1371/journal.pone.0195829

Fu B, Zhou Y, Liu H, Yu X, Tong J (2021) Updated genome assembly of bighead carp (*Hypophthalmichthys nobilis*) and its differences between male and female on genomic, transcriptomic, and methylation level. Front Genet 12:728177. https://doi.org/10.3389/fgene.2021.728177

Gabián M, Morán P, Fernández AI, Villanueva B, Chtioui A, Kent MP et al (2019) Identification of genomic regions regulating sex determination in Atlantic salmon using high density SNP data. BMC Genomics 20(1):764. https://doi.org/10.1186/s12864-019-6104-4

Gandomkar H, Shekarabi SPH, Abdolhay HA, Nazari S, Mehrgan MS (2021) Characterization of novel genotyping-by-sequencing (GBS)-based simple sequence repeats (SSRs) and their application for population genomics of *Capoeta aculeata* (Valenciennes, 1844). Mol Biol Rep 48(9):6471–6480. https://doi.org/10.1007/s11033-021-06653-x

Gao Z, You X, Zhang X, Chen J, Xu T, Huang Y et al (2021) A chromosome-level genome assembly of the striped catfish (*Pangasianodon hypophthalmus*). Genomics 113(5):3349–3356. https://doi.org/10.1016/j.ygeno.2021.07.026

García-Cruz EL, Yamamoto Y, Hattori RS, de Vasconcelos LM, Yokota M, Strüssmann CA (2020) Crowding stress during the period of sex determination causes masculinization in pejerrey *Odontesthes bonariensis*, a fish with temperature-dependent sex determination. Comp Biochem Physiol A Mol Integr Physiol 245:110701. https://doi.org/10.1016/j.cbpa.2020.110701

Geffroy B, Wedekind C (2020) Effects of global warming on sex ratios in fishes. J Fish Biol 97(3):596–606. https://doi.org/10.1111/jfb.14429

Geraerts M, Vangestel C, Artois T, de Oliveira Fernandes JM, Jorissen MWP, Chocha Manda A et al (2022) Population genomics of introduced Nile tilapia *Oreochromis niloticus* (Linnaeus, 1758) in the Democratic Republic of the Congo: repeated introductions since colonial times with multiple sources. Mol Ecol 31(12):3304–3322. https://doi.org/10.1111/mec.16479

Giri S, Chanda A, Maity S, Chakraborty K, Hazra S (2022) Role of tide and lunar phases on the migration pattern of juvenile Hilsa shad (*Tenualosa ilisha*) within a meso-macrotidal estuary. J Fish Biol 100(4):988–996. https://doi.org/10.1111/jfb.15003

Gong J, Li B, Zhao J, Zhou Z, Ke Q, Zhu Q et al (2022) Sex-specific genomic region identification and molecular sex marker development of rock bream (*Oplegnathus fasciatus*). Mar Biotechnol 24(1):163–173. https://doi.org/10.1007/s10126-022-10095-2

Gopi N, Iswarya A, Vijayakumar S, Jayanthi S, Nor SAM, Velusamy P, Vaseeharan B (2022) Protective effects of dietary supplementation of probiotic *Bacillus licheniformis* Dahb1 against ammonia induced immunotoxicity and oxidative stress in *Oreochromis mossambicus*. Comp Biochem Physiol Toxicol Pharmacol 259:109379. https://doi.org/10.1016/j.cbpc.2022.109379

Gotoh RO, Sekimoto H, Chiba SN, Hanzawa N (2009) Peripatric differentiation among adjacent marine lake and lagoon populations of a coastal fish, *Sphaeramia orbicularis* (Apogonidae, Perciformes, Teleostei). Genes Genet Syst 84(4):287–295. https://doi.org/10.1266/ggs.84.287

Gu HR, Wan YF, Yang Y, Ao Q, Cheng WL, Deng SH et al (2019) Genetic and morphology analysis among the pentaploid F1 hybrid fishes (*Schizothorax wangchiachii* ♀ × *Percocypris pingi* ♂) and their parents. Animal 13(12):2755–2764. https://doi.org/10.1017/S1751731119001289

Guo Y-H, Halasan LC, Wang H-Y, Lin H-C (2022) High migratory propensity constitutes a single stock of an exploited cutlassfish species in the Northwest Pacific: a microsatellite approach. PLoS One 17(3):e0265548. https://doi.org/10.1371/journal.pone.0265548

Gutierrez AP, Bean TP, Hooper C, Stenton CA, Sanders MB, Paley RK et al (2018) A genome-wide association study for host resistance to Ostreid herpesvirus in Pacific Oysters (*Crassostrea gigas*). G3 (Bethesda) 8(4):1273–1280. https://doi.org/10.1534/g3.118.200113

Harvey SC, Boonphakdee C, Campos-Ramos R, Ezaz MT, Griffin DK, Bromage NR, Penman P (2003) Analysis of repetitive DNA sequences in the sex chromosomes of *Oreochromis niloticus*. Cytogenet Genome Res 101(3–4):314–319. https://doi.org/10.1159/000074355

He Y, Wu X, Zhu Y, Yang D (2020) Expression profiles of dmrt1 in *Schizothorax kozlovi*, and their relation to CpG methylation of its promoter and temperature. Zool Sci 37(2):140–147. https://doi.org/10.2108/zs190054

Hillestad B, Moghadam HK (2019) Genome-wide association study of piscine myocarditis virus (PMCV) resistance in Atlantic salmon (*Salmo salar*). J Hered 110(6):720–726. https://doi.org/10.1093/jhered/esz040

Hillestad B, Makvandi-Nejad S, Krasnov A, Moghadam HK (2020) Identification of genetic loci associated with higher resistance to pancreas disease (PD) in Atlantic salmon (*Salmo salar* L.). BMC Genomics 21(1):388. https://doi.org/10.1186/s12864-020-06788-4

Hirase S, Takeshima H, Nishida M, Iwasaki W (2016) Parallel mitogenome sequencing alleviates random rooting effect in phylogeography. Genome Biol Evol 8(4):1267–1278. https://doi.org/10.1093/gbe/evw063

Hoekzema K, Sidlauskas BL (2014) Molecular phylogenetics and microsatellite analysis reveal cryptic species of speckled dace (Cyprinidae: *Rhinichthys osculus*) in Oregon's Great Basin. Mol Phylogenet Evol 77:238–250. https://doi.org/10.1016/j.ympev.2014.04.027

Hsu T-H, Chiu Y-T, Lee H-T, Gong H-Y, Huang C-W (2021) Development of EST-molecular markers from RNA sequencing for genetic management and identification of growth traits in potato grouper (*Epinephelus tukula*). Biology 10(1):36. https://doi.org/10.3390/biology10010036

Huang G, Cao J, Chen C, Wang M, Liu Z, Gao F et al (2022a) Genome survey of *Misgurnus anguillicaudatus* to identify genomic information, simple sequence repeat (SSR) markers, and mitochondrial genome. Mol Biol Rep 49(3):2185–2196. https://doi.org/10.1007/s11033-021-07037-x

Huang P, Guo W, Wang Y, Xiong Y, Ge S, Gong G et al (2022b) Genome-wide association study reveals the genetic basis of growth trait in yellow catfish with sexual size dimorphism. Genomics 114(3):110380. https://doi.org/10.1016/j.ygeno.2022.110380

Jasonowicz AJ, Simeon A, Zahm M, Cabau C, Klopp C, Roques C et al (2022) Generation of a chromosome-level genome assembly for Pacific halibut (*Hippoglossus stenolepis*) and characterization of its sex-determining genomic region. Mol Ecol Resour 22(7):2685–2700. https://doi.org/10.1111/1755-0998.13641

Jiang L, Miao L, Yi G, Li X, Xue C, Li MJ et al (2022) Powerful and robust inference of complex phenotypes' causal genes with dependent expression quantitative loci by a median-based Mendelian randomization. Am J Hum Genet 109(5):838–856. https://doi.org/10.1016/j.ajhg.2022.04.004

Johnstone C, Pérez M, Arizcun M, García-Ruiz C, Chaves-Pozo E (2022) Reservoirs of Red-Spotted Grouper Nervous Necrosis Virus (RGNNV) in squid and shrimp species of Northern Alboran Sea. Viruses 14(2):328. https://doi.org/10.3390/v14020328

Khasanah M, Kadir NN, Jompa J (2019) Reproductive biology of three important threatened/near-threatened groupers (*Plectropomus leopardus, Epinephelus polyphekadion* and *Plectropomus areolatus*) in Eastern Indonesia and implications for management. Animals 9(9):643. https://doi.org/10.3390/ani9090643

Klütsch CFC, Maduna SN, Polikarpova N, Forfang K, Beddari B, Gjelland KØ et al (2021) Temporal analysis shows relaxed genetic erosion following improved stocking practices in a subarctic transnational brown trout population. Sci Rep 11(1):17396. https://doi.org/10.1038/s41598-021-96681-1

Kon T, Omori Y, Fukuta K, Wada H, Watanabe M, Chen Z et al (2020) The genetic basis of morphological diversity in domesticated goldfish. Curr Biol 30(12):2260–2274.e6. https://doi.org/10.1016/j.cub.2020.04.034

Kortner TM, Gu J, Krogdahl Å, Bakke AM (2013) Transcriptional regulation of cholesterol and bile acid metabolism after dietary soyabean meal treatment in Atlantic salmon (*Salmo salar* L.). Br J Nutr 109(4):593–604. https://doi.org/10.1017/S0007114512002024

Koubová A, Van Nguyen T, Grabicová K, Burkina V, Aydin FG, Grabic R et al (2022) Metabolome adaptation and oxidative stress response of common carp (*Cyprinus carpio*) to altered water pollution levels. Environ Pollut 303:119117. https://doi.org/10.1016/j.envpol.2022.119117

Kuo H-C, Hsu H-H, Chua CS, Wang T-Y, Chen Y-M, Chen T-Y (2014) Development of pedigree classification using microsatellite and mitochondrial markers for Giant grouper broodstock (*Epinephelus lanceolatus*) management in Taiwan. Mar Drugs 12(5):2397–2407. https://doi.org/10.3390/md12052397

LeBlanc NM, Gahagan BI, Andrews SN, Avery TS, Puncher GN, Reading BJ et al (2020) Genomic population structure of Striped Bass (*Morone saxatilis*) from the Gulf of St.Lawrence to Cape Fear River. Evol Appl 13(6):1468–1486. https://doi.org/10.1111/eva.12990

Lee M-Y, Munroe TA (2021) Unraveling cryptic diversity among shallow-water tonguefishes (Pleuronectiformes: Cynoglossidae: Symphurus) from the Indo-West Pacific region, with descriptions of five new species. Zootaxa 5039(1):1–55. https://doi.org/10.11646/zootaxa.5039.1.1

Lee C-H, Hur S-W, Na O-S, Baek H-J, Noh C-H, Han S-H, Lee Y-D (2014) Induction of primary male in juvenile red spotted grouper *Epinephelus akaara* by immersion of 17-α-methyltestosterone. Balsaeng'gwa Saengsig 18(3):127–131. https://doi.org/10.12717/DR.2014.18.3.127

Leitwein M, Gagnaire P-A, Desmarais E, Berrebi P, Guinand B (2018) Genomic consequences of a recent three-way admixture in supplemented wild brown trout populations revealed by local ancestry tracts. Mol Ecol 27(17):3466–3483. https://doi.org/10.1111/mec.14816

Li L, Wu Y, Zhao C, Miao Y, Cai J, Song L et al (2021) The role of StAR2 gene in testicular differentiation and spermatogenesis in Nile tilapia (*Oreochromis niloticus*). J Steroid Biochem Mol Biol 214:105974. https://doi.org/10.1016/j.jsbmb.2021.105974

Lien S, Koop BF, Sandve SR, Miller JR, Kent MP, Nome T et al (2016) The Atlantic salmon genome provides insights into rediploidization. Nature 533(7602):200–205. https://doi.org/10.1038/nature17164

Lǚ K, Li H, Wang S, Li A, Weng S, He J, Li C (2022) Interferon-induced protein 6-16 (IFI6-16) from *Litopenaeus vannamei* regulate antiviral immunity via apoptosis-related genes. Viruses 14(5):1062. https://doi.org/10.3390/v14051062

Lubieniecki KP, Botwright NA, Taylor RS, Evans BS, Cook MT, Davidson WS (2015a) Expression analysis of sex-determining pathway genes during development in male and female Atlantic salmon (*Salmo salar*). Physiol Genomics 47(12):581–587. https://doi.org/10.1152/physiolgenomics.00013.2015

Lubieniecki KP, Lin S, Cabana EI, Li J, Lai YYY, Davidson WS (2015b) Genomic instability of the sex-determining locus in Atlantic Salmon (*Salmo salar*). G3 (Bethesda) 5(11):2513–2522. https://doi.org/10.1534/g3.115.020115

Lund H, Bakke A, Boysen P, Afanasyev S, Rebl A, Manji F et al (2022) Evaluation of immune status in two cohorts of Atlantic salmon raised in different aquaculture systems (case study). Genes 13(5):736. https://doi.org/10.3390/genes13050736

Luo L, Huang R, Zhang A, Yang C, Chen L, Zhu D et al (2018) Selection of growth-related genes and dominant genotypes in transgenic Yellow River carp *Cyprinus carpio* L. Funct Integr Genom 18(4):425–437. https://doi.org/10.1007/s10142-018-0597-9

Lup SD, Wilson-Sánchez D, Micol JL (2022) Mapping-by-sequencing of point and insertional mutations with easymap. Methods Mol Biol 2484:343–361. https://doi.org/10.1007/978-1-0716-2253-7_23

Lyons J, Piller KR, Artigas-Azas JM, Dominguez-Dominguez O, Gesundheit P, Köck M et al (2019) Distribution and current conservation status of the Mexican Goodeidae (Actinopterygii, Cyprinodontiformes). ZooKeys 885:115–158. https://doi.org/10.3897/zookeys.885.38152

Mahboob S, Al-Ghanim KA, Al-Misned F, Al-Balawi HFA, Ashraf A, Al-Mulhim NMA (2019) Genetic diversity in tilapia populations in a freshwater reservoir assayed by randomly amplified polymorphic DNA markers. Saudi J Biol Sci 26(2):363–367. https://doi.org/10.1016/j.sjbs.2018.11.015

Mar-Silva AF, Arroyave J, Díaz-Jaimes P (2022) The complete mitochondrial genome of the Mexican-endemic cavefish *Ophisternoninfernale* (Synbranchiformes, Synbranchidae): insights on patterns of selection and implications for synbranchiform phylogenetics. ZooKeys 1089:1–23. https://doi.org/10.3897/zookeys.1089.78182

Mehta TK, Koch C, Nash W, Knaack SA, Sudhakar P, Olbei M et al (2021) Evolution of regulatory networks associated with traits under selection in cichlids. Genome Biol 22(1):25. https://doi.org/10.1186/s13059-020-02208-8

Miller-Crews I, Matz MV, Hofmann HA (2021) A 2b-RAD parentage analysis pipeline for complex and mixed DNA samples. Forensic Sci Int Genet 55:102590. https://doi.org/10.1016/j.fsigen.2021.102590

Miya M, Sato Y, Fukunaga T, Sado T, Poulsen JY, Sato K et al (2015) MiFish, a set of universal PCR primers for metabarcoding environmental DNA from fishes: detection of more than 230 subtropical marine species. R Soc Open Sci 2(7):150088. https://doi.org/10.1098/rsos.150088

Mondal D, Dutta S, Chakrabarty U, Mallik A, Mandal N (2019) Development and characterization of white spot disease linked microsatellite DNA markers in Penaeus monodon, and their application to determine the population diversity, cluster and structure. J Invertebr Pathol 168:107275. https://doi.org/10.1016/j.jip.2019.107275

Moustakas-Verho JE, Kurko J, House AH, Erkinaro J, Debes P, Primmer CR (2020) Developmental expression patterns of six6: a gene linked with spawning ecotypes in Atlantic salmon. Gene Expr Patterns 38:119149. https://doi.org/10.1016/j.gep.2020.119149

Mulugeta TD, Nome T, To T-H, Gundappa MK, Macqueen DJ, Våge DI et al (2019) SalMotifDB: a tool for analyzing putative transcription factor binding sites in salmonid genomes. BMC Genomics 20(1):694. https://doi.org/10.1186/s12864-019-6051-0

Munguia-Vega A, Terrazas-Tapia R, Dominguez-Contreras JF, Reyna-Fabian M, Zapata-Morales P (2022) DNA barcoding reveals global and local influences on patterns of mislabeling and substitution in the trade of fish in Mexico. PLoS One 17(4):e0265960. https://doi.org/10.1371/journal.pone.0265960

Mzingirwa FA, Mkare TK, Nyingi DW, Njiru J (2019) Genetic diversity and spatial population structure of a deepwater snapper, *Pristipomoides filamentosus* in the south-west Indian Ocean. Mol Biol Rep 46(5):5079–5088. https://doi.org/10.1007/s11033-019-04962-w

Nagpure NS, Rashid I, Pathak AK, Singh M, Pati R, Singh SP, Sarkar UK (2015) Fmir: a curated resource of mitochondrial DNA information for fish. PLoS One 10(8):e0136711. https://doi.org/10.1371/journal.pone.0136711

Nam S-E, Eom H-J, Park HS, Rhee J-S (2022) Characterization and phylogenetic analysis of the complete mitochondrial genome of the rainbow krib, *Pelvicachromis pulcher* (Perciformes: Cichlidae). Mitochondrial DNA B Resour 7(6):918–920. https://doi.org/10.1080/23802359.2022.2079099

Oh S-R, Kang H-C, Lee C-H, Hur S-W, Lee Y-D (2013) Sex reversal and masculinization according to growth in longtooth grouper *Epinephelus bruneus*. Balsaeng'gwa Saengsig 17(2):79–85. https://doi.org/10.12717/DR.2013.17.2.079

Pang M, Fu B, Yu X, Liu H, Wang X, Yin Z et al (2017) Quantitative trait loci mapping for feed conversion efficiency in crucian carp (*Carassius auratus*). Sci Rep 7(1):16971. https://doi.org/10.1038/s41598-017-17269-2

Pathak AK, Rashid I, Nagpure NS, Kumar R, Pati R, Singh M et al (2019) FisOmics: a portal of fish genomic resources. Genomics 111(6):1923–1928. https://doi.org/10.1016/j.ygeno.2019.01.003

Peñaloza C, Gutierrez AP, Eöry L, Wang S, Guo X, Archibald AL et al (2021) A chromosome-level genome assembly for the Pacific oyster *Crassostrea gigas*. GigaScience 10(3):giab020. https://doi.org/10.1093/gigascience/giab020

Perez-Enriquez R, Robledo D, Houston RD, Llera-Herrera R (2018) SNP markers for the genetic characterization of Mexican shrimp broodstocks. Genomics 110(6):423–429. https://doi.org/10.1016/j.ygeno.2018.10.001

Pla S, Benvenuto C, Capellini I, Piferrer F (2022) Switches, stability and reversals in the evolutionary history of sexual systems in fish. Nat Commun 13(1):3029. https://doi.org/10.1038/s41467-022-30419-z

Podlesnykh AV, Brykov VA, Kukhlevsky AD (2017) Unstable linkage of molecular markers with sex determination gene in pacific salmon (*Oncorhynchus* spp.). J Hered 108(3):328–333. https://doi.org/10.1093/jhered/esx001

Raharjo HM, Budiyansah H, Mursalim MF, Chokmangmeepisarn P, Sakulworakan R, Madyod S et al (2022) Distribution of Vibrionaceae in farmed Asian sea bass, Lates calcarifer in Thailand and their high prevalence of antimicrobial resistance. J Fish Dis 45(9):1355–1371. https://doi.org/10.1111/jfd.13667

Rajendiran P, Jaafar F, Kar S, Sudhakumari C, Senthilkumaran B, Parhar IS (2021) Sex determination and differentiation in teleost: roles of genetics, environment, and brain. Biology 10(10):973. https://doi.org/10.3390/biology10100973

Sahoo L, Das P, Sahoo B, Das G, Meher PK, Udit UK et al (2020) The draft genome of *Labeo catla*. BMC Res Notes 13(1):411. https://doi.org/10.1186/s13104-020-05240-w

Saillant EA, Luque PL, Short E, Antoni L, Reynal L, Pau C et al (2022) Population structure of blackfin tuna (*Thunnus atlanticus*) in the western Atlantic Ocean inferred from microsatellite loci. Sci Rep 12(1):9830. https://doi.org/10.1038/s41598-022-13857-z

Salem M, Al-Tobasei R, Ali A, Lourenco D, Gao G, Palti Y et al (2018) Genome-wide association analysis with a 50K transcribed gene SNP-chip identifies QTL affecting muscle yield in rainbow trout. Front Genet 9:387. https://doi.org/10.3389/fgene.2018.00387

Salisbury SJ, Ruzzante DE (2022) Genetic causes and consequences of sympatric morph divergence in salmonidae: a search for mechanisms. Ann Rev Anim Biosci 10:81–106. https://doi.org/10.1146/annurev-animal-051021-080709

Sandoval-Castillo J, Beheregaray LB, Wellenreuther M (2022) Genomic prediction of growth in a commercially, recreationally, and culturally important marine resource, the Australian snapper (*Chrysophrys auratus*). G3 (Bethesda) 12(3):jkac015. https://doi.org/10.1093/g3journal/jkac015

Seki K (2021) RAD-R scripts: R pipeline for RAD-seq from FASTQ files to linkage maps construction and run R/QTL, operating only at copying and pasting scripts into R console. Breed Sci 71(4):426–434. https://doi.org/10.1270/jsbbs.20159

Shao W-H, Cheng J-L, Zhang E (2021) Eight in one: hidden diversity of the bagrid catfish *Tachysurus albomarginatus* s.l. (Rendhal, 1928) widespread in lowlands of South China. Front Genet 12:713793. https://doi.org/10.3389/fgene.2021.713793

Sinclair-Waters M, Piavchenko N, Ruokolainen A, Aykanat T, Erkinaro J, Primmer CR (2022) Refining the genomic location of single nucleotide polymorphism variation affecting Atlantic salmon maturation timing at a key large-effect locus. Mol Ecol 31(2):562–570. https://doi.org/10.1111/mec.16256

Song W, Xie Y, Sun M, Li X, Fitzpatrick CK, Vaux F et al (2021) A duplicated amh is the master sex-determining gene for Sebastes rockfish in the Northwest Pacific. Open Biol 11(7):210063. https://doi.org/10.1098/rsob.210063

Soyano K, Amagai T, Yamaguchi T, Mushirobira Y, Xu W-G, Phạm NT, Murata R (2022) Endocrine regulation of maturation and sex change in groupers. Cells 11(5):825. https://doi.org/10.3390/cells11050825

Su JJ, Shang XY, Wan QY, Su JG (2018) SNP-based susceptibility-resistance association and mRNA expression regulation analyses of tlr7 to grass carp *Ctenopharyngodon idella* reovirus. J Fish Biol 92(5):1505–1525. https://doi.org/10.1111/jfb.13607

Sudo R, Yada T (2022) Anguillid eels as a model species for understanding endocrinological influences on the onset of spawning migration of fishes. Biology 11(6):934. https://doi.org/10.3390/biology11060934

Sukhavachana S, Tongyoo P, Luengnaruemitchai A, Poompuang S (2021) Optimizing genomic prediction using low-density marker panels for streptococcosis resistance in red tilapia (*Oreochromis* spp.). Anim Genet 52(5):667–674. https://doi.org/10.1111/age.13114

Tan S, Zhou T, Wang W, Jin Y, Wang X, Geng X et al (2018) GWAS analysis using interspecific backcross progenies reveals superior blue catfish alleles responsible for strong resistance against enteric septicemia of catfish. Mol Gen Genomics 293(5):1107–1120. https://doi.org/10.1007/s00438-018-1443-4

Tao W, Cao J, Xiao H, Zhu X, Dong J, Kocher TD et al (2021) A chromosome-level genome assembly of Mozambique tilapia (*Oreochromis mossambicus*) reveals the structure of sex determining regions. Front Genet 12:796211. https://doi.org/10.3389/fgene.2021.796211

Taslima K, Wehner S, Taggart JB, de Verdal H, Benzie JAH, Bekaert M et al (2020) Sex determination in the GIFT strain of tilapia is controlled by a locus in linkage group 23. BMC Genet 21(1):49. https://doi.org/10.1186/s12863-020-00853-3

Tesfaye G, Curto M, Meulenbroek P, Englmaier GK, Tibihika PD, Alemayehu E et al (2021) Genetic diversity of Nile tilapia (*Oreochromis niloticus*) populations in Ethiopia: insights from nuclear DNA microsatellites and implications for conservation. BMC Ecol Evol 21(1):113. https://doi.org/10.1186/s12862-021-01829-2

Tian J-J, Zhang J-M, Yu E-M, Sun J-H, Xia Y, Zhang K et al (2020) Identification and analysis of lipid droplet-related proteome in the adipose tissue of grass carp (*Ctenopharyngodon idella*) under fed and starved conditions. Comp Biochem Physio D Genom Proteom 36:100710. https://doi.org/10.1016/j.cbd.2020.100710

Tian H-F, Hu Q-M, Li Z (2021) Genome-wide identification of simple sequence repeats and development of polymorphic SSR markers in swamp eel (*Monopterus albus*). Sci Prog 104(3):368504211035597. https://doi.org/10.1177/00368504211035597

Valenza-Troubat N, Montanari S, Ritchie P, Wellenreuther M (2022) Unraveling the complex genetic basis of growth in New Zealand silver trevally (*Pseudocaranx georgianus*). G3 (Bethesda) 12(3):jkac016. https://doi.org/10.1093/g3journal/jkac016

Vela-Avitúa S, Thorland I, Bakopoulos V, Papanna K, Dimitroglou A, Kottaras E et al (2022) Genetic basis for resistance against viral nervous necrosis: GWAS and potential of genomic prediction explored in farmed European sea bass (*Dicentrarchus labrax*). Front Genet 13: 804584. https://doi.org/10.3389/fgene.2022.804584

Verta J-P, Debes PV, Piavchenko N, Ruokolainen A, Ovaskainen O, Moustakas-Verho JE et al (2020) Cis-regulatory differences in isoform expression associate with life history strategy variation in Atlantic salmon. PLoS Genet 16(9):e1009055. https://doi.org/10.1371/journal.pgen.1009055

Wallace EN, Reed EMX, Aguilar A, Burford Reiskind MO (2022) Resolving the phylogenetic relationship among recently diverged members of the rockfish subgenus *Sebastosomus*. Mol Phylogenet Evol 173:107515. https://doi.org/10.1016/j.ympev.2022.107515

Wallis DJ, La Du J, Thunga P, Elson D, Truong L, Kolluri SK et al (2022) Leveraging a high-throughput screening method to identify mechanisms of individual susceptibility differences in a genetically diverse zebrafish model. Front Toxicol 4:846221. https://doi.org/10.3389/ftox.2022.846221

Wang Q, Liu Y, Peng C, Wang X, Xiao L, Wang D et al (2017a) Molecular regulation of sex change induced by methyltestosterone -feeding and methyltestosterone -feeding withdrawal in the protogynous orange-spotted grouper. Biol Reprod 97(2):324–333. https://doi.org/10.1093/biolre/iox085

Wang Q, Qi X, Tang H, Guo Y, Li S, Li G et al (2017b) Molecular identification of StAR and 3βHSD1 and characterization in response to GnIH stimulation in protogynous hermaphroditic grouper (*Epinephelus coioides*). Comp Biochem Physiol B Biochem Mol Biol 206:26–34. https://doi.org/10.1016/j.cbpb.2017.01.001

Wang W, Ma C, Ouyang L, Chen W, Zhao M, Zhang F et al (2021) Genetic diversity and population structure analysis of *Lateolabrax maculatus* from Chinese coastal waters using polymorphic microsatellite markers. Sci Rep 11(1):15260. https://doi.org/10.1038/s41598-021-93000-6

Wang J, Yu J, Lipka AE, Zhang Z (2022) Interpretation of Manhattan plots and other outputs of genome-wide association studies. Methods Mol Biol 2481:63–80. https://doi.org/10.1007/978-1-0716-2237-7_5

Weiss SJ, Kopun T, Bajec SS (2013) Assessing natural and disturbed population structure in European grayling *Thymallus thymallus*: melding phylogeographic, population genetic and jurisdictional perspectives for conservation planning. J Fish Biol 82(2):505–521. https://doi.org/10.1111/jfb.12007

Weng Z, Yang Y, Wang X, Wu L, Hua S, Zhang H, Meng Z (2021) Parentage analysis in giant grouper (*Epinephelus lanceolatus*) using microsatellite and SNP markers from genotyping-by-sequencing data. Genes 12(7):1042. https://doi.org/10.3390/genes12071042

White SL, Eackles MS, Wagner T, Schall M, Smith G, Avery J, Kazyak DC (2021) Optimization of a suite of flathead catfish (*Pylodictis olivaris*) microsatellite markers for understanding the population genetics of introduced populations in the northeast United States. BMC Res Notes 14(1):314. https://doi.org/10.1186/s13104-021-05725-2

Wu C-S, Ma Z-Y, Zheng G-D, Zou S-M, Zhang X-J, Zhang Y-A (2022) Chromosome-level genome assembly of grass carp (*Ctenopharyngodon idella*) provides insights into its genome evolution. BMC Genomics 23(1):271. https://doi.org/10.1186/s12864-022-08503-x

Xu P, Zhang X, Wang X, Li J, Liu G, Kuang Y et al (2014) Genome sequence and genetic diversity of the common carp, *Cyprinus carpio*. Nat Gene 46(11):1212–1219. https://doi.org/10.1038/ng.3098

Yagishita N, Kume G (2021) Genetic characteristics of the amphidromous fish Ayu *Plecoglossus altivelis* altivelis (Osmeriformes: Plecoglossidae) on Yaku-shima Island in Japan, the southern-most population of the subspecies. Genetica 149(2):117–128. https://doi.org/10.1007/s10709-021-00117-7

Yang M, Wang Q, Chen J, Wang Y, Zhang Y, Qin Q (2021) Identification of candidate SNPs and genes associated with anti-RGNNV using GWAS in the red-spotted grouper, *Epinephelus akaara*. Fish Shellfish Immunol 112:31–37. https://doi.org/10.1016/j.fsi.2021.02.010

Yu X, Wu L, Xie L, Yang S, Charkraborty T, Shi H et al (2014) Characterization of two paralogous StAR genes in a teleost, Nile tilapia (*Oreochromis niloticus*). Mol Cell Endocrinol 392(1–2):152–162. https://doi.org/10.1016/j.mce.2014.05.013

Yüncü E, Açan SC, Onar V, Karakulak FS, Gökoğlu M, Alıçlı TZ et al (2021) Demography of swordfish (*Xiphias gladius* Linneus) populations from the coasts of Turkey, based on mitochondrial DNA and microsatellites. J Fish Biol 99(1):37–48. https://doi.org/10.1111/jfb.14696

Zabidi A, Yusoff FM, Amin N, Yaminudin NJM, Puvanasundram P, Karim MMA (2021) Effects of probiotics on growth, survival, water quality and disease resistance of red hybrid tilapia (*Oreochromis* spp.) fingerlings in a biofloc system. Animals 11(12):3514. https://doi.org/10.3390/ani11123514

Zainal Abidin DH, Mohd Nor SA, Lavoué S, Rahim MA, Jamaludin NA, Mohammed Akib NA (2021) DNA-based taxonomy of a mangrove-associated community of fishes in Southeast Asia. Sci Rep 11(1):17800. https://doi.org/10.1038/s41598-021-97324-1

Zhang X, Yuan J, Sun Y, Li S, Gao Y, Yu Y et al (2019) Penaeid shrimp genome provides insights into benthic adaptation and frequent molting. Nat Commun 10(1):356. https://doi.org/10.1038/s41467-018-08197-4

Zhang X, Luan P, Cao D, Hu G (2021) A high-density genetic linkage map and fine mapping of QTL For feed conversion efficiency in common carp (*Cyprinus carpio*). Front Genet 12: 778487. https://doi.org/10.3389/fgene.2021.778487

Zhao Y, Peng W, Guo H, Chen B, Zhou Z, Xu J et al (2018) Population genomics reveals genetic divergence and adaptive differentiation of Chinese sea bass (*Lateolabrax maculatus*). Mar Biotechnol 20(1):45–59. https://doi.org/10.1007/s10126-017-9786-0

Zheng S, Tao W, Yang H, Kocher TD, Wang Z, Peng Z et al (2022) Identification of sex chromosome and sex-determining gene of southern catfish (*Silurus meridionalis*) based on XX, XY and YY genome sequencing. Proc Biol Sci/R Soc 289(1971):20212645. https://doi.org/10.1098/rspb.2021.2645

Zhou Y, Qin W, Zhong H, Zhang H, Zhou L (2021) Chromosome-level assembly of the *Hypophthalmichthys molitrix* (Cypriniformes: Cyprinidae) genome provides insights into its ecological adaptation. Genomics 113(5):2944–2952. https://doi.org/10.1016/j.ygeno.2021.06.024

Zhu Y, Cheng Q, Rogers SM (2016) Genetic structure of *Scomber japonicus* (Perciformes: Scombridae) along the coast of China revealed by complete mitochondrial cytochrome b sequences. Mitochondrial DNA A DNA Mapp Seq Anal 27(6):3828–3836. https://doi.org/10.3109/19401736.2014.958671

Zhu Z-X, Lin Y-L, Ai C-H, Xiong Y-Y, Huang D-D, Yao Y-Y et al (2022) First identification of two co-existing genome-wide significant sex quantitative trait loci (QTL) in red tilapia using integrative QTL mapping. Dong Wu Xue Yan Jiu = Zool Res/"Dong Wu Xue Yan Jiu" Bian Ji We Yuan Hui Bian Ji 43(2):205–216. https://doi.org/10.24272/j.issn.2095-8137.2021.402

Ziarati M, Zorriehzahra MJ, Hassantabar F, Mehrabi Z, Dhawan M, Sharun K et al (2022) Zoonotic diseases of fish and their prevention and control. Vet Q 42:1–42. https://doi.org/10.1080/01652176.2022.2080298

5 Aquaculture and Applications of Green Seaweeds of the Genus *Caulerpa* J.V. Lamouroux, 1809

Wahidatul Husna Zuldin

Abstract

The genus *Caulerpa* is highly distributed within tropical and subtropical regions. Some species are found around the Mediterranean Sea and temperate regions of Australia, with the highest species diversity occurring in southern Australia. The consumption of this genus has been booming recently in Asia, with more species having been established to be edible as common local foods. Most of the species are wildly harvested, and the cultivation of several species is still underexplored. Nevertheless, the applications of this genus are massively studied for their nutritional and medicinal properties. Therefore, this chapter highlights the details on the cultivation status and applications of several known species under *Caulerpa* that are beneficial as baseline data for the developing seaweed industry.

5.1 Introduction

Seaweeds are members of the Plantae kingdom, also known as benthic marine macroalgae, that become the most important primary producers in the aquatic ecosystem when they emerge from the intertidal and subtidal zones of the sea, also distributed along the coasts from the tropical to the polar region (Salehi et al. 2019). They also have photosynthetic pigments that aid in food production through photosynthesis (Chen et al. 2017). Marine macroalgae are divided into three divisions with different evolutionary histories, comprising Rhodophyta (red macroalgae), Ochrophyta (brown or dark macroalgae, previously classified as Phaeophyta and Heterokontophyta), and Chlorophyta (green macroalgae), whereby each division is

W. H. Zuldin (✉)
Borneo Marine Research Institute, Universiti Malaysia Sabah, Kota Kinabalu, Sabah, Malaysia
e-mail: wahidatul@ums.edu.my

M. D. Shah et al. (eds.), *Marine Biotechnology: Applications in Food, Drugs and Energy*, https://doi.org/10.1007/978-981-99-0624-6_5

composed of thousands of species (García-Poza et al. 2020). The seaweed divisions differed primarily in the pigment composition involved in photosynthesis, as well as in a variety of ultrastructural and biochemical characteristics such as storage compounds, cell wall composition, and chloroplast fine structure (Ibañez et al. 2012; Koushalya et al. 2021).

Marine macroalgae, or seaweeds, vary in size, ranging from a few centimetres up to 100 m in length, and more than 15,000 species have been described worldwide. The differences between macroalgae as marine plants and terrestrial plants are the absence of conductive tissues, adsorption of nutrients throughout their whole surface, and lack of real roots, whereby some seaweeds may have rhizoids or basal discs as a support system, but not for nourishment purposes like a real plant (Gomez-Zavaglia et al. 2019). Seaweeds also contain a variety of biochemical compounds that are useful for the industries involved in agriculture, nutraceutical, pharmaceutical, and biotechnological applications (Pereira 2018). Prior to history, seaweeds have been recognised as valuable nutritional algae around Asian countries that have spread to other continents such as Europe, South America, North America, and Australia, where edible seaweeds are found to be high in proteins, lipids, and dietary fibres (García-Poza et al. 2020).

The utilisation of brown macroalgae for goitre by Chinese people, red macroalgae such as *Gelidium* J.V. Lamouroux, 1813, for intestinal afflictions, and dehydrated brown macroalgae such as *Laminaria* J.V. Lamouroux, 1813, for cervix dilation in difficult childbirths is early example of seaweed used for medicinal purposes (Buschmann et al. 2017). Furthermore, green macroalgae also contain many bioactive compounds that can be used in various industrial applications. Thus, extensive research has been prompted on useful compounds in seaweeds and aquaculture potential which differs from onshore to offshore cultivation methods (Pereira 2018; Ferrara 2020).

Other than that, the algae population (microalgae and macroalgae) is claimed to be among the largest aquaculture crops in the world after freshwater fish, molluscs, and seaweed production for human consumption, and industrial application is the largest sector in the aquaculture industry (Nagappan and Vairappan 2014; Wikfors and Ohno 2002). Red and brown seaweeds are currently the most extensively consumed macroalgae. The green macroalgae are comparatively uncommon species for consumption, except for the *Monostroma* genus, which has been directly consumed as fresh seaweed (Nagappan and Vairappan 2014). Recent reports indicate that certain green seaweeds, such as the genus *Caulerpa*, are widely consumed as fresh salads in the Indo-Pacific region, as well as the genus *Ulva*, which is used as a dietary supplement and condiment in Japan, China, Chile, France, and the United States (Phang et al. 2019; Peña-Rodríguez et al. 2011; Zubia et al. 2020). In addition, certain *Ulva* species, primarily *U. lactuca*, often known as sea lettuce, are consumed as 'aonori', a form of dried seaweed used in numerous Japanese recipes (de Gaillande et al. 2017; Zubia et al. 2020).

Malaysia's seaweed sector is understudied and little known, with only three types of red seaweed being cultivated commercially (*Kappaphycus alvarezii* (Doty) L.M. Liao, 1996, *K. striatum* (F. Schmitz) L.M.Liao, 1996, and *Eucheuma denticulatum* (N.L. Burman) Collins & Hervey, 1917) for carrageenan extraction and processing (Nor et al. 2019). Malaysia is divided into Peninsular or West Malaysia and East Malaysia, with an average coastal depth of less than 200 m (Phang et al. 2019). Almost all of Malaysia's islands have been discovered with seaweed varsity, particularly red and brown seaweed (Phang et al. 2005). Malaysia is also located within the Coral Triangle area, one of the most important reef systems in the world that supports a tremendous amount of marine biodiversity (Hussin and Khoso 2017). Thus, the strategic location and tropical conditions around Malaysian coastal waters have created favourable habitats for high seaweed biodiversity. According to Phang (2006), the eastern coast of Sabah (East Malaysia) provides an ideal habitat for the growth of high-value seaweeds, such as *K. alvarezii* and *Caulerpa lentillifera* J. Agardh, 1837.

To date, Sabah has become the only and highest seaweed-producing state in Malaysia due to its good geography, being situated below the monsoon and typhoon belt, which facilitated commercial seaweed farming (Hussin and Khoso 2017). Therefore, increased global demand for seaweed and its derivatives provided an opportunity for Malaysia to further develop the seaweed industry, and more seaweed species needed to be explored for future commercialisation (Nor et al. 2019). The rich seaweed resources in Malaysia are an opportunity for all researchers, farmers, and industry players to explore and utilise. Indeed, seaweed resource identification has supported the research and development of new areas, such as the identification of industrial biomaterials, potential therapeutics, environmental chemistry, and genomics (Phang et al. 2019).

5.2 Taxonomy, Ecology, and Distribution of Genus *Caulerpa*

The most distinctive edible green macroalgae are those in the family Caulerpaceae and class Bryopsidophyceae of the division Chlorophyta as shown in Table 5.1, which contain more than 100 species recognised on a global scale (Zubia et al. 2020). These species are identifiable solely based on their growth form and internal morphology (Kumar et al. 2019). However, due to phenotypic plasticity in physical characteristics as well as polymorphism, *Caulerpa* species have encountered some

Table 5.1 Taxonomical hierarchy of *Caulerpa* J. V. F. Lamouroux 1809

Kingdom	Plantae
Subkingdom	Viridiplantae
Infrakingdom	Chlorophyta
Division	Chlorophyta
Subdivision	Chlorophytina
Class	Ulvophyceae
Order	Bryopsidales
Family	Caulerpaceae
Genus	*Caulerpa*

Source: (ITIS Report 2020)

difficulties in taxonomic identification, particularly at the species level, as well as unstable classification of varieties and forms (Zubia et al. 2020; Belton et al. 2014; Prud'homme Van Reine et al. 1996).To date, a total of 173 *Caulerpa* species names have been described in the world algae database, as well as 232 infraspecific names, and only 104 species names are taxonomically valid (Guiry and Guiry 2020). *Caulerpa prolifera* (Forsskål) J.V. Lamouroux was the first species of the genus *Caulerpa* to be discovered in 1809 (Zubia et al. 2020).

Besides, these species are called siphonous macroalgae due to the lack of transverse cell walls that also display complexity with the thallus differentiating into rhizoids, stolons, and upright assimilators (fronds) that usually bear branchlets or ramuli (Belton et al. 2014). The morphology of these various structures of *Caulerpa* has been historically used for species delimitation. Environmentally controlled phenotypic plasticity in all of these characters, on the other hand, had resulted in a plethora of synonyms and a classification scheme that included subspecies, varieties, types, and 'ecads' (de Senerpont Domis et al. 2013). Coppejans and Prid'homem van Reine (1992) applied the term 'ecad' to describe intermediate growth forms of *Caulerpa* species, in addition to many varieties, subspecies, and forms. The sectional division among the taxa was predominantly supported by differences in the morphology of the fronds or assimilators (Benzie et al. 2008). These fronds, however, can be highly plastic and are strongly influenced by the environment (Belton et al. 2014).

The taxa currently associated with *C. racemosa* (Forsskål) J. Agardh and *C. peltata* J. V. Lamouroux are claimed to be the most taxonomically problematic. They are commonly referred to as the *C. racemosa–peltata* complex, which consists of six distinct species-level entities involving more than 30 identified varieties and forms (Guiry and Guiry 2020). The issue arises when the *Caulerpa* encounters a problem that necessitates additional species identification and confirmation, such as *C. racemosa* var. cylindracea (Sonder), which has recently received a lot of attention due to its invasion in the Mediterranean Sea and South Australia (Jongma et al. 2013). Many of the varieties and forms within the complex were first described as distinct species, such as *C. cylindracea* and *C. macrodisca*. Nevertheless, the later discovery of intermediate thalli within *C. racemosa* and *C. peltata* type morphologies has caused these taxa to be described as varieties or forms and not species, mostly associated with *C. racemosa*, for instance, *C. racemosa* var. *macrodisca* and *C. peltata* var. *macrodisca* (Belton et al. 2014). Thus, more than 250 sequences in GenBank remained simply labelled as either *C. racemosa* or *C. peltata*, despite the fact that some sequences included multiple varieties. This has raised concerns regarding the inaccuracy of the species identification of *Caulerpa*, despite the availability of molecular sequence data, as this species continued to expand beyond their native ranges (Perez-Estrada et al. 2013; Amat et al. 2008; Lapointe et al. 2005; Jongma et al. 2013).

The thallus (sometimes called thalli in the plural form) of multicellular macroalgae or seaweed often refers to the entire organism. The organisation of the *Caulerpa* thallus is always siphonous whereby each species consists of a single giant multinucleate cell that usually takes the form of a branching tube or siphon containing one to several central vacuoles and a thin peripheral layer of protoplasm (Barsanti and Gualtieri 2014). In addition, the chloroplasts and amyloplasts of *Caulerpa* are heteroplastidic and lie in the peripheral protoplasm together with the nuclei (Cremen et al. 2019). The *Caulerpa* species exhibit a complex external morphology with the thalli differentiated into several parts that consist of green cylindrical, prostate, and creeping tube-like stolons, anchored by bundles of downward growing colourless rhizoids, where upright photosynthetic 'shoots' (assimilators or fronds) arise from the stolons bearing distinctive branchlets termed ramuli, and these units are called metameres (de Senerpont Domis et al. 2013).

Caulerpa species' stolons can be glabrous (smooth) or squamiform (covered with scale-like appendages), while the assimilators or fronds can be ligulate (leaf-like) or rachis (consisting of a central axis) and bear lateral branches or ramuli in a variety of arrangements, including distich, pinnate, irregular, and vertical (Zubia et al. 2020). Ramuli on a stem can be club-shaped (clavate), cylindrical (terete), trumpet-shaped (turbinate), sickle-shaped (falcate), disc-shaped (peltate), or spherical (vesiculate). The entire *Caulerpa* thallus can be more than 1 m long, and the rachis of some species can have regular constrictions (annulations). The ramuli are the vegetative part of the *Caulerpa* thallus, and they are commonly used to identify the species. Some *Caulerpa* species are capable of displaying a high degree of morphological plasticity. This is due to environmental and habitat changes, which have resulted in incredible taxonomic confusion and a massive description of approximately 400 species, varieties, forms, and ecads (Belton et al. 2019). The previous study also demonstrated that assimilates on the same stolon can have distinct morphologies and that distinct assimilates can be identified as different subspecific taxa or species (Ohba and Enomoto 1987).

There are 173 *Caulerpa* species and 232 infraspecific names that have been documented and are widely distributed in tropical to subtropical regions, with some species reaching as far as the Mediterranean and temperate regions of Australia, outside their original range (Guiry and Guiry 2020). The greatest *Caulerpa* diversity is found in the Caribbean, Indo-Malay Archipelago, and the temperate waters of southern Australia (Zubia et al. 2020). *Caulerpa* species occur only in the Indo-Malay region, with the majority of species native to South Australia and 28 records of the species' existence (Belton et al. 2019). *Caulerpa* species also occur in other parts of the world, including north-west Australia, southern Japan, and East and West Africa. Due to challenges in defining morphological species boundaries, reliable estimates of species richness for each region are uncertain, and the diversity of some regions has only been explored using DNA-based species identifications. For example, several studies found 20 species in north-western Australia, 15 in the Caribbean and Gulf of Mexico, 10 in north-western and south-eastern India, and only six in Central America (Kazi et al. 2013; Fernandez-Garcia et al. 2016; Famà et al. 2002; Sauvage et al. 2013; Belton et al. 2014). *Caulerpa* species are currently identified in Southeast Asian waters. In Malaysian coastal areas, about 13 species and seven variants of *Caulerpa* have been identified and reported (Phang et al. 2016).

The *Caulerpa* species are benthic organisms that are widely distributed around the intertidal and subtidal zones of coastal waters and mostly inhabit vertical ranges including the intertidal area with at least 50 m of depth where the substrates are mostly unconsolidated sand of seagrass meadows and hard surfaces such as coral rocks (Lisette et al. 2003). In many *Caulerpa* species, the stolons grow through the surface layers of sandy or muddy sediment, forming dense swards on the bottom of tropical lagoons (Katsanevakis et al. 2014a). These species can also be found on shady rocks, under-surfaces of overhanging rocks, in hollows and among branches of corals in shallow water, or on unshaded rocks below a depth of 5 m (Diaz-Pulido and McCook 2008). These species are often discovered to be distributed in patches, however; some species are claimed to have the ability to dominate large areas, especially the ones with invasive characteristics as what has been observed within the Mediterranean Sea (Inderjit et al. 2006). Most of these species can inhabit a range of habitats with variable climatic conditions, such as temperature, light intensity, water velocity, grazing pressure, depth, and benthic substrate (Zubia et al. 2020). These *Caulerpa* species contain a number of invasive taxa, including *C. taxifolia* and *C. cylindracea* (Anderson 2005; Verlaque et al. 2003). The invasive potential of several *Caulerpa* species has been linked to their inherent characteristics that facilitate their adaptation to different environmental niches. For example, the native tropical *C. taxifolia* can tolerate a wide range of environmental conditions and appears to be able to survive and become invasive in temperate estuaries (Glasby and Gibson 2007).

Aside from that, these species are known to survive in a wide variety of habitats due to their phenotypic plasticity, asexual reproduction, production of surface metabolites, and ability to nutrient uptake through the rhizoids (Crocket and Keough 2014). The phenotypic flexibility of several *Caulerpa* species provides an additional physiological advantage in adapting to changes in light intensity and temperature. For example, the morphological changes of *C. racemosa* in terms of ramule organisation occur under different climatic conditions (Ohba et al. 1992). *C. cylindracea* is also able to change its pigment composition to protect itself from high radiation intensity and maximise photosynthetic capacity in low light (Bernardeau-Esteller et al. 2015). According to Burfeind and Udy (2009), seasonal and temporal variations in temperature and light influence the growth and morphology of the vegetative structures of *C. taxifolia*. The ability of these species to reproduce asexually through rhizoid expansion and fragmentation increases the resistance of populations to predators and unfavourable environmental conditions, which could be a factor contributing to the spread of *Caulerpa* outside its original habitat. In addition, species such as *C. sertularioides* have high persistence in sunlight, so thallus fragments can drift for many days before settling on a suitable substrate (Fernandez-Garcia et al. 2016).

Furthermore, the production of bioactive surface metabolites by *Caulerpa* species has become an additional deterrent against predators. *C. ashmeadi*, for example, produces a variety of sesquiterpenoids with antibacterial activity against certain marine bacteria, making the species unattractive to grazers and dangerous to certain fish predators (Zubia et al. 2020). Numerous *Caulerpa* species produce chemicals with antifouling properties against microorganisms to inhibit larval settlement and defend against predators (Nagaraj and Osborne 2014). However, the amounts of surface metabolites vary among species and are inconsistent. In addition, *Caulerpa* species are known to invade seagrass beds, and the presence of seagrass can influence the synthesis of surface metabolites within species if production is regulated and boosted to improve competitiveness (Katsanevakis et al. 2014b). In addition, *Caulerpa* species can affect the benthic assemblages of neighbouring macrophytes by extracting nutrients from sediments, making them particularly aggressive and invasive (Garcia et al. 2015). These invasive *Caulerpa* species exploit degraded habitats and cause an ecological shift by preventing the return of native benthic ecosystems (Bulleri et al. 2010).

5.3 Nutritional Content

Seaweeds in general produce organic compounds such as secondary metabolites as part of their defence mechanisms and homeostasis maintenance tools through changes in biological activity, water solubility increment, and chemical stability improvement (Mashjoor et al. 2015; Christobel et al. 2011). More than 3000 novel chemical compounds with therapeutic applications were discovered in marine creatures (Martins et al. 2014). Numerous marine compounds have been shown to be useful biological instruments for investigating cellular processes at the molecular level (Kiuru et al. 2014). Depending on their size, the varied and heterogeneous group of organisms known as algae can be separated into multicellular macroalgae and unicellular microalgae (Mashjoor et al. 2016). These algae are frequently encountered in unfavourable circumstances of temperature, salinity, and light. To adapt to these harsh environmental circumstances, the majority of algae synthesise a wide array of secondary metabolites, many of which possess physiologically active qualities (Ibañez et al. 2012). Seaweed, also known as marine macroalgae, is one of the most crucial natural resources for maintaining the chemical and biological equilibrium of oceanic ecosystems (Gomez-Zavaglia et al. 2019). Due to its abundance of bioactive components, seaweed is employed as a source of unsaturated fatty acids, alginates, proteins, agar, carrageenan, minerals, and vitamins (Pereira 2018). Numerous research investigations have demonstrated that the bioactive chemicals obtained from macroalgae have a wide range of biological activities, including antibiotics, antioxidants, antivirals, anticancer activity, cytotoxicity, and the potential to induce apoptosis in cancer cells (Kim 2012; Ibtissam et al. 2009).

Aside from that, seaweeds contain a wide range of polysaccharides that have been shown to have a variety of pharmacological properties, including anticancer, anti-inflammatory, and great antioxidant properties (Rengasamy et al. 2020). The structural polysaccharides, mucopolysaccharides, and storage polysaccharides in seaweed are known to be carbohydrate polymers that are made up of simple sugars connected by glycoside linkages (Kraan 2012). Because of their various biological features and low toxicity, high molecular weight polysaccharides and their breakdown products of low molecular weight oligosaccharides are economically very important for new drug discovery (Rengasamy et al. 2020). Within the food industry, polysaccharides are commonly employed as stabilisers, thickeners, and emulsifiers. Among the significant polysaccharides in seaweed are the alginates, agarans, and carrageenan which are derived from red seaweeds such as Kappaphycus, Gracilaria, and Eucheuma species; fucoidans and laminarin which are derived from brown seaweeds such as Sargassum and Laminaria species, and ulvans that are derived from the green seaweeds, mostly from Ulva species (Fernando et al. 2018). Other than that, some seaweeds are also known to have phlorotannins, commonly known as algal polyphenols, which are the polymers of phloroglucinols (Aminina et al. 2020). Brown seaweeds are known to have the most abundant phlorotannin content. Numerous biological and pharmacological activities such as antibacterial, antioxidant, antiproliferative, anticancer, anti-inflammatory and antidiabetic effects have been identified in phlorotannins (Cotas et al. 2020).

Another type of macroalgae that has been widely studied in terms of its bioactive compounds is the Caulerpaceae family. They produce several secondary metabolites including sesquiterpenoids and diterpenoids that act as protection tools from grazers, whereby caulerpenyne, a type of sesquiterpene that is present in *Caulerpa* species of the same family, plays a major role in chemical defence (Movahhedin et al. 2014). Besides, the genus *Caulerpa* of the family Caulerpaceae contains several active metabolites such as caulerpin, caulerpicin, caulerpenyne, palmitic acid, ß-sitosterol, taraxerol, caulerpol, flexilin, and trifarin whereby the first compound to be identified from the genus was caulerpicin (Handley 2003). In addition, triterpenes; squalene, squalene epoxides, sterols, di-indole pigments; caulerpin and its analogues, caulersin, a mixture of ceramide derivatives; and caulerpicin are the other secondary metabolites that have been isolated from several *Caulerpa* species (Movahhedin et al. 2014). Earlier studies had shown anti-herpes simplex virus 1, antibacterial activity, antitumor activity, and plant regulatory effects for caulerpin extracted from *Caulerpa* species (Macedo et al. 2012). In many reports, the linear terpenoid secondary metabolites, especially caulerpenyne, have been associated with anticancer, antiproliferative, antimicrobial, antiherpetic, and antiviral properties (Cavas and Pohnert 2010). In addition, extracts from various *Caulerpa* species are used as animal feed and in traditional medicine to cure rheumatism and lower blood pressure (Novaczek 2001; Prud'homme van Reine and Trono 2001).

Caulerpin is a type of bis-indolic alkaloid with antinociceptive and anti-inflammatory properties. It is commonly produced by certain *Caulerpa* species, including *C. racemosa*, *C. peltate*, *C. cylindracea,* and *C. mexicana* (Kumar et al. 2019). Alkaloid is one of the major components utilised in the development of new drugs due to its various chemical structures that exhibit several pharmacological properties (Herrero et al. 2013). *C. racemosa*, a popular edible and widely cultivated plant in the Indo-Pacific region, contains caulerpin as one of its key phytochemical constituents, attracting more researchers to investigate the bioactive ingredient of *Caulerpa* species with pharmaceutical potential (Cavalcante-Silva et al. 2016; Ornano et al. 2014). Some *Caulerpa* species also contain Caulerpenyne which is a type of sesquiterpene with a bis-enol acetate functional group that has been shown to exhibit antibiotic and antioxidant activity. However, this compound can also exhibit cytotoxicity in mammalian cells (Sfecci et al. 2017).

Caulerpa species also have a substantial proportion of proximate components (Table 5.2). The protein content of the edible Caulerpa species ranged from 0.6% to 20.8% in dry matter, which is low compared to protein-rich legumes such as soybeans. The moisture content of Caulerpa species ranged from 75% to 94% on a wet-weight basis (de Gaillande et al. 2017). The protein content of seaweed varies according to preparation technique, season, and location (Wells et al. 2017). Indeed, the protein content also differs based on species, where the amount is low for brown seaweed (3–15% of dry weight), moderate for green seaweeds (9–26% of dry weight), and high for red seaweeds (maximum 47% of dry weight) (Fleurence et al. 2018). In addition, the presence of important amino acids influences the protein quality of *Caulerpa* species (Table 5.3), with aspartic acid and glutamic acid being the most abundant and histidine, lysine, and methionine the least abundant (Matanjun et al. 2009; Ratana-arporn and Chirapart 2006).

Carbohydrates make up between 3.6% and 83.2% of the dry mass of Caulerpa species, followed by proteins. Seaweeds contain a significant amount of polysaccharides, a type of polymeric carbohydrate widely recognised as an alternative source of dietary fibre for human consumption (Fleurence et al. 2012). Although the human body is unable to digest dietary fibres, they are still essential for human health to aid in increasing the faecal mass, promoting colonic fermentation, and lowering postprandial glucose and preprandial cholesterol levels (Elleuch et al. 2011). Fibre, which makes up the bulk of seaweed biomass and can be ingested in sufficient amounts to avoid metabolic syndrome, is connected to obesity, type II diabetes, and cardiovascular problems (Jakobsdottir et al. 2014; Mayakrishnan et al. 2013). According to some studies, the polysaccharides contained in *Caulerpa* species, such as arabinose, and xylose, fucose, and rhamnose, have antiviral, anticancer, and immunostimulant properties (de Gaillande et al. 2017). For example, *C. cupressoides* var. lycopodium contains a non-anticoagulant sulphated polysaccharide, which is a new potential source of the analgesic chemical that needs to be investigated in biomedical studies (Rodrigues et al. 2019). *C. racemosa* also contains sulphated polysaccharides which, according to a separate study, have antinociceptive and anti-inflammatory properties (Ribeiro et al. 2014). The lipid content of *Caulerpa* species, which ranges from 0.1 to 7.2% of dry matter, is thought to be low. However, the presence of polyunsaturated fatty acids (PUFAs) in this species has attracted the interest of scientists (Table 5.4). Most edible *Caulerpa* species contain the polyunsaturated fatty acid–linolenic acid (C18:3w3), with *C. racemosa* having the highest proportion of PUFAs, accounting for 60.8% of the total fatty acid content (de Gaillande et al. 2017).

In addition, *Caulerpa* species have a significant ash content of up to 55% of dry matter, suggesting that they may contain significant amounts of critical minerals and trace elements required for human nutrition (Matanjun et al. 2009). The high content of ash might also reflect a high concentration of macronutrients (phosphorus, sodium, magnesium, potassium, chloride, sulphur, and calcium), microminerals, and trace elements (zinc, iron, selenium, cobalt, copper, molybdenum, iodine,

Table 5.2 Proximate composition of edible *Caulerpa* species

Species	Moisture (% WW)	Ash (% DW)	Crude lipid (% DW)	Crude protein (% DW)	Crude fibre (% DW)	Carbohydrate (% DW)	References
C. cupressoides	n.d	11.3	3.8	20.8	n.d	47.4	Carneiro et al. (2014)
C. lentillifera	74.69	24.21	0.86	12.49	3.17	59.27	Ratana-arporn and Chirapart (2006)
	89.24	37.15	1.11	10.41	1.91	38.66	Matanjun et al. (2009)
	90.84	14.10	0.17	13.24	19.40	53.08	Ahmad et al. (2012)
	87.05	29.61	2.87	19.38	4.12	44.02	Nagappan and Vairappan (2014)
	95.95	27.36	2.32	12.50	8.60	44.82	Zhang et al. (2019)
	95.01	3.41	0.79	0.43	14.38	n.d	Nofiani et al. (2019)
C. racemosa	92.00	10.64	0.15	10.52	11.29	67.40	Ahmad et al. (2012)
	91.36	23.81	2.21	17.36	3.11	52.81	Nagappan and Vairappan (2014)
	15.37	12.15	7.65	19.72	11.51	48.97	Bhuiyan et al. (2016)
	13.85 ± 0.93	28.25 ± 0.27	4.20 ± 0.32	20.27 ± 0.14	n.d	33.42	Aroyehun et al. (2020)

DW stands for dry weight, WW stands for wet weight, and n.d stands for not determined.

Table 5.3 Amino acid content of edible *Caulerpa* species

Species	Essential amino acids (EAA), mg g^{-1} of dry sample								Nonessential amino acids (NEAA), mg g^{-1} of dry sample							
	Valine	Lysine	Histidine	Leucine	Methionine	Isoleucine	Phenylalanine	Threonine	Alanine	Proline	Serine	Glycine	Arginine	Glutamine	Tyrosine	Asparagine
C. racemosa var. turbinata	0.68	0.58	0.20	0.86	0.18	0.51	0.62	0.60	0.67	0.47	0.55	0.70	0.57	1.39	0.45	1.20
C. lentillifera	0.62–0.87	0.12–0.82	0.08–0.14	0.78–0.99	0.16	0.51–0.62	0.61–1.99	0.58–0.79	0.69–0.85	0.43–0.57	0.55–0.76	0.51–0.85	0.57–0.87	1.38–1.78	0.33–0.48	0.83–1.43
C. racemosa	0.57	0.33	0.11	0.69	0.17	0.43	0.64	0.51	0.49	0.57	0.45	0.61	0.49	1.13	0.44	0.94

Table 5.4 Fatty acid content (% of total fatty acids, TFA) of edible *Caulerpa* species

	C. cupressoides	*C. lentillifera*	*C. racemosa*
C12:0	1.30	0.13–0.69	0.38–1.03
C14:0	2.30	1.64–4.70	2.20–1.03
C16:0	33.0	8.74–67.83	27.40–87.98
C18:0	1.20	2.10–11.10	0.70–2.14
C20:0	n.d	0.20–1.98	0.03–0.04
SAFs	37.80	40.70–82.69	33.40–93.06
C16:1	0.80	0.10–7.49	4.40–8.50
C18:1ω9	2.00	0.93–32.49	1.90–4.75
C20:1ω9	n.d	0.17–1.36	0.40–1.04
MUFAs	7.95	8.37–36.83	5.70–14.10
C18:2ω6	10.20	4.50–11.85	1.90–10.30
C18:3ω6	0.80	0.31–5.99	0.60–2.13
C18:3ω3	13.70	5.15–14.71	0.78–18.84
C20:2ω6	0.80	0.07–4.27	1.06
C20:3ω6	0.40	0.50–3.30	n.d
C20:4ω6	4.50	0.84–6.70	2.40–6.70
C20:5ω3	3.10	0.83–1.60	1.70–9.54
C22:6ω3	4.60	0.83–3.64	0.81
PUFAs	50.80	9.49–47.30	27.17–60.80
PUFAs ω6	19.70	8.04–25.01	16.00–16.01
PUFAs ω3	30.30	7.55–28.52	11.25–31.11
Ratio ω6/ω3	1.54	1.07–2.85	0.20–2.00

n.d = not determined.

manganese, nickel, and boron) in the species (Table 5.5). Ash contained in seaweed is often regarded as a criterion of seaweed quality to examine its nutritional and bifunctional properties (Reka et al. 2017). *Caulerpa* species are good candidates for functional foods because they are naturally rich in essential nutrients as well as a variety of health-promoting compounds. Furthermore, algae have emerged as one of today's most important foods, particularly among vegetarians, vegans, and health-food consumers. To meet the increasing demand in the field of healthy nutrition, the cultivation of *Caulerpa* species is being expanded worldwide.

5.4 Aquaculture

In the Philippines, Indonesia, Malaysia, and Japan, fresh salads from *C. lentillifera* are considered a delicacy. The species was virtually extinct in the wild due to intensive natural harvesting, which has led to a huge production of this species in captivity (Aroyehun et al. 2020). In addition, the increasing demand for *Caulerpa* in Southeast Asian countries has led to the establishment or assessment of several seaweed farms to generate more income for local communities and develop a new

Table 5.5 Macroelement and microelement composition g (100 g)$^{-1}$, dry weight in different edible *Caulerpa* species

	Microelements						Macroelements				
	Fe	Zn	Mn	Cu	Se	Mo	Na/K	Mg	K	Ca	Na
C. lentillifera	0.02–21.4	0.002–3.5	0.001–7.9	0–0.1	0.4–1.07	<0.01	7.8–22.0	0.6–1.6	0.7–1.1	0.6–1.9	0.02–9.0
C. racemosa	0.02–8.13	0–2.5	0.5.8	0–0.8	0.1–12.3	0–0.3	0.3 $\geq$ 10	0.4–4.1	0.3–3.9	0.6–5.9	0.02–7.0
C. sertularioides	0.41	0.03	n.d	2.5	n.d	n.d	2.3	0.4	0.03	1.2	0.1

form of commercial aquaculture products (Zubia et al. 2020; Buschmann et al. 2017; Titlyanov and Titlyanova 2010; Trono 1999; Trono 1988). On the island of Mactan, Cebu, massive cultivation of this species in small tidal ponds began more than 30 years ago in response to increasing demand. Since then, it has spread throughout the Philippines and is exported to Japan and Taiwan (Nagappan and Vairappan 2014). Previously, fishermen in the Philippines discovered *C. lentillifera* growing in abandoned ponds previously used for milkfish aquaculture in an area cleared of mangroves, which eventually led to the start of production of *Caulerpa* species (Trono 1988). The pond culture applied a bottom planting method where the *C. lentillifera* rhizoidal thalli were buried in the mud for around 2 cm or the young thalli were spread on the bottom of the pond. The dams of the ponds were built of dried black mud and coral cairns and filled to a depth of 0.5 to 0.6 m, depending on the tidal range. Sluice gates regulated the water level and the filling of the ponds, which ranged in area from 1000 square metres to about 1 hectare. After 45 days of cultivation, C. lentillifera is harvested by removing part of the plants from the mud and leaving part of the plants to promote new development (Zubia et al. 2020).

The farming system has evolved via constant innovation, particularly in cultural techniques. The species can be farmed in a single step by connecting the fragments to a substrate like ropes, nets, cages, or trays, which requires little in the way of expensive infrastructure support or specialised knowledge. This makes the species' cultivation highly viable (de Gaillande et al. 2017; Bast 2013). A number of cultivation techniques, such as off-bottom cultivation, land-based racetracks, floating longlines, and bottom planting, have been adopted by numerous farms across Asia (Zubia et al. 2020). Nevertheless, all cultivation methods used are varied and adapted differently depending on the country and site conditions (Table 5.6). To date, cultivation of *C. lentillifera* and *C. racemosa* has been carried out massively in the Philippines, Thailand, and Japan to cope with the increasing demand (Nagappan and Vairappan 2014). *C. lentillifera* is commonly cultivated as part of the water treatment indicator in shrimp ponds within Thailand and is used as animal feed (Ratana-arporn and Chirapart 2006). The propagation and cultivation methods of *C. lentillifera* vary across Southeast Asia's regions; for instance, in the Philippines, *C. lentillifera* is cultivated through sea farming using a bamboo raft method. However, commercial production is limited through farming due to the vulnerability of the *Caulerpa* when cultured in the open sea with unpredictable tidal patterns. Thus, research on the cultivation of *C. lentillifera* using a tank culture system was performed in Australia, and the culture method was also patented in 2010 (Paul et al. 2014). Nevertheless, mass cultivation of *Caulerpa* species in the open sea should also be considered as one of the high-potential seaweed farming activities with a proper investigation of the limiting survival factor that can be overcome through thorough research.

C. lentillifera production has decreased as a result of the Philippines' ban on turning mangrove areas into cultivation ponds due to the negative effects on the ecosystem (Tanduyan et al. 2013). Due to extensive wild harvesting in various Asian nations, including the Philippines and Malaysia, the *Caulerpa* species has nearly completely disappeared from the natural world. Thus, advances in growing methods

Table 5.6 Cultivation of *Caulerpa* species within the Asian Region

Species	Techniques	Status	Location
Caulerpa racemosa var. *occidentalis* *C. lentillifera*	Bottom-placing method	Prohibited Commercial	Philippines Trono (1988) Tanduyan et al. (2013)
C. lentillifera	Off-bottom cages method	Experimental farm	
C. racemosa	Floating long lines	Experimental farm	Bangladesh Zafar (2005)
C. racemosa var. *turbinata*	Off-bottom trays method	Experimental farm	French Polynesia de Gaillande et al. (2017)
C. lentillifera	Land-based raceway Off-bottom cages method	Commercial	Japan Titlyanov and Titlyanova (2010)
C. lentillifera	Land-based raceway	Commercial	Thailand Bambaranda et al. (2019)
C. lentillifera *C. racemosa*	Recirculating aquaculture system	Experimental farm	Malaysia The Star (2019)
C. lentillifera	Land-based raceway	Experimental farm	India Mary et al. (2009)
C. racemosa	Off-bottom trays method	Experimental farm	Samoa Morris et al. (2014)
C. lentillifera	Off-bottom trays method	Experimental farm	Vietnam Titlyanov and Titlyanova (2010)

have led to a resurgence of *Caulerpa* farming. The Nha Trang Oceanography Institute and Tri Tin Company Limited created a method for producing *C. lentillifera* in Vietnam utilising bamboo frames and plastic trays in response to *Caulerpa* cultures' success in the Philippines. The plastic trays containing the *C. lentillifera* seeds were fixed on the bamboo soil and the bamboo frames were positioned 0.5 m above the bottom of the culture tanks, with seawater filtered through the drains to avoid organic contamination. Two weeks later, *C. lentillifera* was removed from the bamboo raft (Tadokoro et al. 2000).

Apart from this, an off-bottom method is used in the Philippines, Vietnam, Samoa, and Japan, where *Caulerpa* is cultivated a few metres above the ground in different systems, e.g. cage and tray cultures (de Gaillande et al. 2017). In the off-bottom method, the seaweed is attached either on trays or cages and then placed a few metres above the bottom of sandy substrates where poles with appropriate heights are selected such that the nets are submerged at high tides and exposed at low tides (de Gaillande et al. 2017). Similar to red seaweed farming, the longline technique is used in Bangladesh to cultivate *Caulerpa* on suspended lines, ropes, or nets above or below the water surface (Zafar 2005). In addition, farmers in Japan, India, and Thailand use land-based troughs to overcome environmental variations in the cultivation of *Caulerpa* species (de Gaillande et al. 2017). Several cultivation methods have been patented in China. Patent CN106973780A proposes a method for growing *C. lentillifera* in ponds, while patent CN201042155Y describes a strategy for cultivating *C. racemosa* using land-based breeding facilities (Zubia et al. 2020). Paul and de Nys (2008) proposed integrated multitrophic aquaculture for the widespread propagation of *Caulerpa* species for sustainability reasons. Research on the propagation of *C. racemosa* was started on the island of St. Martin, using three types of culture set-ups: longline, net method, and hanging rope method (Zafar 2005). The results of the study suggest that *C. racemosa* grows better with the longline and net methods.

C. lentillifera is grown in Japan using the net method to a depth of at least 0.5 m at the cultivation site (Titlyanov and Titlyanova 2010). In addition, a number of trials have been started in Samoa to grow *C. racemosa* using different growing systems such as submerged, with or without substrate, and floating (Morris et al. 2014). In French Polynesia, Tubuai, a pilot farm for the cultivation of *C. chemnitzia var. turbinata* has been established in a lagoon using submerged trays made of perforated plastic (de Gaillande et al. 2017). *C. lentillifera* is the most commonly available species on the local market in Malaysia. However, the information on commercial cultivation is still limited. Besides, *Caulerpa* also farmed in land-based raceways in a particular country where this farming method can control environmental fluctuations such as light, salinity, silt, current, and nutrients. According to the country and location conditions, all these farming techniques differ and are adjusted in different ways (Titlyanov and Titlyanova 2010). *Caulerpa* is grown in open-tidal ponds in Vietnam and the Philippines (Zubia et al. 2020; de Gaillande et al. 2017), where seaweed cuttings are planted 1 m apart in the pond bottom. After it begins to grow, frequent water exchange is required to maintain good water quality and supply nutrients (Rabia 2016). In China and Japan, the demand for *Caulerpa* is incredibly high, and land-based farming has already been done to some degree (Zubia et al. 2020). A land-based cultivation system for *C. lentillifera* has also been applied in India using raceways under controlled conditions, allowing optimisation of seaweed growth. The raceways are made of fibreglass tanks with muddy sand as the substrate. The raceway-cultured *C. lentillifera* resulted in a 12.4 kg biomass gain with 16 g of initial weight after a 6-month cultivation period (Mary et al. 2009).

5.5 Applications of *Caulerpa* Species

Caulerpa is highly diverse and widely distributed throughout tropical and subtropical regions (Lewmanomont 2008; Phang et al. 2019). Many species in the *Caulerpa* genus have become economically important species (Zubia et al. 2020). For instance, *C. lentillifera* is commonly consumed as a fresh salad in Japan. Other than that, *C. lentillifera* is widely used in pond bioremediation associated with shrimp aquaculture in Thailand. Besides, the invasive *C. taxifolia* is commonly sold for use in home aquaria throughout the world, mostly in the Mediterranean region, Australia, and California, USA (Wiedenmann et al. 2001). The species under this genus have also gained a considerable amount of attention for their antioxidant and antibacterial properties, for instance, the commercial *C. racemosa* and *C. lentillifera*, also known as 'green caviar' and 'sea grapes', respectively, in the Asian region such as the Philippines, China, Indonesia, Malaysia, Thailand, Japan, Vietnam, and Singapore (Aroyehun et al. 2020; Yap et al. 2019). Several *Caulerpa* species are popular as human foods in the Indo-Pacific region due to their delicious taste, crunchy texture, and health benefits. Meanwhile, locals have a long history in the Far East and Asian Pacific region of eating green seaweed as part of their diet (Wong and Cheung 2000). *Caulerpa* species in general contain various nutritional compounds such as proteins, fibres, essential fatty acids, vitamins, minerals, and antioxidants that are good for human health (Darmawan et al. 2020). Therefore, various research has been done on the nutritional properties and biochemical composition of the *Caulerpa* species that are significant for human and other animal consumption (Nofiani et al. 2019; Ratana-arporn and Chirapart 2006; Van Nguyen et al. 2011). Nevertheless, more profound research is required to further comprehend the distribution, ecology, and biology of this highly potential commercial species with beneficial biochemical composition (Zubia et al. 2020).

Caulerpa is claimed to be beneficial as a natural food and a source of fibre without a toxic or harmful effect on humans (Movahhedin et al. 2014). In addition, the nutritional properties of *Caulerpa* species have been documented by various researchers, particularly those of *C. lentillifera* and *C. racemosa*, which were reported to have high antioxidants, vitamins, minerals, and protein as well as balanced amino acids and essential fatty acid content (Aroyehun et al. 2020; Daud et al. 2016; Hafting et al. 2015). Cavas and Pohnert (2010) also claimed that some *Caulerpa* species contain several active metabolites such as caulerpin, caulerpicin, caulerpenyne, palmitic acid, ß-sitosterol, taraxerol, caulerpol, flexilin, and trifarin that are good for health benefits. For instance, one of the metabolites named caulerpenyne is claimed to have antimicrobial, anticancer, antiherpetic, antiproliferative, and antiviral properties (Barzkar et al. 2019). Indeed, most edible macroalgae have been linked to a variety of health benefits (Cengiz et al. 2010). The highly reported *Caulerpa* species for their nutritional values are mostly *C. lentillifera* and *C. racemosa* (Zubia et al. 2020). The information on the nutritional values and secondary metabolite composition of other high-potential species is still lacking, and further research is required to better understand their potential. Thus, in-depth research on the nutritional values of this species would be extremely important for a better understanding of their potential as functional foods and marine vegetables (Nagappan and Vairappan 2014).

Currently, the *Caulerpa* species are becoming increasingly popular and consumed. *C. lentillifera, C. racemosa var. clavifera f. macrophysa,* and *C. racemosa var. laetevirens* are frequently consumed as a delicacy and fresh salads in Malaysia (Nagappan and Vairappan 2014; Phang 2006). *C. lentillifera* is the most favoured species due to its appealing colour and good texture. *C. racemosa var. laetevirens* and *C. racemosa var. clavifera f. macrophysa*, on the other hand, are light green with radial fronds carrying a clear spherical grape-like structure of ramuli. The common marketable species in Malaysia is *C. lentillifera*, also known as Latok, generally consumed as fresh salads by the local people in Sabah, as shown in Figs. 5.1 and 5.2.

Fig. 5.1 Wild-harvested *C. lentillifera* sold in the Local Market Tuaran, Sabah, Malaysia

Fig. 5.2 Wild *C. lentillifera* found in Sepanggar Bay during one of the site surveys for *Caulerpa* sp. within West Malaysia (Sabah) coastal area

Fig. 5.3 Wild *C. macrodisca* was found to have accidentally grown inside a local farmer's cage in Menumbok, Sabah, Malaysia

Nevertheless, *C. lentillifera* is yet to be massively cultivated commercially, and most of these species have been harvested from the wild for consumption by the locals. Several local farmers and agencies in Malaysia have recently begun exploring *Caulerpa* cultivation, primarily on *C. lentillifera* and *C. racemosa* (Aroyehun et al. 2020).

C. macrodisca (Decaisne) Weber-van Bosse, a fascinating *Caulerpa* species, was recently discovered in large numbers on the west coast of Sabah and recognised as shown in Fig. 5.3. *C. macrodisca*, formerly known as *C. racemosa var. macrodisca* and *C. peltata var. macrodisca*, was introduced by Decaisne in 1842 and is native to the region around Anambas Island in Indonesia (Belton et al. 2014). This species, known locally as eaba-eaba, is commonly consumed as a fresh salad in the Philippines, where divers collect it from the wild and sell it seasonally. In the Philippines, *C. macrodisca* is similarly thought to be locally extinct due to extensive land reclamation efforts that have wiped out the algal populations (Belleza and Liao 2007). The Republic of Palau, Sri Lanka, Australia, New Zealand, the Philippines, Singapore, the South China Sea, Indonesia, and the Samoan Archipelago are all known locations where C. macrodisca is known to be found (Guiry and Guiry 2020). According to Phang et al. (2016), *C. macrodisca* is exclusively present in Singapore's, Thailand's, and Vietnam's coastal regions.

5.6 Conclusion

In summary, the widely distributed green seaweed species that come from the genus *Caulerpa* exhibit various benefits for the seaweed industry, especially for food, pharmaceutical, nutraceutical, and industrial applications. The secondary metabolites that exist in the species are crucial for various applications, including in the biotechnology field. Thus, the cultivation of the genus is important in ensuring a sufficient raw material supply for the industry and preventing over-exploitation of the species that are wildly harvested.

References

Ahmad F, Sulaiman MR, Saimon W, Chye FY, Matanjun P (2012) Proximate compositions and total phenolic contents of selected edible seaweed from Semporna, Sabah, Malaysia. Borneo Sci 3:85–96

Amat JN, Cardigos F, Santos RS (2008) The recent northern introduction of the seaweed *Caulerpa webbiana* (Caulerpales, Chlorophyta) in Faial, Azores Islands (North-Eastern Atlantic). Aquat Invasions 3(4):417–422

Aminina NM, Karaulova EP, Vishnevskaya TI, Yakush EV, Kim YK, Nam KH, Son KT (2020) Characteristics of polyphenolic content in brown algae of the pacific coast of Russia. Molecules 25:3909

Anderson LWJ (2005) California's reaction to *Caulerpa taxifolia*: a model for invasive species rapid response. Biol Invasions 7(6):1003–1016

Aroyehun AQ, Abdul Razak S, Palaniveloo K, Nagappan T, Rahmah NSN, Gan WJ, Chelappan DK, Chellian J, Kunnath AP (2020) Bioprospecting cultivated tropical green algae, *Caulerpa racemosa* (Forsskal) J. Agardh: a perspective on nutritional properties, antioxidative capacity and anti-diabetic potential. Foods 9(9):1313–1333

Bambaranda BVSAM, Tsusaka TW, Chirapart A, Salin KR, Sasaki N (2019) Capacity of Caulerpa lentillifera in the removal of fish culture effluent in a recirculating aquaculture system. Processes 7(440):1–15

Barsanti L, Gualtieri P (2014) Chlorophyta. In: Algae: anatomy, biochemistry, and biotechnology, 2nd edn. CRC Press, Boca Raton

Barzkar N, Tamadoni Jahromi S, Poorsaheli HB, Vianello F (2019) Metabolites from marine microorganisms, micro, and macroalgae: immense scope for pharmacology. Mar Drugs 17(8): 464

Bast F (2013) Agronomy and cultivation methods for edible seaweeds. Int J Agric Food Sc Technol 4(7):661–666

Belleza DFC, Liao LM (2007) Taxonomic inventory of the marine green algal genus *Caulerpa* (Chlorophyta, Bryopsidales) at the University of San Carlos (Cebu) herbarium. Philipp Sci 44: 71–104

Belton GS, Draisma SGA, Prud'homme van Reine WF, Huisman JM, Gurgel CFD (2019) A taxonomic reassessment of Caulerpa (Chlorophyta, Caulerpaceae) in southern Australia, based on *tuf*A and *rbc*L sequence data. Phycologia 58(3):234–253

Belton GS, Prud'homme Van Reine WF, Huisman JM, Draisma SGA, Gurgel CFD (2014) Resolving phenotypic plasticity and species designation in the morphologically challenging *Caulerpa racemosa-peltata* Complex (Caulerpaceae, Chlorophyta). J Phycol 50(1):32–54

Benzie J, Price IR, Ballment E (2008) Population genetics and taxonomy of *Caulerpa* (Chlorophyta) from the Great Barrier Reef, Australia. J Phycol 33(3):491–504

Bernardeau-Esteller J, Ruiz JM, Tomas F, Sandoval-Gil JM, Marin-Guirao L (2015) Photoacclimation of Caulerpa cylindracea: light as as limiting factor in the invasion of native Mediterranean grass meadows. J Exp Mar Biol Ecol 465:130–141

Bhuiyan MKA, Qureshi S, Kamal AHM, Uddin SA, Siddique AM (2016) Proximate chemical composition of sea grapes *Caulerpa racemosa* (J. Agardh, 1873) collected from a sub-tropical coast. Virol Mycol 5(2):1–6

Bulleri F, Balata D, Bertocci I, Tamburello L, Benedetti-Cecchi L (2010) The seaweed *Caulerpa racemosa* on Mediterranean rocky reefs: from passenger to driver of ecological change. Ecology 91(8):2205–2212

Burfeind DD, Udy J (2009) The effects of light and nutrients on *Caulerpa taxifolia* and growth. Aquat Bot 90(2):105–109

Buschmann AH, Camus C, Infante J, Neori A, Israel A, Hernández-González MC, Pereda SV, Gomez-Pinchetti JL, Golberg A, Tadmor-Shalev N, Critchley AT (2017) Seaweed production: overview of the global state of exploitation, farming and emerging research activity. Eur J Phycol 52(4):391–406

Carneiro JG, Rodrigues JAG, Teles FB, Belfort A, Cavalcante D, Benevides NMB (2014) Analysis of some chemical nutrients in four Brazilian tropical seaweeds. Acta Sci Biol Sci 36(2):137–145

Cavalcante-Silva LHA, Correia ACC, Sousa JCF, Barbosa-Filho JM, Santos BVO, de Miranda GEC, Alexandre-Moreira MS, Cavalcante FA (2016) Involvement of β adrenergic receptors in spasmolytic effect of caulerpin on Guinea pig ileum. Nat Prod Res 30(22):2605–2610

Cavas, L. & Pohnert, G. 2010. The potential Of *Caulerpa* Spp. for biotechnological and pharmacological applications. In Seaweeds and their role in globally changing environments 15 (2009), 386–397, Springer, Dordrecht

Cengiz S, Cavaz L, Yurdakoc K (2010) Alpha-amylase inhibition kinetics by Caulerpenyne. Mediterr Mar Sci 11:93–103

Chen K, Ríos JJ, Pérez-Gálvez A, Roca M (2017) Comprehensive chlorophyll composition in the main edible seaweeds. Food Chem 228:625–633

Christobel GJ, Lipton AP, Aishwarya MS, Sarika AR, Udayakumar A (2011) Antibacterial activity of aqueous extract from selected macroalgae of southwest coast of India. Seaweed Res Utilisation 33(1&2):67–75

Coppejans E, Prid'homem van Reine WF (1992) Seaweeds of the Snellius-II expedition (Indonesia): the genus *Caulerpa* (Chlorophyta-Caulerpales). Bulletin des Séances de l'Académie Royale des Sciences d'Outre-Mer 37:667–712

Cotas J, Leandro A, Monteiro P, Pacheco D, Figueirinha A, Goncalves AMM, de Silva GJ, Pereira L (2020) Seaweed phenolics: from extraction to applications. Mar Drugs 18:384

Cremen MCM, Leliart F, West J, Lam DW, Shimada S, Lopez-Bautista JM, Verbruggen H (2019) Reassessment of the classification of bryopsidales (Chlorophyta) based on chloroplast *Phylogenomic analyses*. Mol Phylogenet Evol 130:397–405

Crocket P, Keough M (2014) Ecological niches of three abundant species in port Philip Bay, Southeast Australia. Aquat Bot 119:120–131

Darmawan M, Fajarningsih ND, Sihono & Irianto, H.E. (2020) *Caulerpa*: ecology, nutraceutical and pharmaceutical potential. In: Nathani NM, Mootapally C, Gadhvi IR, Maitreya B, Joshi CG (eds) Marine niche: applications in pharmaceutical sciences. Springer Nature, Singapore, pp 299–318

Daud D, Arsad NMA, Ismail A, Tawang A (2016) Anti-pyretic action of *Caulerpa lentillifera*, *Hibiscus Rosa-sinensis* and *Piper sarmentosum* aqueous extract in mice. Asian J Pharm Clin Res 9(1):145–147

de Gaillande C, Payri C, Remoissenet G, Zubia M (2017) *Caulerpa* consumption, nutritional value and farming in the Indo-Pacific region. J Appl Phycol 29:2249–2266

de Senerpont Domis LN, Elser JJ, Gsell AS, Huszar VLM, Ibelings BW, Jeppesen E (2013) Plankton dynamics under different climatic conditions in space and time. Freshw Biol 58: 463–482

Diaz-Pulido G, McCook L (2008) Macroalgae (seaweeds). In: Chin A (ed) The state of the great barrier reef on-line. Great Barrier Reef Marine Park Authority, Townsville, pp 1–44

Elleuch M, Bedigian D, Roiseux O, Besbes S, Blecker C, Attia H (2011) Dietary fibre and fibre-rich by-products of food processing: characterisation, technological functionality and commercial applications: A review. Food Chem 124(2):411–421

Famà P, Wysor B, Kooistra WHCF, Zuccarello GC (2002) Molecular phylogeny of the Genus Caulerpa (Caulerpales, Chlorophyta) inferred from Chloroplast tufA gene. J Phycol 38:1040–1050

Fernandez-Garcia C, Riosmena-Rodriguez R, Wysor B, Pena-Salamanca E, Verbruggen H (2016) DNA-assisted identification of *Caulerpa* (Caulerpaceae, Chlorophyta) reduces species richness estimates for the Eastern Tropical Pacific. Phytotaxa 252:185–204

Fernando S, Kim KN, Kim D, Jeon YJ (2018) Algal polysaccharides: potential bioactive substances for cosmeceutical applications. Crit Rev Biotechnol 39(2):1–15

Ferrara L (2020) Seaweeds: A food for our future. J Food Chem Nanotechnol 6(2):56–64

Fleurence J, Morancais M, Dumay J (2018) Seaweed Proteins. In: Yada RY (ed) Proteins in food processing. Elsevier, Cambridge, pp 245–262

Fleurence J, Morançais M, Dumay J, Decottignies P, Turpin V, Munier M, Garcia-Bueno N, Jaouen P (2012) What are the prospects for using seaweed in human nutrition and for marine animals raised through aquaculture? Trends Food Sci Technol 27(1):57–61

Garcia M, Weitzmann B, Pinedo S, Cebrian E, Bellesteros E (2015) First report on the distribution and impact of marine alien species in Coastal Benthoc assemblages along the Catalan Coast. In: Munne A, Gonebreda A, Prat N (eds) Experiences from ground, coastal and transitional water quality monitoring, The handbook chemistry. Springer Nature, Cham, pp 249–270

García-Poza S, Leandro A, Cotas C, Cotas J, Marques JC, Pereira L, Goncalves AMM (2020) The evolution road of seaweed aquaculture: cultivation technologies and the Industry 4.0. Int J Environ Res Public Health 17(6528):1–42

Glasby TM, Gibson PT (2007) Limited evidence for increased cold-tolerance of invasive versus native *Caulerpa taxifolia*. Mar Biol 152:255–263

Gomez-Zavaglia A, Lage MAP, Jimenez-Lopez C, Mejuto JC, Simal-Gandara J (2019) The potential of seaweeds, as a source of functional ingredients of prebiotic and antioxidant value. Antioxidants 8(406):1–30

Guiry MD, Guiry GM (2020) AlgaeBase. World-wide electronic publication, National University of Ireland, Galway. http://www.algaebase.org. Accessed 10 Nov 2020

Hafting JT, Craigie JS, Stengel DB, Loureiro RR, Buschmann AH, Yarish C, Edwards MD, Critchley AT (2015) Prospects and challenges for industrial production of seaweed bioactives. J Phycol 51(5):821–837

Handley JT (2003) Studies of the secondary metabolite content of Tasmanian Caulerpa species. Doctoral Dissertation. University of Tasmania, Tasmania, Australia. https://core.ac.uk/download/pdf/33330668.pdf

Herrero M, Mendiola JA, Plaza M, Ibañez E (2013) Screening for bioactive compounds from algae. In: Lee J (ed) Advanced biofuels and bioproducts. Springer, New York, pp 833–872

Hussin H, Khoso A (2017) Seaweed cultivation and coastal communities in Malaysia: an overview. Asian Fish Sci 30(2017):87–100

Ibañez E, Herrero M, Mendiola JA, Castro-Puyana M (2012) Extraction and characterization of bioactive compounds with health benefits from marine resources: macro and micro algae, cyanobacteria, and invertebrates. In: Hayes M (ed) Marine bioactive compounds. Springer, Boston, pp 55–98

Ibtissam C, Hassane R, Jose ML, Francisco DSJ, Antonio GVJ, Hassan B, Mohamed K (2009) Screening of antibacterial activity in marine green and brown macroalgae from the Coast of Morocco. Afr J Biotechnol 8(7):1258–1262

Inderjit, Callaway RM, Vivanco JM (2006) Can plant biochemistry contribute to understanding of invasion ecology? Trends Plant Sci 11:574–580

ITIS Report (2020). http://www.itis.gov. Accessed 15 Oct 2020

Jakobsdottir G, Nyman M, Fak F (2014) Designing future prebiotic fiber to target metabolic syndrome. Nutrition 30:497–502

Jongma DN, Campo D, Dattolo E, D'esposito D, Duchi A, Grewe P, Huisman J, Verlaque M, Yokes MB, Procaccini G (2013) Identity of a slender *Caulerpa taxifolia* strain introduced into the Mediterranean Sea. Bot Mar 56(1):27–39

Katsanevakis S, Coll M, Piroddi C, Steenbeek J, Lasram FBR, Zenetos A, Cardoso AC (2014a) Invading the Mediterranean Sea: biodiversity patterns shaped by Huan activities. Front Mar Sci 1(32):1–11

Katsanevakis S, Wallentinus I, Zenetos A, Leppakoski E, Cinar ME, Ozturk B, Grabowski M, Golani D, Cardoso AC (2014b) Impacts of invasive alien marine species on ecosystem services and the biodiversity: a pan-European view. Aquat Invasions 9(4):391–423

Kazi MA, Reddy CRK, Jha B (2013) Molecular phylogeny and barcoding of *Caulerpa* (Bryopsidales) based on the tufA, rbcL, 18S rDNA and ITS rDNA genes. PLoS One 8(12): e82438

Kim KS (2012) Handbook of marine macroalgae, biotechnology and applied phycology. Wiley-Blackwell, Oxford

Kiuru P, D'Auria MV, Muller CD, Tammela P, Vuorela H, Yli-Kauhaluoma J (2014) Exploring marine resources for bioactive compounds. Planta Med 80(14):1234–1246

Koushalya S, Vishwakarma R, Malik A (2021) Unraveling the diversity of algae and its biomacromolecules. In: Surajit D, Hirak RD (eds) Microbial and natural macromolecules. Academic Press, London, pp 179–204

Kraan S (2012) Algal polysaccharides, novel applications and outlook. In: Chuan-Fa C (ed) Carbohydrates – comprehensive studies on glycobiology and glycotechnology. London, IntechOpen, pp 489–531

Kumar JGS, Umamaheswari S, Kavimani S, Ilavarasan R (2019) Pharmacological potential of green algae *Caulerpa*: a review. Int J Pharm Sci Res 5:1014–1024

Lapointe BE, Barile P, Littler MM, Littler DS (2005) Macroalgal blooms on southeast Florida coral reefs: II. Cross-shelf discrimination of nitrogen sources indicates widespread assimilation of sewage nitrogen. Harmful Algae 4:1106–1122

Lewmanomont K (2008) In: Phang SM, Lewmanomont K, Lim PE (eds) Taxonomy of Southeast Asian seaweeds. Some Caulerpa and Caulerpella (Bryopsidales, Chlorophyta) from Thailand. Institute of Ocean and Earth Science (IOES), University of Malaya, Kuala Lumpur, pp 27–32

Lisette N, de Senerport D, Fama P, Bartlett AJ, Prud'homme van Reine WF, Espinosa CA, Trono GC (2003) Defining taxon boundaries in members of the morphologically and genetically plastic genes *Caulerpa* (Caulerpales, Chlorophyta). J Phycol 39(6):1019–1037

Macedo NRPV, Ribeiro MS, Villaça RC, Ferreira W, Pinto AMV, Teixeira V, Cirne-Santos CC, Paixão I, Giongo V (2012) Caulerpin as a potential antiviral drug against herpes simplex virus type 1. Rev Bras 22(4):861–867

Martins A, Vieira H, Gaspar H, Santos S (2014) Marketed marine natural products in the pharmaceutical and cosmeceutical industries: tips for success. Mar Drugs 12(2):1066–1101

Mary A, Mary V, Lorella A, Matias J (2009) Rediscovery of naturally occurring seagrape *Caulerpa lentillifera* from the Gulf of Mannar and its mariculture. Curr Sci 97(10):1418–1420

Mashjoor S, Yousefzadi M, Esmaeili MA, Rafiee R (2016) Cytotoxicity and antimicrobial activity of marine macro algae (Dictyotaceae and Ulvaceae) from the Persian Gulf. Cytotechnology 68(5):1717–1726

Matanjun P, Mohamed S, Mustapha NM, Muhammad K (2009) Nutrient content of tropical edible seaweeds, *Eucheuma cottonii*, *Caulerpa lentillifera* and *Sargassum polycystum*. J Appl Phycol 21:75–80

Mayakrishnan V, Kannappan P, Abdullah N, Ahmed ABA (2013) Cardioprotective activity of polysaccharides derived from marine algae: an overview. Trends Food Sci Technol 30:98–104

Morris C, Bala S, South GR (2014) Supply chain and marketing of sea grapes, *Caulerpa racemosa* (Forsskål) J. Agardh (Chlorophyta: Caulerpaceae) in Fiji, Samoa and Tonga. J Appl Phycol 26: 783–789

Movahhedin N, Barar J, Fathi FA, Barzegari A, Nazemiyeh H (2014) Phytochemistry and biologic activities of *Caulerpa peltata* native to Oman Sea. Iran J Pharm Res 13(2):515–521

Nagappan T, Vairappan C (2014) Nutritional and bioactive properties of three edible species of green algae, genus *Caulerpa* (Caulerpaceae). J Appl Phycol 26:1019–1027

Nagaraj S, Osborne J (2014) Bioactive compounds from *Caulerpa racemosa* as a potent larvicidal and antibacterial agent. Front Biol 9(4):300–305

Van Nguyen T, Ueng JP, Tsai GJ (2011) Proximate composition, total phenolic content and antioxidant activity of sea grape (*Caulerpa lentillifera*). J Food Sci 76:950–958

Nofiani R, Hertanto S, Zaharah TA, Gafur S (2019) Proximate composition and biological activities of *Caulerpa lentillifera*. Mol Ther 13(2):141–147

Nor AM, Gray TS, Caldwell GS, Stead SM (2019) A value chain analysis of Malaysia's seaweed industry. J Appl Phycol 32:2161–2171

Novaczek I (2001) A guide to the common edible and medicinal sea plants of the Pacific Islands, vol 3. University of the South Pacific, Suva, p 40

Ohba H, Enomoto S (1987) Culture studies on *Caulerpa* (Caulerpales, Chlorophyceae) II. Morphological variation of *C. racemosa* var. *laetevirens* under various culture conditions. Jpn J Phycol 35:178–188

Ohba H, Nashima H, Enomoto S (1992) Culture studies on *Caulerpa* (Caulerpales, Chlorophyceae) III. Reproduction, development and morphological variation of laboratory-cultured *C. racemosa* var. *peltata*. Bot Mag Tokyo 105:589–600

Ornano L, Donno Y, Sanna C, Ballero M, Serafini M, Bianco A (2014) Phytochemical study of Caulerpa racemosa (Forsk.) J. Agarth, an invading alga in the habitat of La Maddalena Archipelago. Nat Prod Res 28:1795–1799

Paul NA, de Nys R (2008) Promise and pitfalls of locally abundant seaweeds as biofilters for integrated aquaculture. Aquaculture 281:49–55

Paul NA, Neveux N, Magnusson M (2014) Comparative production and nutritional value of "sea grapes"—the tropical green seaweeds *Caulerpa lentillifera* and *C. racemosa*. J Appl Phycol 26: 1833–1844

Peña-Rodríguez A, Mawhiney TP, Ricque-Marie D, Cruz-Suárez LE (2011) Chemical composition of cultivated seaweed *Ulva clathrata* (Roth) C. Agardh. Food Chem 129(20122):491–498

Pereira L (2018) Seaweeds as source of bioactive substances and skin care therapy-cosmeceuticals, algotherapy and thalassotherapy. Cosmetics 5(68):1–41

Perez-Estrada CJ, Rodriguez-Estrella R, Palacios-Salgado DS, Paz-Garcia DA (2013) Initial spread of an invasive green alga *Caulerpa verticillata* over coral reef communities in the Gulf of California. Coral Reefs 32:865

Phang SM (2006) Seaweed resources in Malaysia: current status and future prospects. Aquat Ecosyst Health Manag 9(2):185–202

Phang SM, Wong CL, Lim PE, Yeong HY, Chan CX (2005) Seaweed diversity of the Langkawi Islands with emphasis on the northeastern region. Malaysian J Sci 24:77–94

Phang SM, Yeong HY, Lim PE (2019) The seaweed resources of Malaysia. Bot Mar 62:265–273

Phang SM, Yeong HY, Ganzon-Fortes ET, Lewmanomont K, Prathep A, Hau LN, Gerung GS, Tan KS (2016) Marine algae of the South China sea bordered by Indonesia, Malaysia, Philippines, Singapore, Thailand and Vietnam. Raffles Bull Zool 34:13–59

Prud'homme Van Reine WF, Verheij E, Coppejans E (1996) Species and ecads of *Caulerpa* (Ulvophyceae, Chlorophyta) in Malesia (South-East Asia): taxonomy, biogeography and biodiversity. Netherland J Aquat Ecosyst 30(2–3):83–98

Prud'homme van Reine WF, Trono GC Jr (2001) Plant resources of south-east Asia 15 (1), in cryptogams: algae. Backhuys Publishers, Leiden

Rabia MDS (2016) Cultivation of *Caulerpa lentillifera* using tray and sowing methods in Brackish water pond. Environ Sci 4(1):23–29

Ratana-arporn P, Chirapart A (2006) Nutritional evaluation of tropical green seaweeds *Caulerpa lentillifera* and *Ulva reticulata*. Kasetsart J Nat Sci 40:75–83

Reka P, Banu T, Seethalakshmi M (2017) Elemental composition of selected edible seaweeds using SEM-energy dispersive spectroscopic analysis. Int Food Res J 24(2):600–606

Rengasamy KRR, Mahomoodally MF, Aumeeruddy MZ, Zengin G, Xiao J, Kim DH (2020) Bioactive compounds in seaweeds: an overview of their biological properties and safety. Food Chem Toxicol 135(2020):111013

Ribeiro NA, Abreu TM, Chaves HV, Bezerra MM, Monteiro HSA, Jorge RJB, Benevides NMB (2014) Sulfated polysaccharides isolated from the green seaweed *Caulerpa racemosa* plays antinociceptive and anti-inflammatory activities in a way dependent on HO-1 pathway activation. Inflamm Res 63(7):569–580

Rodrigues JAG, Benevides NMB, Tovar AMF, Mourao PAS (2019) Molecular characteristics of polysulfated fractions from green seaweed *Caulerpa cupressoides* and actions on thrombin generation in vivo. Acta Sci Biol Sci 41:e35655

Salehi B, Sharifi-Rad J, Seca A, Pinto D, Michalak I, Trincone A, Mishra AP, Nigam M, Zam W, Martins N (2019) Current trends on seaweeds: looking at chemical composition, phytopharmacology, and cosmetic applications. Molecules 24(22):4182

Sauvage T, Payri C, Draisma SDG, Prud'homme van Reine WF, Verbruggen H, Belton GS, Gurgel CFD, Gabriel D, Sherwood AR, Fredericq S (2013) Molecular diversity of the *Caulerpa racemosa-Caulerpa peltata* complex (Caulerpaceae, Bryopsidales) in New Caledonia with new Australasian records for *C. racemosa* var. *cylindracea*. Phycologia 52(1):6–13

Sfecci E, Le Quemener CL, Lacour T, Massi L, Amade P, Audo G, Mehiri M (2017) Caulerpenyne from *Caulerpa taxifolia*: a comparative study between CPC and classical chromatographic techniques. Phytochem Lett 20:406–409

Tadokoro Y, Sulit VT, Abastillas RB (2000) Technologies in mangrove-friendly aquaculture. Final Report of and papers presented to the on-site training on mangrove-friendly aquaculture Hai Phong City, Socialist Republic of Vietnam. SEAFDEC. https://repository.seafdec.org.ph. Accessed 15 Oct 2020

Tanduyan SN, Gonzaga RB, Bensig VD (2013) Off-bottom culture of *Caulerpa lentillifera* in three different water levels in the marine waters of San Francisco, Cebu, Philippines. J Coral Reef Stud 15:123–132

The Star (2019) Langkawi FRI studying commercial viability of cultivating 'lattut' served in local salad. 6th March. https://www.thestar.com.my

Titlyanov EA, Titlyanova TV (2010) Seaweed cultivation: methods and problems. Algology 36: 227–242

Trono GC (1999) Diversity of the seaweed flora of the Philippines and its utilization. Hydrobiologia 398:1–6

Trono GCJ (1988) Pond culture of *Caulerpa*. Manual on seaweed culture. ASEAN/SF/88/manual no. 3. ASEAN/UNDP/FAO regional small-scale coastal fisheries development project, Manila, Philippines

Verlaque M, Durand C, Husiman JM, Boudouresque CF, Parco YL (2003) On the identity and origin of the Mediterranean invasive *Caulerpa racemosa* (Caulerpales, Chlorophyta). Eur J Phycol 38(4):325–339

Wells ML, Potin P, Craigie JS, Raven JA, Merchant SS, Helliwell KE, Smith AG, Camire ME, Brawley SH (2017) Algae as nutritional and functional food sources: revisiting our understanding. J Appl Phycol 29(2):949–982

Wiedenmann J, Baumstark A, Pillen TL, Meinesz A, Vogel W (2001) DNA fingerprints of *Caulerpa taxifolia* provide evidence for the introduction of an aquarium strain into the Mediterranean Sea and its close relationship to an Australian population. Mar Biol 138:229–234

Wikfors GH, Ohno M (2002) Impact of algal research in aquaculture. J Phycol 37(6):968–974

Wong KK, Cheung PCK (2000) Nutritional evaluation of some subtropical red and green seaweeds. Part I-proximate composition, amino acid profiles and some physico-chemical properties. Food Chem 71(2000):475–482

Yap WF, Tay V, Tan SH, Yow YY, Chew J (2019) Decoding antioxidant and antibacterial potentials of Malaysian green seaweeds: *Caulerpa racemosa* and *Caulerpa lentillifera*. Antibiotics 8(152):1–18

Zafar M (2005) Seaweed culture in Bangladesh holds promise. INFOFISH Int 1:8–10

Zhang M, Ma Y, Che X, Huang Z, Chen P, Xia G (2019) Comparative analysis of nutrient composition of *Caulerpa lentillifera* from different regions. J Ocean Univ China 19(2):439–445

Zubia M, Draisma SGA, Morissey KL, Varela-Alvarez E, De Clerck O (2020) Concise review of the genus *Caulerpa* J.V. Lamoroux. J Appl Phycol 32:23–39

Potential and Challenges of Sea Cucumber *Holothuria scabra* Mariculture in Sabah, Malaysia

6

Sitti Raehanah Muhamad Shaleh, Rafidah Othman, and Fui Fui Ching

Abstract

Sea cucumbers in Sabah, Malaysia, appear to be heavily fished, especially the high-value tropical sea cucumber *Holotuira scabra, often known as sandfish.* Due to its high protein and low-fat contents as well as the presence of various bioactive compounds with antibacterial, antifungal, and anticancer activities, *H. scabra* is in high demand on a global scale. Despite overfishing, there is no government regulation to prevent the situation from worsening. Farming activities still rely on wild seeds and this situation needs to be discontinued for sustainability reasons. Sandfish seed production in hatcheries is being developed to meet the needs of the aquaculture industry. The established technique of hatchery production will enable a constant supply of sandfish juveniles. Artificial spawning using several methods, including heat shock, dryness, and diet stimulants has been well established. However, larvae and juvenile rearing remain the major constraint in the production due to very low survival rates. More research is urged to establish the larval-rearing technique as well as the grow-out system. The sea ranching programme of *H. scabra* for stock enhancement not only depends on the success of hatchery seed production, but it also requires the cooperation of community members to enable proper fisheries management that could be done to ensure the sustainability of the resources.

S. R. Muhamad Shaleh (✉) · R. Othman · F. F. Ching
Borneo Marine Research Institute, Universiti Malaysia Sabah, Kota Kinabalu, Sabah, Malaysia
e-mail: sittirae@ums.edu.my

M. D. Shah et al. (eds.), *Marine Biotechnology: Applications in Food, Drugs and Energy*, https://doi.org/10.1007/978-981-99-0624-6_6

6.1 Introduction

Sea cucumbers are marine invertebrates that belong to the phylum Echinodermata and are also referred to as holothurians. The unique feature of sea cucumbers compared to other echinoderm is the worm-shaped and soft body with calcareous ossicles. The mouth has tentacles located at the anterior end, while the anus is at the posterior end. The tentacles' shape and the types of calcareous rings and spicules on the body were used in species identification (Kamarudin et al. 2010). Sea cucumbers are a sessile animal found in the sandy reefs, and as a benthic feeder, they play an important role in maintaining a clean and healthy benthic environment (Plotieau et al. 2013). In terms of fishery value, sea cucumber is one of the most valuable fishery commodities in the world (Pietrak et al. 2014).

In tropical and subtropical Asian countries, sea cucumber fisheries consist of various species of holothurians, while in temperate countries, *Apostichopus japonicus* is usually the only species harvested. *Holothuria scabra, H. nobilis, H. edulis, and H. atra* are among the species that have been fished, traded, and provide a livelihood for the fishermen in Indonesia, the Philippines, India, Madagascar, and some other countries in Oceania. In Malaysia, Sabah is an important state for sea cucumber fisheries, where the landing ports can be found in the districts of Kudat, Semporna, Sandakan, and Kota Belud. Based on the FAO (2006) report, it was indicated that Indonesia is the top harvester of wild sea cucumbers with an increasing trend. However, the Philippines demonstrated a dramatic reduction from 1998 to the present, due to the low quality of catches. For Malaysia, no data are available since 1989, despite local statistics reporting Sabah as the most significant state in Malaysia that contributes to sea cucumber landings.

In Malaysia, sea cucumbers were harvested for traditional medicine by the Malays, while among the Chinese, sea cucumbers are one of the most luxurious foods and are culturally believed to bring good luck and fortune (Vaitilingon et al. 2016). The genus *Stichopus* locally known as *gamat* was processed in Langkawi, Malaysia, for traditional tonic and ointment (Fig. 6.1). In Sabah, the genus of *Holothuria* locally known as *balat* was processed for food. Decreased production of sea cucumber in Malaysia has been reported by Kamarudin et al. (2010) with a tremendous reduction from 862 tonnes in 2008 to only 177 tonnes in 2009. It was directly related to the increased demand for dried sea cucumber in the international market that worsened the overfishing of sea cucumbers. Excessive exploitation and the capture of sea cucumbers regardless of size have resulted in the decline of the wild population (Kumara and Dissanayake 2015). Lack of resource management in most countries is also one of the reasons for the population decline. Therefore, many studies have been conducted on sea cucumber aquaculture, including farming techniques, seed production, and stock enhancement programmes.

Fig. 6.1 *Stichopus variegatus herrmanni* (**a**) and *Stichopus chloronotus* (**b**) were harvested from the wild and processed for traditional medicine

6.2 The Importance of Sea Cucumbers to the Local Fishing Community

The coastal communities that have long been active in sea cucumber fishing have found that sea cucumbers are becoming difficult to obtain and the numbers are also declining. This situation indeed shows that sea cucumbers are at risk of extinction. Although sea cucumbers are not listed in the Convention on International Trade in Endangered Species of Wild Fauna and Flora (CITES), *H. scabra* has been listed as endangered species by the International Union for Conservation of Nature and Natural Resources (IUCN) and the population declined at least 50% over the past 50 years (Hamel et al. 2013).

Sea cucumbers are versatile marine animals, as they can be eaten raw or cooked. It was high in protein, low in fat, and contained fine chemicals and bioactive compounds. It has long been used in traditional medicine in China, where it is known as sea ginseng (Bahrami et al. 2014). Due to the high value of the dried sea cucumber, which is known as *beche-de-mer* or *hai-som* (Fig. 6.2), it has been

Fig. 6.2 Salting, cooking, and sun-drying are part of the various stages in the traditional process of producing *beche-de-mer* or *hai-som*

exported mainly to China and the Middle East markets (Watanabe et al. 2014). However, only certain species can be processed for the premium grade of dried sea cucumber such as *H. edulis, H. rigida, H. nobilis, H. fuscogilva,* and *H. scabra,* at a range price of USD 22-112 per individual (Perez and Brown 2012). Apart from different species factors, size, and quality of processing, the festive season factor also influences the price of the *beche-de-mer* (Agudo 2012).

The sea cucumber market is becoming increasingly popular as most restaurants have replaced the shark fin soup menu with sea cucumber. This is due to the pressure and support from various parties who are urging restaurants to stop selling shark fins to protect endangered species. In Malaysia, dried sea cucumber production is actively carried out in Sabah, particularly in the areas of Kudat, Sandakan, Kunak, and Semporna. However, the sector relies solely on wild sea cucumbers caught from the sea. Apart from meeting local needs, sea cucumbers also contribute to state income and are exported mainly to several states in Peninsular Malaysia, Sarawak, Singapore, Thailand, Hong Kong, Taiwan, and also to China (Lotavelli et al. 2004). Meanwhile, according to the local fishermen in the Kota Belud district who processed *beche-de-mer,* larger and high-quality sea cucumbers are often exported outside Malaysia while the low-grade and small-sized sea cucumbers are sold for the local market.

According to Uthicke (2004), continuous catching of wild sea cucumbers will reduce the density of the wild population, and in turn, this will result in unbalanced larval production of the population. Therefore, overexploitation will eventually weaken the population's ability to produce sufficient juveniles for population growth in that particular area. Overexploitation can be prevented or reduced through active management measures to maintain the remaining stocks and promote the recovery of the wild population. Until now, there have been no regulations on the size limit or species that can be caught or processed for dry product in Malaysia (Choo 2012). Due to the declining lading pattern of sea cucumbers, the Malaysian Fisheries Department has started conducting studies on sea cucumbers.

Previous studies suggested a complete ban on sea cucumber landings for a set time, such as 5 years and a ban on the use of 'bombs' to allow the persistence of a deeper population to replenish stocks. A sea cucumber fishery management plan is important to ensure that the community can gain maximum economic benefits with minimal impact on the marine and coastal environment. Several methods for sea cucumber fishery management could be done to ensure the sustainability of the resources.

(i) Enforcement and improvement of the existing fisheries laws.
(ii) Introduction of permits and a fisheries cooperation system.
(iii) Introduction of the catching and closing seasons for sea cucumber.
(iv) Implementation of a Joint Management Area between the community and government.
(v) Creating a Marine Reserve Area (19–40 ha).
  - as the breading nucleus of the sea cucumber population.
  - as a mechanism for monitoring
(vi) Implementation of a catching quota system.
(vii) Implementation of catch size limits for live sea cucumber.
(viii) Introducing import tariffs.
(ix) Temporary closure for catching activity.
(x) Prohibition of night-catching activity.
(xi) Prohibition on the use of certain equipment.
(xii) Awareness campaign on sea cucumber conservation and marine environment.

6.2.1 Holothuria scabra

Holothuria scabra, commonly known as sandfish, is one of the sea cucumbers that occurs in all of the major oceans and seas of the world (Purcell et al. 2016). Sandfish is mainly found in the Indo-Pacific region, from East Africa to the eastern Pacific. According to Agudo (2006), sandfish is normally found between the latitudes of 30°N and 30°S. Morphologically, sandfish has a cylindrical, elongated, and robust body with a wide range of colours from olive green to black. It has a convex dorsal side, while the ventral body surface is flat and covered by tiny scattered tube feet that facilitate locomotion. The thick, slimy, and gritty body wall covers almost 56% of its total body weight. However, Baskar (1994) stated the variation in the weight of the sandfish was influenced by the amount of sediment and coelomic water present in the alimentary canal. The mouth is located antero-ventrally and surrounded by 20 tentacles, which help in the feeding activity and the anus is in the postero-dorsal area. (Agudo 2006). The internal anatomy of sandfish comprises the digestive system, respiratory tree, gonads, water vascular system, and haemal system (Mary Bai 1994). The respiratory trees can be found lying at the posterior of the body and open to the cloaca while the gonadal tubules of sandfish appeared in one tuft and open dorsally at the anterior end through a single gonopore. The gonopore is located parallel to the foregut in the mid-dorsal region and close to the mouth (Agudo 2006). The species of sea cucumber can be identified using microscopic examination of the

Fig. 6.3 *Holothuria scabra* body covered by fine sand is naturally found in a seagrass bed

calcareous plates called spicules. The spicules of sandfish are normally present in the shape of tables and knobbed buttons (Agudo 2006).

Sandfish are usually found on the inner reef flats in shallow tropical water less than 20 m in depth and near the estuaries (Ramofafia et al. 2003). Sometimes, sandfish can also be found in brackish water, as they are tolerant of lower salinity levels down to 20 ppt for short periods (Agudo 2006). This species preferred sheltered areas with high nutrient levels such as seagrass beds and muddy substrata. The association of sandfish with seagrass were also reported in several other places and sandfish was not the only sea cucumber species that inhabit the seagrass bed. Other species of sea cucumbers such as *Actinopyga mauritiana*, *A. echinites*, *Bohadschia marmorata*, *B. argus*, *Holothuria leucospilota*, *H. hilla*, *H. atra*, *H. aculeate*, *Stichopus variegatus,* and *Theleonota ananas* also can be found in seagrass beds (Fortes 1995). Sandfish play a significant role in the seagrass bed ecosystem (Fig. 6.3) that act as the suspension feeders, detritivores, and prey (Anderson et al. 2010). Widely known as bioturbators, the deposit feeder sandfish were capable to alter the stratification and stability of muddy and sandy bottoms via bioturbation. Subsequently, this species facilitates the recycling of nutrient and significantly reduce the microalgal biomass trapped in the sediments (Uthicke 2001).

6.2.2 Spawning Behaviour of *H. scabra*

Most of the holothurian species exist as male or female (Kuganathan 2014) even though a few of the species are hermaphroditic (Purcell et al. 2010). Holothurians are mainly broadcast spawners that release their sperm and oocytes into the water body (Purcell et al. 2010) and they reproduce both sexually and asexually (Hoareau and Conand 2001). For *H. scabra* species, sexual reproduction happened without mating and fertilization occurred externally in the water column (Kuganathan 2014).

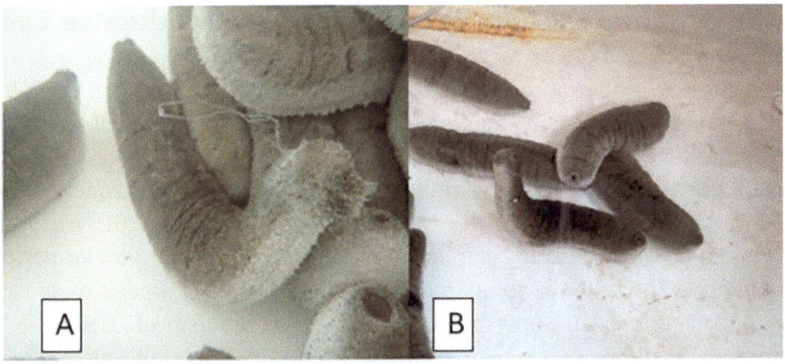

Fig. 6.4 Spawning behaviour of male *H. scabra* releasing sperm, white-thread like through the gonophore (**a**). Female *H. scabra* has a bulge at the anterior end and is ready to spray the eggs (**b**)

Holothuria atra, in contrast, is an example of a species that forms asexual reproduction through fission (Rowe 1982).

Unlike fish and shrimp, sex identification in holothurian is almost impossible to be distinguished externally. The sex can only be determined via microscopic examination and also through the observation of spawning behaviour (Fig. 6.4), either releasing sperm or eggs (Agudo 2006). In a microscopic examination, in the ripe gonad of a male, the testis is milky white, whereas, in the ripe gonad of a female, the ovary is translucent. In addition, the biopsy can also be used to determine the sex (Agudo 2006). The gonad of *H. scabra* is in the anterior and dorsal parts of the animal. Some of the holothurians may have gonads on both sides of the mesentery, but in *H. scabra*, they occur only on the left side of the animal (Demeuldre and Eeckhaut 2012). Based on the spawning behaviour, the male will rise first and sway the anterior end to release the sperm from the gonophore situated on the dorsal side near the oral region. It keeps on releasing the sperm for several hours, depending on the size. The female will be triggered to release the egg by the sperm released into the water. Females typically start spawning a few hours after the first male releases the sperm. The anterior end of the female gets bulged due to the pressure built up inside (Agudo 2006), and then, the anterior part will rear up to spur the eggs.

Gametogenesis in sea cucumbers can be classified as synchronous and asynchronous. Synchronous gametogenesis occurs when both males and females release mature gametes simultaneously. Asynchronous reproduction, on the other hand, occurred when males and females release gametes at different times (Ramofafia et al. 2003; Morgan 2000). *H. scabra* performed asynchronous gametogenesis; hence, gametes are constantly available throughout the year. Due to this, the aquaculture of *H. scabra* is of great advantage, as the sandfish breeding programme can be done year-round.

6.3 Environmental Conditions and the Reproductive Cycle

Numerous studies on the reproductive cycle of this species have been done from the Middle East to Southeast Asia and New Caledonia (Rasolofonirina et al. 2005). This is because the life cycles of tropical and temperate sea cucumbers are varied. Some of the species are ripe at most times of the year, with one or two seasonal peaks. Most of the tropical species have a peak of spawning activity near the early summer months and some of the species can be spawn several times per year or periodically every month. Meanwhile, the temperate species tend to have a peak of spawning in spring or early summer and only occur once a year (Purcell et al. 2010).

There were many studies of the *H. scabra* reproductive cycle in several places and the reports showed that sandfish spawning peaks vary with location (Pangkey et al. 2012). It is important to know the peak spawning activity of *H. scabra* as it can help to determine the right time to collect the broodstock into the hatchery for the high chance of successfully induced spawning since the broodstock has a ripe gonad and is ready to release the gametes (Keshavarz et al. 2015). The pattern of *H. scabra* maturity in Sabah, Malaysia, studied by Arsad et al. (2020), can be used as a guide in broodstock procurement for hatchery breeding (Fig. 6.5). In Kudat, the Gonad Index (GI) was high from July to November 2015 and can be associated with the maturation of *H. scabra* in that particular month. The GI dropped to the lowest level in December 2015 and was continuously low until January 2016. GI then increased in February 2016 and rose a bit in April 2016. High GI was recorded in August 2016, as 40% of the gonad were at Stage 5. In general, there are three peaks of high GI recorded in Kudat.

Over the world, *H. scabra* shows two major reproductive patterns that are seasonal spawning at high latitudes and low latitudes (Ramofafia et al. 2003). *H. scabra* situated in countries near the equator can spawn throughout the year (Choo 2008; Agudo 2006), whereas in higher latitudes, annual or biannual reproductive periodicity is more common. In the Solomon Islands (09°S), the

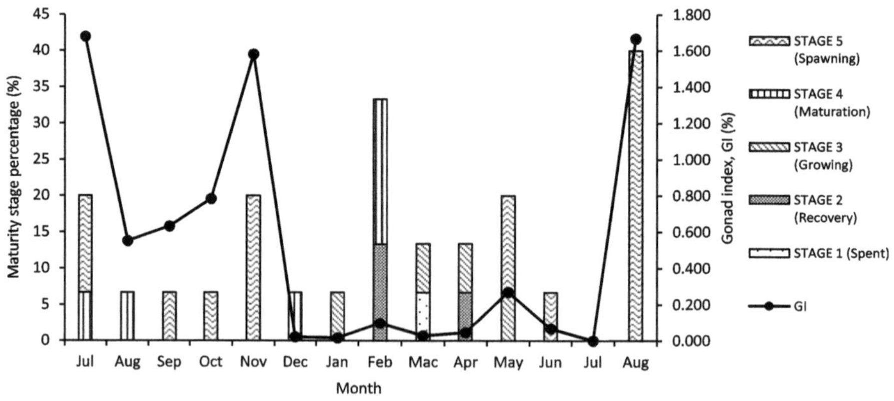

Fig. 6.5 Maturity stage of *H. scabra* from July 2015 to August 2016 in Kudat, Malaysia

reproductive pattern in *H. scabra* is characterized by continuous, asynchronous gametogenesis with mature gametes that can be found year-round (Ramofafia et al. 2003). Choo (2008) and Pitt (2001) noted that in a study by Ong Che and Gomez in 1985, the reproduction of *H. scabra* in the Philippines has two spawning events occurring from May to June with high temperatures and from October to November with cooler water.

Holothuroids in the temperate zone are reported to have seasonal breeding cycles, whereas tropical species can breed for longer periods. Earlier works done in the tropics but at higher latitudes show a seasonal breeding pattern for sea cucumbers (Muthiga and Kawaka 2008). The release of gametes, sperm, and oocytes by mature adults is caused by the surrounding situation (Purcell et al. 2010). Environmental changes perceived by sea cucumbers may explain the interannual variation. Therefore, it triggers reactions that lead to changes and modifications in reproductive metabolism that are possible through gene activation or hormone synthesis (Hamel and Mercier 2004).

Light intensity, photoperiod, temperature, salinity, tidal changes, food availability, and food type are believed to have an effect on the spawning and gametogenesis in holothurians (Kumara et al. 2013; Muthiga et al. 2009; Battaglene et al. 2002). Kazanidis et al. (2014) reported that echinoderms spawning correlates with algae blooms and water temperature. Apart from that, the lunar phase can affect not only the echinoids, and crinoids but also the holothuroids (Mercier et al. 2007). Spawning stimulation studies are being conducted for certain species; for example, *H. pulla* and *H. coluber* show that temperature, monsoon, lunar cycle, and chemicals produced by males and females are crucial factors for spawning (Purwati et al. 2003).

Dabbagh and Sedaghat (2012) described the seasonal and non-seasonal breeding of *H. scabra* based on its distribution globally. They showed that breeding patterns can be influenced by seasonally predictable factors such as water temperature and daytime, particularly at high latitudes where the annual breeding patterns have been recorded. A peak breeding pattern of biennial and continuous double spawning has been observed in the Indo-Pacific region. This may be due to the effects of environmental conditions such as temperature, salinity, and photoperiod as reported in Indonesia, the Philippines, New Caledonia, and India. Purwati et al. (2003) mentioned that the spawning behaviour of *H. scabra* is stimulated by changes in salinity.

6.4 Aquaculture and Seed Production

Aquaculture is globally known as one of the fastest-growing sectors. In Malaysia, the sea cucumber breeding project towards aquaculture farming is implemented based on the rapid development of the sea cucumber industry in Southeast Asia and southern China. The production of sea cucumber seeds in captivity for aquaculture purposes and sea ranching programmes is becoming increasingly important. In addressing the issue of the declining sea cucumber population and the threat of extinction, efforts must be continued through more systematic farming development. According to Kumara and Dissanayake (2015), to avoid the continued decline in the population, marine programmes and stock enhancement through seed production from hatcheries should be focused on and developed.

In Southeast Asia, sea cucumbers have been widely farmed in sea pens, ponds, and hatcheries. China has dominated sea cucumber aquaculture and is the major producer (Sicuro and Levine 2011). *H. scabra* is one of the tropical sea cucumber species that is commercially produced. Several studies have been done on the artificial spawning of holothurians, including *H. scabra.* According to Agudo (2006), spawning often occurs before full and new moons but can occur at other times as well. Battaglene et al. (2002) reported that in the Solomon Islands and Papua New Guinea, *H. scabra* spawns during a full moon. Thermal stimulation or heat shock method is claimed by many as the most successful method to induce spawning in *H. scabra* (Kumara et al. 2013). Other physical stressors such as drying, water jetting, UV irradiation, and high concentrations of food are alternative techniques for spawning induction (Pitt and Duy 2003).

Breeding and cultivation of holothuroids have started a long time ago, when *Apostichopus japonicus* was first produced in the 1950s by the Japanese and a few decades later by China (Battaglene 1999). In 1988, *H. scabra* was first produced in India using thermal shock, following a similar technique used in the breeding of *A. japonicus* in China (James et al. 1994). India is the first country to successfully produce seeds in the hatchery (Kumara and Dissanayake 2015) and go along with other countries such as Australia, Indonesia, the Philippines, the Republic of Maldives, Fiji, and Solomon Island. Sea cucumber production in hatcheries is not only for aquaculture but also as a source of seedlings for stock enhancement or population rehabilitation of declining wild stock (Hamel et al. 2001).

Juvenile production of sandfish in hatcheries is by natural spawning or induction. Broodstock of above 300 g could be purchased from sea cucumber farmers or by catching from their natural areas. Healthy broodstocks with no injuries or lesions on their bodies were selected as candidates for breeding. Broodstocks transported to the hatchery required special handling using a plastic bag filled with oxygen and water at 3:1 ratio. The plastic bags could be packed in a polystyrene box and maintained at a temperature of 26 to 27 °C using some ice cubes (Kumara and Dissanayake 2015). Upon arrival at the hatchery, sandfish broodstocks were acclimatized in a tank containing treated sand for a minimum of one week before spawning induction. A flow-through water system equipped with a sand filter was installed on the broodstock tank to maintain the water quality. The water parameters such as temperature, salinity, dissolved oxygen, and pH should be monitored daily to prevent extreme fluctuations in the water parameters, which might cause mortality in the broodstocks. *Sargassum* sp. was given to the broodstocks either in the form of fresh seaweed paste, juice, or dried powder twice daily at the rate of 3% of body weight.

Broodstock should be held at a low density in the tank, as high stocking density may enhance stress and promote severe loss of body weight. Rearing the broodstock in the hatchery is an advantage to the farmers, as they can reduce the cost of transporting it from the wild into the hatchery and it is easier to monitor the broodstock's condition. Besides, it also helps to prevent stress caused by handling

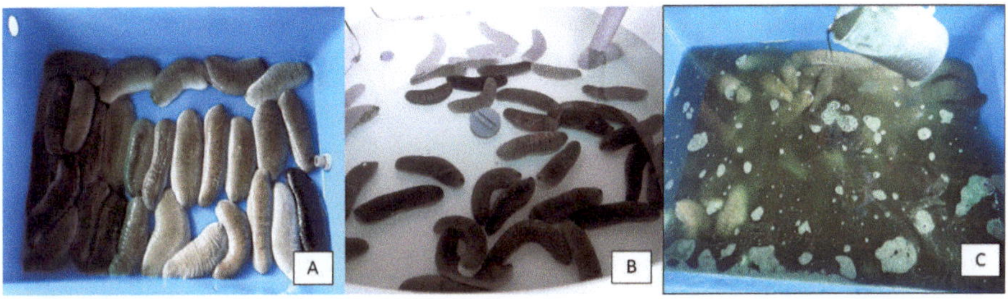

Fig. 6.6 Spawning induction protocol for sandfish consists of three spawning induction methods: (**a**) air stimulant, (**b**) thermal stimulant, and (**c**) feed stimulant

and transporting that may affect the spawning of the broodstock (Agudo 2006). Rearing broodstock in the hatchery could be a success if the optimum condition can be maintained. Kuganathan (2014) listed the optimum range of environmental parameters, as these parameters should be highlighted before initiating commercially successful aquaculture. The parameters are as follows: water temperature of 28–32 °C; salinity of 5–6 ppt; DO (mg/L) of 5–6; and pH of 6.8–9.3 The artificial spawning of sandfish is considered well-established. Mazlan and Hashim (2015) have reported that thermal stimulation is a better method for spawning induction. However, based on Abidin et al. (2016), using a combination of three methods, air stimulant, heat, stimulant and feed stimulant give a higher success rate in spawning (Fig. 6.6). The air stimulant or desiccation protocol was conducted by exposing the broodstocks to dry conditions. The broodstocks were placed in a clean container before being transferred into the spawning tank, which contained ambient temperature seawater. Thermal induction was performed by increasing the water temperature in the spawning tank by 3 to 5 °C above the ambient temperature. The seawater could be heated up either using aquarium heaters or by adding hot water into the tank. The thermal stimulant protocol was carried out for 60 min, and during that time, spawning behaviour should be monitored. The third induction method, which is feed stimulation, is necessary if heat induction fails. *Spirulina* powder was used as the feed stimulant at a concentration of 30 g/500 L. Spirulina bath was first prepared by dissolving *Spirulina* powder in a clean container using filtered seawater. Then broodstocks were transferred into the bath and left for 30–45 min. Finally, the broodstocks were thoroughly rinsed and transferred into the spawning tank for spawning observation.

Fertilized eggs were aspirated or siphoned out of the spawning tank and transferred to the larval-rearing tank for larval rearing. The larval-rearing tank must be covered with a fine mesh net to prevent any foreign substances and infestation of bloodworms (*Chironomid*). In 2019, Abidin et al. again reported that at the larval stage, mixed microalgae were given at different concentrations to improve growth and the metamorphosis of the larvae. For the pentactula, settlement plates are prepared using *Spirulina* powder (1–2 g m^{-2}) (Agudo 2006) or by using plates covered with *Navicula* sp. biofilm. Biofilm was introduced once the first doliolaria

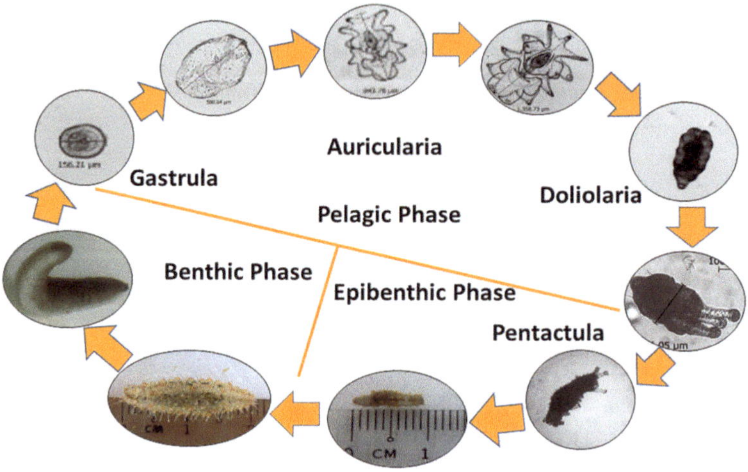

Fig. 6.7 Life cycle of *H. scabra* that includes pelagic phase (or larval stage), epibenthic phase, and benthic phase

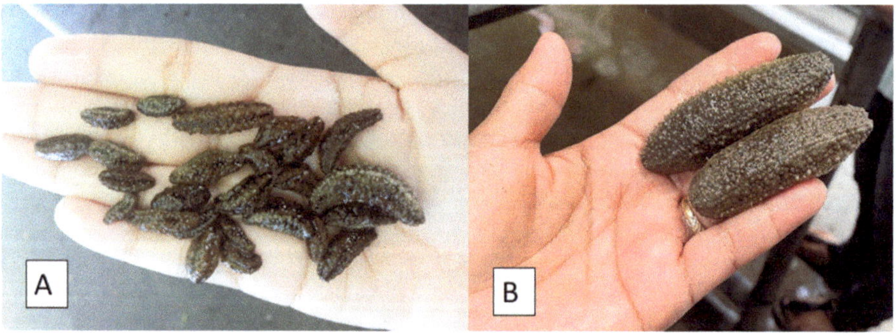

Fig. 6.8 Juvenile sandfish produced in the hatchery are ready to be released into the nursery (**a**) and sea pens for grow-out or for sea ranching (**b**)

was observed. Larval rearing is normally between 2–3 weeks until the pentactula stage (Fig. 6.7).

Sandfish juveniles produced in the hatchery were transferred to the nursery stage. At the nursery, juvenile sandfish were allowed to grow until they reached a size that was suitable for growing out. Shallow tanks or hapa, a fine mesh net set inside the rearing pond or in a floating cage were used for the nursery. Feeds were given at a minimal rate since natural biofilms were readily available in the tank and the hapa was to be grazed. Larval and juvenile stages are considered the most critical stage that requires extra care and proper handling. Thus, effective management is important to produce fast growth and healthy seeds.

From the nursery, sandfish juveniles (2–3 inches) were transferred to the farms to grow out at the rate of 1 individual per square metre (Fig. 6.8). A sandfish farm for

Fig. 6.9 Different designs of sea pens for the grow-out stage were built using a net, wooden pole, rock, and dead corals

growing out is either an earthen pond with sandy bottom or a sea pen (enclosure) with fine sand and rich organic matter. Site selection for a sea pen is important because the location should be safe from strong waves, and high organic matter, far from freshwater sources and toxic pollution and easy to access. The sea pens were built using wooden stakes and nets to prevent the sea cucumbers from escaping (Fig. 6.9). Monitoring and net cleaning should be done regularly to ensure predators such as crabs could be controlled. Feeding is not necessary for extensive farming as the sandfish obtained their food from the substrate and the sea floor.

6.5 Sandfish Farming

H. scabra is farmed by the coastal communities in Sabah as a source of livelihood. Sandfish are usually picked or harvested by fishermen during low tide along the tidal zone as well as using trawlers in the deep sea. For coastal fishermen, harvesting wild sea cucumber using a small boat and simple equipment is one of their major activities for living. Sea cucumbers harvested were sold to small-scale sea cucumber farmers either to be processed for dried sea cucumber or to be released in their sea pens for farming.

Sandfish farming among the coastal community in Sabah has been practised since the 1980s. However, an activity that still relies on wild seed supplies has been unable to meet the growing demand in international markets. Seed production from hatcheries is in urgent need of sustainable and competitive activities. The following are some of the issues, challenges, and potentials that need to be faced in the development of sea cucumber farming in Sabah.

Securing the supply of seeds is very important to ensure the sustainability of the sea cucumber enterprises. Seed production requires orderly and organized work, and in this regard, hatchery techniques and sea ranching need to be introduced to the farmers. According to Mills et al. (2012), sea ranching can be defined as the release of juvenile sandfish into the sea pen or marine estuaries to be harvested at larger sizes. The released sea cucumbers are not expected to contribute to population spawning and not all the released juveniles will be able to be harvested. Nevertheless, studies in India showed positive results when local fishermen reported an increased stock population after the implementation of an effective sea ranching programme (Asha et al. 2015).

Currently, the sea cucumber breeding technology in Malaysia is still in early development. Manuals and protocols for spawning, larviculture, juvenile rearing, and grow-out systems should be provided for references and training among potential entrepreneurs. An adequate supply of seeds will certainly be able to support the development of the sea cucumber industry as well as conservation programmes.

Cooperation between the community, institutions of higher learning, the Department of Fisheries, and private commercial companies play an important role in improving the production technology and the quality of sea cucumbers via research and development projects. For instance, a smart partnership between institutions of higher learning and private companies to develop patented sea cucumber seed production technology to be supplied to the fishermen through a contract farming system. Through this system, the harvested sea cucumber will be resold back to the seed suppliers for product processing. This method will be able to guarantee the quality, a better price, and stability. From this framework of collaboration, local communities will benefit in terms of sales potential, be able to reach the global markets and avoid manipulation by middlemen.

The practice of sustainable aquaculture methods through this smart partnership will ensure the production of quality products that can penetrate the international market. Aquaculture management will also be more organized, the product is safe to eat, and production is consistent, disease free, and has zero effects on the environment. In Madagascar, community-based sea cucumber farming project that are using juveniles from commercial hatcheries has successfully become a source of livelihood and at the same time has created a trading company that focused on sea cucumber aquaculture (Robinson and Pascal 2009).

6.6 Sea Ranching and Stock Enhancement Programme

Intensification of sea cucumber farming without proper management can hurt the environment. It should be noted that different capacities require the implementation of different techniques. Among the communities and entrepreneurs, an increase in responsibility is important to ensure the sustainable use of the resources. Some factors determine the success of sea cucumber aquaculture including the unpolluted marine environment, well-planned ecological care, the marine traditional care system, and basic knowledge of sea cucumber rearing. Alteration or destruction of the natural environment to meet the needs of sea cucumber farming may harm the resilience and productivity of the ecosystem. Indeed, one of the major challenges in any type of aquaculture is to minimize the negative impacts on the natural environment that include natural processes, natural productivity, and biodiversity (Troell 2009).

Sea ranching is the practice of releasing cultured juveniles into their natural habitat and harvesting them when they reach commercial size (Mills et al. 2012). This activity can help to restock the diminished populations in the wild due to overfishing (Hartati et al. 2017). However, several requirements have to be considered while selecting a cultural site. The survival of sea cucumbers in their natural habitat depends on the environmental surroundings. Biotic and abiotic factors are important for the survival of sea cucumbers. Abiotic environment, including the existence of seaweed and sea grass, may provide shelter and food for the juveniles and broodstock that leading to better survival. The abiotic factors such as temperature, salinity, dissolved oxygen, and pH could affect sea cucumbers as their growth and reproductive development would be better in favourable conditions (Agudo 2006). The sediment type also plays an important role since sea cucumber has a daily burrowing cycle and can be influenced by the size of the sediment particle (Mercier et al. 1999). Sediment with higher organic matter content will be preferable to enhance the growth and promote better survival of sea cucumbers. According to Plotieau et al. (2013), fine sediment may have the richest organic matter compared to coarse sediment.

Gazetting some of the areas for Marine Parks is one of the effective approaches in the efforts to protect the sea cucumbers and other marine life. The Sabah government has begun to expand the marine protected areas and the implementation of integrated marine management. Tun Mustapha Marine Park (TMP) which was located in northern Sabah has been announced in 2016 as the largest marine protected area with a total area of 1.02 million hectares. The multi-purpose parks that include activities for wildlife conservation, aquaculture, tourism, and sustainable fisheries have been a source of income for most of the coastal communities. Guidelines or work procedures related to the development of marine areas for sea cucumber farming shall be developed. According to Bell et al. (2008), sea cucumber farming requires a lagoon area for sea pens or sea ranching. Activities related to sea cucumber rearing could be an alternative for the economic activity of the coastal communities as well as to replenish the declining wild stocks of sea cucumbers.

6.7 Conclusion

Sea cucumbers in Sabah appear to be heavily fished, especially the high-value *Holotuira scabra*. Despite overfishing, there is no regulation by the government to prevent the condition from worsening. Hatchery seed production can secure a continuous supply and sustainable aquaculture of *H. scabra*. Seed production techniques and aquaculture skills are also important factors and are necessary to ensure the success of a stock enhancement programme for this species in this country. Continuous work is necessary for a better understanding of the embryonic and larval development of *H. scabra* as well as the larval rearing for higher survival. Sea ranching of *H. scabra* not only depends on the success of hatchery seed production but also requires the cooperation of community members and enables the released juveniles to survive and grow to harvest size. Thus, for the stock enhancement programme, it is expected that soon we will be able to release large numbers of sandfish juveniles at selected locations, particularly the Sabah gazetted marine parks (i)Tun Mustapha, (ii) Tun Sakaran and (iii) Tunku Abdul Rahman Marine Park.

Funding Information This project was financially supported by the Ministry of Higher Education, Malaysia, under the Niche Research Grant Scheme (NRGS0002).

References

Abidin NA, Sitti Raehanah MS, Ching F, Othman R, Manjaji-Mutsumoto M, Senoo S, Mustafa S (2016) Larval development of induced spawned sandfish (*Holothuria scabra*) in hatchery. Int J Aquat Sci 7(2):107–110

Abidin NA, Muhamad Shaleh SR, Ching FF, Othman R, Manjaji-Matsumoto M, Mustafa S, Senoo S (2019) Appropriate diet and stocking density for sea cucumber (*Holothuria scabra)* larvae rearing. J Phys Conf Ser 1358:012015

Agudo NS (2006) Sandfish hatchery techniques. Australian Centre for International Agricultural Research (ACIAR), Canberra

Agudo NS (2012) Pond grow-out trials for sandfish (Holothuria scabra) in New Caledonia. In: Hair CA, Pickering TD, Mills DJ (eds) Asia-pacific tropical sea cucumber aquaculture. Canprint Communications Pty Ltd, Canberra, pp 104–112

Anderson SC, Flemming JM, Watson R, Lotze HK (2010) Serial exploitation of global sea cucumber fisheries. In: Fish and fisheries. Blackwell, Hoboken

Arsad NA, Othman R, Sitti Raehanah MS, Ching FF, Manjaji-Matsumoto M (2020) Effects of physicochemical parameters on the reproductive pattern of sea cucumber Holothuria scabra in Sabah. Songklanakarin J Sci Technol 42(1):109–116

Asha PS, Diwakar K, Santhanavalli G, Manissery MK (2015) Comparative distribution and habitat preference of the sea cucumber Holothuria atra Jaeger at protected and unprotected sites in Thoothukudi region of Gulf of Mannar, southeast coast of India. Indian J Fish 62(1):52–57

Bahrami Y, Zhang W, Franco C (2014) Discovery of novel saponins from the viscera of the sea cucumber *Holothuria lessoni*. Mar Drugs 12(5):2633–2667

Baskar BK (1994) Some observations on the biology of the Holothurian *Holothuria (Metriatyla) scabra* (Jaeger). Bull Centre Mar Fish Res Inst 46:39–43

Battaglene SC (1999) Culture of tropical sea cucumbers for stock restoration and enhancement. ICLARM Q 22(4):4–11

Battaglene SC, Syemour JE, Ramofafia C, Lane I (2002) Spawning induction of three tropical sea cucumbers, *Holothuria scabra, H. fuscogilva* and *Actinopyga mauritiana*. Aquaculture 207:29–47

Bell JD, Purcell SW, Nash WJ (2008) Restoring small-scale fisheries for tropical sea-cucumber. Ocean Coast Manag 51:589–593

Choo PS (2008) Population status, fisheries and trade of sea cucumbers in Asia. In: Toral-Granda V, Lovatelli A, Vasconcellos M (eds) Sea cucumbers. A global review of fisheries and trade. FAO Fisheries and Aquaculture Technical Paper. No. 516. FAO, Rome, pp 81–118

Choo PS (2012) The sea cucumber fishery in Semporna, Sabah, Malaysia. SPC Trepang Inf Bull 32: 43–48

Dabbagh AR, Sedaghat MR (2012) Breeding and rearing of the sea cucumber *Holothuria scabra* in Iran. SPC Beche-de-mer Inf Bull 32:49–52

Demeuldre M, Eeckhaut I (2012) Gonad development in the sea cucumber *Holothuria scabra* Jaeger, 1833. SPC Beche-de-mer Inf Bull 32:15–23

FAO (2006) State of world aquaculture 2006. FAO fisheries technical paper no 500. FAO, Rome

Fortes MD (1995) Seagrasses: a resource unknown in the ASEAN region. ICLARM Education Series 5. International Center for Living Aquatic Resources Management, Manila, p 46

Hamel JF, Conand C, Pawson DL, Mercier A (2001) The sea cucumber Holothuria scabra (Holothuroidea: Echinodermata): its biology and exploitation as Beche-de-Mer. In: Southward AJ, Young CM, Tyler PA, Fuiman LA (eds) Advance in marine biology, vol 41. Academic Press, London, pp 129–132

Hamel J-H, Mercier A (2004) Synchronous gamete maturation and reliable spawning induction method in Holothurians. Aquacult Adv:359–370

Hamel J-F, Mercier A, Conand C, Purcell S, Toral-Granda T-G, Gamboa R (2013) Holothuria scabra. The IUCN Red List of Threatened Species 2013:e.T180257A1606648. https://doi.org/10.2305/IUCN.UK.2013-1.RLTS.T180257A1606648.en

Hartati R, Trianto A, Widianingsih (2017) Habitat characteristics of two selected location for sea cucumber ranching purposes

Hoareau T, Conand C (2001) Sexual reproduction of *Stichopus chloronotus*, a Fissiparous Sea Cucumber, on Reunion Island, Indian Ocean. SPC Beche-de-mer Inf Bull 15:4–12

James DB, Gandhi AD, Palaniswamy N, Rodrigo JX (1994) Hatchery techniques and culture of the sea-cucumber *Holothuria scabra*. Paico Printing Press, Cochin

Kamarudin KR, Rehan AM, Hashim R, Usup G (2010) An update on diversity of sea cucumbers (Echinodermata: Holothuroidea) in Malaysia. Malay Nat J 62(3):315–334

Kazanidis G, Lolas A, Vafidis D (2014) Reproductive cycle of the traditionally exploited sea cucumber *Holothuria tubulosa* (Holothuroidea: Aspidochirotida) in Pagasitikos Gulf, Western Aegean Sea, Greece. Turkish J Zool 38:306–315

Keshavarz M, Mohammadikia D, Dabbagh AR, Kamrani E (2015) Reproductive biology of the sea cucumbers for successful breeding: a review. J Anim Prod Adv 2(5):208–213

Kuganathan S (2014) Sea cucumbers status and culture potential in the Jaffna Lagoon, Sri Lanka. Department of Fisheries University of Jaffna, Jaffna

Kumara A, Dissanayake C (2015) Preliminary study on broodstock rearing, induced breeding and grow-out culture of the sea cucumber *Holothuria scabra* in Sri Lanka. Aquac Res 48:1058–1069

Kumara PADA, Jayanatha JS, Pushpakumara J, Bandara W, Dissanayake DCT (2013) Artificial breeding and larval rearing of three tropical sea cucumber species – *Holothuria scabra, Pseudocolochirus violaceus* and *Colochirus quadrangularis* – in Sri Lanka. SPC Trepang Inf Bull 33:30–37

Lotavelli A, Conand C, Purcell SW, Uthicke S, Hamel F, Mercier A (2004) Advances in sea cucumber aquaculture and management. FAO, fisheries technical paper no 463. FAO, Rome

Mary Bai M (1994) Studies on regeneration in the holothurian *Holothuria* (Meriatyla) *scabra* Jaeger. In: Rengarajan K, James DB (eds) Proceedings of the National Workshop on *Beche-de-*

mer. Central Marine Fisheries Research Institute, Indian Council of Agricultural Research, Cochin, pp 44–50
Mazlan N, Hashim R (2015) Spawning induction and larval rearing of the sea cucumber Holothuria scabra in Malaysia. SPC Beche-de-mer information Bulletin #35
Mercier A, Battagleme SC, Hamel JF (1999) Daily burrowing cycle and feeding activity of juvenile sea cucumbers *Holothuria scabra* in response to environmental factors. J Exp Mar Biol Ecol 239:125–156
Mercier A, Ycaza RH, Hamel J-F (2007) Long-term study of gamete release in a broadcast-spawning Holothurian: predictable lunar and diel periodicities. Mar Ecol Prog Ser 329:179–189
Mills DJ, Duy NDQ, Juinio-Menez MA, Raison CM, Zarate JM (2012) Overview of sea cucumber aquaculture and sea-ranching research in the south-east Asian region. In: Hair CA, Pickering TD, Mills DJ (eds) Proceeding of an international symposium held in Noumea, New Caledonia. Australian Centre for International Agricultural Research, Canberra, pp 22–31
Morgan AD (2000) Induction of spawning in the sea cucumber *Holothuria scabra* (Echinodermata: Holothuroidea). J World Aquacult Soc 31(2):186–194
Muthiga NA, Kawaka J (2008) The effects of temperature and light on the gametogenesis and spawning of four sea urchin and one sea cucumber species on coral reefs in Kenya. In: Proceedings of the 11th International Coral Reef Symposium, Ft. Lauderdale, Florida, pp 356–360
Muthiga NA, Kawaka JA, Ndirangu S (2009) The timing and reproductive output of the commercial sea cucumber *Holothuria scabra* on the Kenyan Coast. Estuar Coast Shelf Sci 84:353–360
Pangkey H, Lantu S, Manuand L, Mokolensang J (2012) Prospect of sea cucumber culture in Indonesia as potential food sources. J Coast Dev 15(2):114–124
Perez ML, Brown EO (2012) Market potential and challenges for expanding the production of sea cucumber in South-East Asia. In: Hair CA, Pickering TD, Mills DJ (eds) Asia-Pacific tropical sea cucumber aquaculture. Australian Centre for International Agricultural Research, Canberra, pp 177–188
Pietrak M, Kim JK, Redmond S, Kim Y-D, Yarish C, Bricknell I (2014) Culture of sea cucumbers in Korea: a guide to Korean methods and the local sea cucumber in the northeast U.S. Maine Sea Grant College Program, Orono, Seagrant.umaine.edu/extension/korea-aquaculture
Pitt R (2001) Review of sandfish breeding and rearing methods. SPC Beche-de-mer Inf Bull 14:14–21
Pitt R, Duy NDQ (2003) Breeding and culture of the sea cucumber *Holothuria scabra* in Vietnam. Aquacult Asia 8(1):36–39
Plotieau T, Baele J-M, Vaucher R, Hasler C-A, Koudad D, Eeckhaut I (2013) Analysis of the impact of *Holothuria scabra* intensive farming on sediment. Cah Biol Mar 54:703–711
Purcell SW, Lovatelli A, Vasconcellos M, Ye Y (2010) Managing sea cucumber fisheries with an ecosystem approach. Food and Agriculture Organization of the United Nations, Rome
Purcell SW, Conand C, Uthicke S, Byrne M (2016) Ecological roles of exploited sea cucumbers. Oceanogr Mar Biol Annu Rev 54:367–386
Purwati E, Radu S, Ismail A, Kqueen CY, Maurice L (2003) Characterization of Listeria monocytogenes isolated from chicken meat: evidence of conjugal transfer of plasmid-mediated resistance to antibiotic. J Anim Vet Adv
Ramofafia C, Byrne M, Battaglene CS (2003) Reproduction of the commercial sea cucumber *Holothuria scabra* (Echinodermata: Holothuroidea) in the Solomon Islands. Mar Biol 142:281–288
Rasolofonirina R, Vaitilingon D, Eeckhaut I, Jangoux M (2005) Reproductive cycle of Edible Echinoderms from the South-Western Indian Ocean. Western Indian Ocean J Mar Sci 4(1):61–75
Robinson G, Pascal B (2009) From hatchery to community – Madagascar's first village-based holothurian mariculture programme. SPC Inf Bull 29:38–43
Rowe FWE (1982) Papers from the Echinoderm Conference. Australian Museum, Sydney

Sicuro B, Levine J (2011) Sea cucumber in the Mediterranean: a potential species for aquaculture in the Mediterranean. Rev Fish Sci 19(3):299–304

Troell M (2009) Integrated marine and brackish water aquaculture in tropical regions: research, implementation and prospects. In: Soto D (ed) Integrated mariculture: a global review. FAO fisheries and aquaculture technical paper no 529. FAO, Rome, pp 47–131

Uthicke S (2001) Nutrient regeneration by abundant coral reef holothurians. J Exp Mar Biol Ecol 265:153–170

Uthicke S (2004) Overfishing of holothurians: lesson from the Great Barrier Reef. In: Lovatelli A, Conand C, Purcell S, Uthicke S, Hamel JH, Mercier A (eds) Advances in sea cucumber aquaculture and management. Food and Agriculture Organization of the United Nations, Rome, pp 193–205

Vaitilingon D, Smith S, Watson G, Miller T, Alattas S, Hock KO, Zainoddin J, Zaidnuddin I, Azhar H (2016) Sea cucumber hatchery seed production in Malaysia: from research and development, to pilot-scale production of the sandfish *Holothuria scabra*. SPC Trepang Inf Bull 36:67–75

Watanabe S, Sumbing JG, Lebata-Ramos MJH (2014) Growth pattern of the tropical sea cucumber, *Holothuria scabra*, under captivity. JARQ 48(4):457–464

Sea Cucumber (Echinodermata: Holothuroidea) Species Diversity on the West Coast of Sabah, Malaysia

7

Nazia Binti Abdul Kadar, Nur Suhaila Binti Rosli, and Siti Syahirah Binti Muzli

Abstract

The sea cucumbers are primitive marine animals that are widely distributed in the ocean habitats, such as coral reefs, seagrass, rocky, sandy-muddy and sandy habitats. Holothurians are significant nutritional food and therefore many commercial species are harvested across the world. Malaysia is one of the major producers of Holothuroidea that exports both fresh and dried sea cucumbers, "beche-de-mer," with significant economic value worldwide, providing income to the local fishermen and community living near the coastline. Sea cucumber is a local delicacy that is consumed for its healing properties and general well-being. Holothuroidea are deposit feeders that consume organic matter and microalgae in the sediment using their tentacles. Sea cucumbers are known as ecosystem engineers that play important parts in the marine ecosystem, especially in the mineralization of sediments and nutrient cycling. Their diversity has been described by several researchers from a taxonomic view. In recent years, the rising demand for sea cucumbers in the international market has caused a decline in the wild stocks, so many species are being endangered. The chapter focuses on the general biology and diversity of holothurian species on the west coast of Sabah.

N. B. Abdul Kadar (✉) · N. S. B. Rosli · S. S. B. Muzli
Borneo Marine Research Institute, Universiti Malaysia Sabah, Kota Kinabalu, Sabah, Malaysia
e-mail: nazia@ums.edu.my

M. D. Shah et al. (eds.), *Marine Biotechnology: Applications in Food, Drugs and Energy*, https://doi.org/10.1007/978-981-99-0624-6_7

7.1 Sea Cucumber and Its Production in Malaysia

Sea cucumbers (Class: Holothuroidea) are marine animals that belong to the Phylum Echinodermata. Sea cucumbers are known as "timun laut," "gamat," "bat," "balat" and "brunok" by local community. Approximately 80 species of sea cucumbers can be found inhabiting the Malaysian waters. Of this, 38 species were recorded in Sabah and 24 species in Peninsular Malaysia. Only 10 species were found in both regions whereby *Holothuria (Mertensiothuria) leucospilota* is the most common species (Kamarudin et al. 2015). The main landing ports for Holothuroidea in Sabah are Kota Kinabalu, Kota Marudu, Kudat, Kota Belud, Sandakan and Semporna (Choo 2008). The most common species landed are *Stichopus herrmanni* (curryfish), *Actinopyga lecanora* (stonefish), *Holothuria coluber* (snakefish) and *Actinopyga echinites* (deep-water redfish) (Choo 2008). Prices for fresh and dried sea cucumber vary. In Tawau, dried tiger fish *Bohadschia argus* was sold at USD 192/kg (Choo 2012). The commercial value of the sandfish *Holothuria scabra* can exceed USD 100/kg dry weight (Tuwo et al. 2021). Currently, dried sea cucumbers are sold at USD 28/g in Kota Kinabalu, Sabah.

Malaysia is one of the major producers of holothurians in the world, which brings significant profits to the economy and important income to the fishermen and villagers living along the coastline. Sea cucumber fishing is mostly artisanal with high demand and a high price. Overexploitation has led to the extinction of some species, thus many attempts to restore the populations are being made through sea ranching and restocking in some countries. Few states such as Perak, Kedah and Terengganu in Peninsular Malaysia are engaged in sea cucumber harvesting. Sea cucumber harvesting in Sabah (139 tonnes) is superior compared to other states in Malaysia; however, inferior in comparison to countries like China (5500 tonnes) due to artisanal fishing methods (Choo 2008). Commercial sea cucumber species are usually overexploited by exceeding the maximum sustainable yield (Anderson et al. 2011). Holothurians are vulnerable to overharvesting due to their characteristics, such as slow growth rate, long lifespan, late age at maturity and others.

7.2 Biological Characteristics and Life Cycle

Holothurians can be found in shallow habitats such as coral reefs, seagrass meadows and sandy areas. The body is usually cylindrical and elongated with leathery skin. The body wall thickness shows the market value of the species as it is processed for consumers (Purcell et al. 2012). Sea cucumbers are benthic deposit feeders that can boost the nutrients in the sediments by their feeding and burrowing actions. This ocean dweller feeds using its tentacles by filtering seawater or waste material on the seafloor. Holothurians breathe from the anus through their respiratory tree and eject their internal organs when threatened. The orders of sea cucumbers are generally based on the morphology of calcareous rings and tentacles, the appearance of an internal respiratory tree, the existence of tentacular retractor muscles and the distribution of podia (Hasmah et al. 2012). Sea cucumbers can tolerate temperatures ranging from 22 to 32 °C and salinities ranging from 26 to 33 ppt. Meanwhile, the suitable pH for sea cucumbers is between 7.5 and 8.6 (Minami et al. 2019; Rohayat et al. 2021). Holothurians can breed asexually or sexually by releasing eggs and sperm into the water that will be fertilized when in contact. A juvenile takes approximately 5 years to grow into an adult.

7.3 Nutritional and Medicinal Value

The exploitation of sea cucumbers is more towards the food industry and at the same time contributes to the economy of Sabah (Kamarudin et al. 2015). Holothurians has been used in making liniment oil, facial wash, soap and many more. Chinese community refers to sea cucumbers as "hoi sum" or "hai shen" meaning sea ginseng for their healing ability. Sea cucumber is crucial as a source of income for Malaysia especially in producing traditional medicine such as "minyak gamat" (lipid extracts), "air gamat" (body fluid extracts) and food supplement, which is based in Langkawi and Pangkor islands. Meanwhile, in Sabah, East Malaysia, holothurians are an essential food supply for the food industry and there are light uses of *Holothuria atra* as fishing poison (Kamarudin et al. 2009). The processed sea cucumber as a food source is usually called "beche-der-mer," dry tunics or "trepang." "Trepang" is also recognized as a dried sea cucumber that has undergone a series of processes including removals of the organs, boiling process to soften the tissues and drying process to allow the processed sea cucumbers to be kept for long term during transportation and marketing. Holothurians especially "gamat" also become the traditional medicine for wound treatment, high blood pressure, white spot disease as well as aches in the joints (Sunmugam et al. 2021). Holothurians are important tonic that is rich in nutrients such as Vitamin A, Vitamin B and other essential minerals (Xing et al. 2021).

7.4 Morphology and Taxonomy of Sea Cucumber

The classification of sea cucumbers is mostly based on their morphological characteristics and habitat. Identification at the species is a more perplexing work because of close resemblances in the morphological traits (Massin 2007). In the current study, live holothurians were obtained from the intertidal zone covering shallow rocky shores, coral reefs and sandy beaches with seagrass patch areas (Table 7.1). Fresh specimens were measured and identified on the spot based on the external morphological traits such as body colour and structure of the tentacles and based on identification keys from various reference materials (Purcell et al. 2012; Kamarudin et al. 2015).

Table 7.1 Comparison of external morphological characteristics of sea cucumber species documented

Species	Colour of the body	Body size	Distribution of podia and papillae	Cuvierian tubules	Habitat
Holothuria (Mertensiothuria) leucospilota (Brandt, 1835)	Entirely black	Length: 17.0 cm Width: 3.2 cm	Randomly distributed	Present	Shallow rocky shore area and sandy bottom with coral rubbles and seagrass patch
Holothuria (Metriatyla) scabra Jaeger, 1833	Black with grey transverse lines (dorsal) White (ventral)	Length: 15.0 cm Width: 3.5 cm	Short papillae and deep wrinkles	Absent	Shallow waters with seagrass beds and muddy sandy substrates
Holothuria (Mertensiothuria) hilla Lesson, 1830	Reddish-brown (dorsal) Brown (ventral)	Length: 11.6 cm Width: 2.0 cm	Bulky and pointed papillae. Scattered podia	Present	Under coral rubbles and rocks within the rocky shore area
Holothuria impatiens (Forsskål, 1775)	Light brown with dark brown transverse band (dorsal) Beige (ventral)	Length: 13.2 cm Width: 2.3 cm	Thin podia	Present	Under rocks in shallow water within the rocky shore area and coral reef area with sandy bottom
Holothuria (Lessonothuria) pardalis Selenka, 1867	Beige with darker spots (dorsal) White (ventral)	Length: 17.7 cm Width: 2.7 cm	Scattered short and pointed papillae with curve	Absent	Rocks in shallow water within the rocky shore area and coral reef area with sandy bottom
Pearsonothuria graeffei (Semper, 1868)	Bronze with many large brown blotches (dorsal)	Length: 28.0 cm Width: 10.5 cm	Scattered long pointed papillae with white tips. Bands of many, lengthy, brown podia	Present	Hard surfaces of the wrecked dead coral. Inhabits the shallow water on the reef slopes with live coral

(continued)

Table 7.1 (continued)

Species	Colour of the body	Body size	Distribution of podia and papillae	Cuvierian tubules	Habitat
Stichopus ocellatus Massin, Zulfigar, Hwai & Boss, 2002	Brown (dorsal) Whitish brown (ventral)	Length: 10.8 cm Width: 9.0 cm	Protruding, huge, rounded, green to grey, wart-like papillae with white base settled in a zig-zag pattern. Green to brown podia	Absent	Inhabits sandy or muddy-sand bedrocks in a seagrass bed and coral reef area

Kingdom:	Animalia
Phylum:	Echinodermata
Class:	Holothuroidea
Order:	Aspidochirotida
Family:	Holothuriidae
Genus:	*Holothuria*
Species:	*Holothuria (Mertensiothuria) leucospilota* (Brandt, 1835)

Classification: *Holothuria (Mertensiothuria) leucospilota*, Order Aspidochirotida, Family Holothuriidae, commonly known as white thread fish or the local name is "bat puntil."

Description: It has an elongated and entirely black body. The body size is also broader in the posterior half. With both anterior and posterior ends, the body of this species tapers abstemiously. On the dorsal side of the body, long podia and papillae are randomly distributed. It has many ventral podia. Fine sediment and mucus sometimes covered the tegument. The mouth is ventral with big black tentacles where the anus is terminal. It ejects Cuvierian tubules as its defence mechanism (Purcell et al. 2012). The juvenile appearance of this species is similar to adults. The average observed length and width of this species were 17.0 cm and 3.2 cm, respectively (Fig. 7.1).

Habitat and distribution: *H. leucospilota* inhabits the shallow and rocky area at Outdoor Development Centre (ODEC) beach, Kota Kinabalu (6^{0}02′40″N, 116^{0}06′ 39″E), and sandy and muddy bottoms with coral rubble and seagrass patches at Kg. Limau-limauan, Kudat (6°49′17″N, 116°51′39″E) and Gaya Island, Kota Kinabalu, Sabah (5°59′43″N, 116°03′48″E).

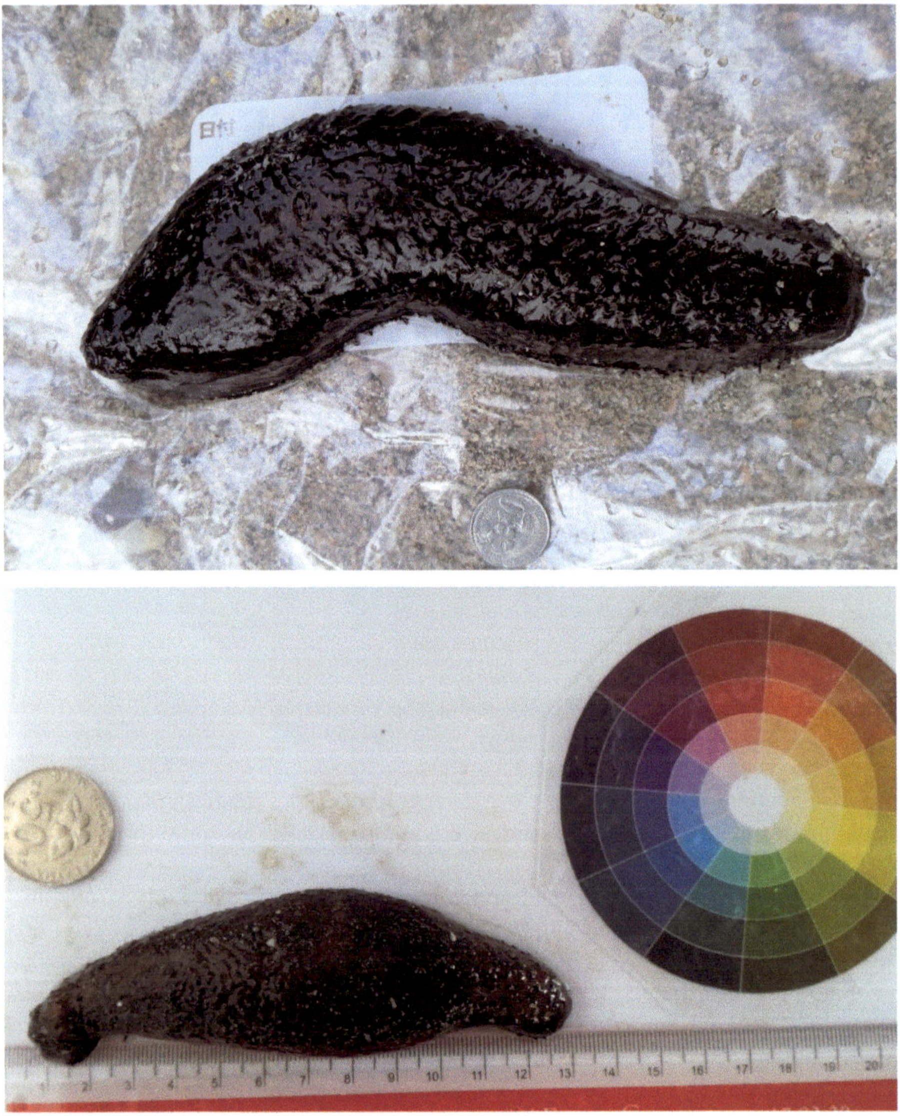

Fig. 7.1 *Holothuria (Mertensiothuria) leucospilota* (Brandt, 1835). A = dorsal view, B = ventral view

Kingdom:	Animalia
Phylum:	Echinodermata
Class:	Holothuroidea
Order:	Aspidochirotida
Family:	Holothuriidae
Genus:	*Holothuria*
Species:	*Holothuria (Metriatyla) scabra* Jaeger, 1833

Classification: *Holothuria (Metriatyla) scabra*, Order Aspidochirotida, Family Holothuriidae, commonly known as sandfish or the local name is "bat putih."

Description: It is observed as black, along with the transverse lines that are greyish black. The ventral side is white, along with smooth, dark spots. It has an elliptical body where the dorsal is more rounded and the ventral is moderately flattened. It has short papillae and deep wrinkles on the dorsal. It has a ventral mouth along with small tentacles that are grey. The anus has no teeth, and it is terminal (Purcell et al. 2012). *H. scabra* does not eject Cuvierian tubules. The observed length was 15.0 cm, and the width was 3.5 cm (Fig. 7.2).

Fig. 7.2 *Holothuria (Metriatyla) scabra* Jaeger, 1833. A = dorsal view, B = ventral view

Habitat and distribution: *H. scabra* was found in shallow waters with seagrass beds and muddy sandy substrates. It is found buried with its body in fine muddy sand at Outdoor Development Centre (ODEC) beach, Kota Kinabalu, Sabah (6°02′40"N, 116°06′39″E).

Kingdom:	Animalia
Phylum:	Echinodermata
Class:	Holothuroidea
Order:	Aspidochirotida
Family:	Holothuriidae
Genus:	*Holothuria*
Species:	*Holothuria (Mertensiothuria) hilla* Lesson, 1830

Classification: *Holothuria (Mertensiothuria) hilla*, Order Aspidochirotida, Family Holothuriidae, commonly known as tiger tail or the local name is "bat."

Description: It is reddish-brown on the dorsal side along with bulky, pointed, yellowish papillae. The ventral side is brown with spots consistent with big podia that are scattered in rows of three to four. *H. hilla* has a tubular body with a flat body wall. The mouth is ventral and encircled by small tentacles. The anus is terminal (Purcell et al. 2012). The Cuvierian tubules are present in this species but never ejected. The observed length was 11.6 cm and the width was 2.0 cm (Fig. 7.3).

Habitat and distribution: *H. hilla* was found under coral rubble and rocks within the rocky shore area at Outdoor Development Centre (ODEC) beach, Kota Kinabalu, Sabah (6°02′40″N, 116°06′39″E).

Classification: *Holothuria impatiens*, Order Aspidochirotida, Family Holothuriidae, commonly known as brown-spotted sea cucumber or the local name is "bat."

Description*:* It was observed in light brown on the dorsal side with five or more transverse bands that are dark brown which posteriorly becomes spots. The ventral side is beige in colour. It has been described as the shape of the bottle and rough. The podia are rather thin. It has a ventral mouth along with tentacles (Purcell et al. 2012). The Cuvierian tubules are long, white and thick. The average observed length and width of this species were 13.2 cm and 2.3 cm, respectively (Fig. 7.4).

Habitat and distribution: *H. impatiens* was found under rocks in shallow water within the rocky shore area at Outdoor Development Centre (ODEC) beach, Kota Kinabalu (6°02′40″N, 116°06′39″E), and coral reef area with a sandy bottom at Gaya Island, Kota Kinabalu, Sabah (5°59′43″N, 116°03′48″E).

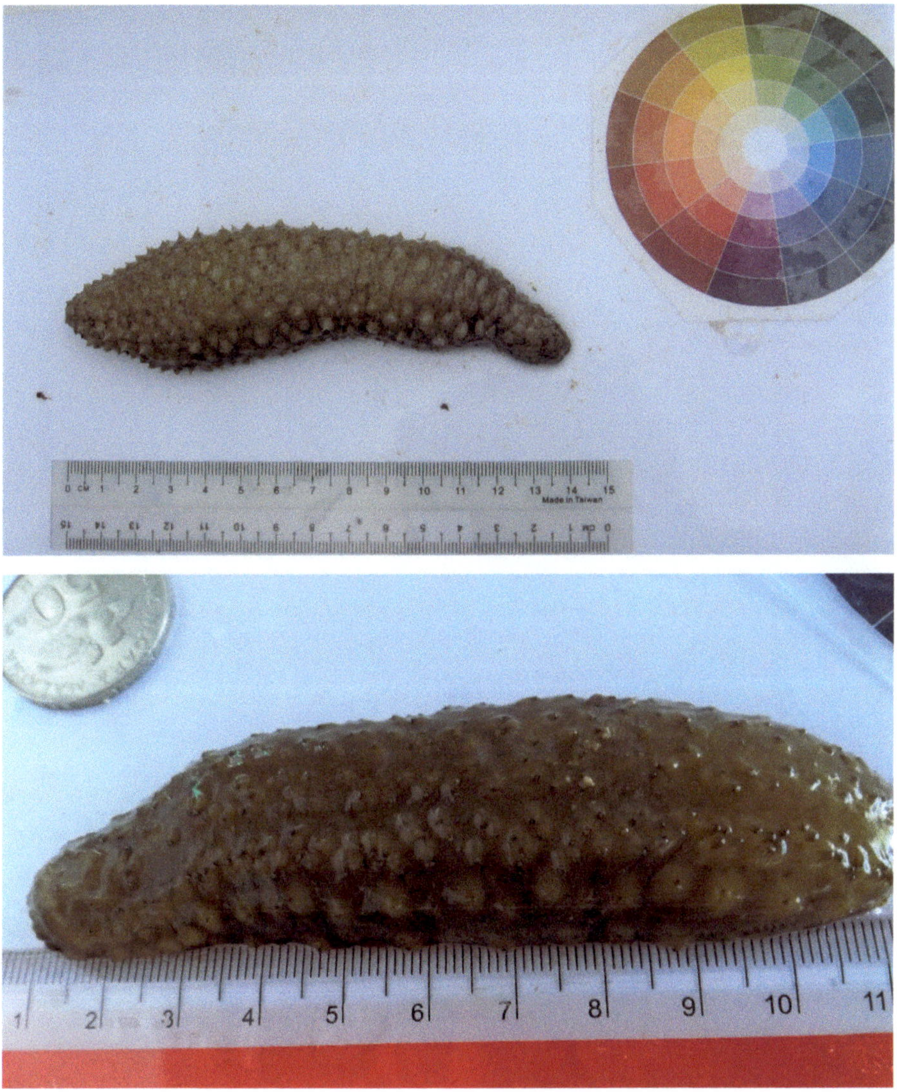

Fig. 7.3 *Holothuria (Mertensiothuria) hilla* Lesson, 1830. A = dorsal view, B = ventral view

Kingdom:	Animalia
Phylum:	Echinodermata
Class:	Holothuroidea
Order:	Aspidochirotida
Family:	Holothuriidae
Genus:	*Holothuria*
Species:	*Holothuria impatiends* Forsskål, 1775

Fig. 7.4 *Holothuria impatiens* (Forsskål, 1775). A = dorsal view, B = ventral view

Kingdom:	Animalia
Phylum:	Echinodermata
Class:	Holothuroidea
Order:	Aspidochirotida
Family:	Holothuriidae

(continued)

Genus:	*Holothuria*
Species:	*Holothuria (Lessonothuria) pardalis* Selenka, 1867

Classification: *Holothuria (Lessonothuria) pardalis*, Order Aspidochirotida, Family Holothuriidae local name is "bat."

Description: It is beige on the dorsal side along with two rows of large spots that are darker in colour and several tiny dark spots. The dorsal side of the body is covered with many short, pointed papillae with curved or somewhat conical tips which are dark brown or black and it is scattered throughout the side. The ventral side is white. The body is tubular and elongated, whereas the posterior end is wider. *H. pardalis* has numerous short and stout ventral podia. The mouth is ventral, encircled by a dual circle of papillae as well as tentacles. The anus is terminal and enclosed by tapering papillae. There are no Cuvierian tubules present in *H. pardalis* (Purcell et al. 2012). The average observed length and width of this species were 17.7 cm and 2.7 cm, respectively (Fig. 7.5).

Habitat and distribution: *H. pardalis* was found under rocks in shallow water within the rocky shore at Outdoor Development Centre (ODEC) beach, Kota Kinabalu (6°02′40″N, 116°06′39″E), and coral reef area with a sandy bottom at Gaya Island, Kota Kinabalu, Sabah (5°59′43″N, 116°03′48″E).

Fig. 7.5 *Holothuria (Lessonothuria) pardalis* Selenka, 1867. A = dorsal view, B = ventral view

Classification: *Pearsonothuria graeffei*, Order Aspidochirotida, Family Holothuriidae, commonly known as flowerfish or the local name is “bat.”

Description: It is bronze in colour with many large brown blotches along with fine dark speckling on the dorsal side. It has an extended and tubular body with many oblique folds. It has a more flattened ventral side, and the dorsal side of the body is

Kingdom:	Animalia
Phylum:	Echinodermata
Class:	Holothuroidea
Order:	Aspidochirotida
Family:	Holothuriidae
Genus:	*Holothuria*
Species:	*Pearsonothuria graeffei* (Semper, 1868)

Fig. 7.6 *Pearsonothuria graeffei* (Semper, 1868). A = dorsal view, B = ventral view

scattered with long pointed papillae with white tips and the ventral side has three bands of many, lengthy, brown podia. The tentacles of *P. graeffei* have a distinctive white edge and black encircled the ventral mouth. The anus is terminal and papillae is absent. It has Cuvierian tubules that is not ejected (Purcell et al. 2012). The average observed length and width of this species were 28.0 cm and 10.5 cm, respectively (Fig. 7.6).

Habitat and distribution: *P. graeffei* was found on the hard surfaces of coral rubbles and the shallow water on the reef slopes with live coral at Gaya Island, Kota Kinabalu, Sabah (5°59′43″N, 116°03′48″E).

Kingdom:	Animalia
Phylum:	Echinodermata
Class:	Holothuroidea
Order:	Aspidochirotida
Family:	Stichopodidae
Genus:	*Stichopus*
Species:	*Stichopus ocellatus* Massin, Zulfigar, Hwai & Boss, 2002

Classification: *Stichopus ocellatus*, Order Aspidochirotida, Family Stichopodidae, commonly known as curry fish or the local name "gamat."

Description: It has a dorsal side brown along with protruding, huge, rounded, green to grey, wart-like papillae with a white base. The huge papillae are settled in a zig-zag pattern and occur in four rows. *S. ocellatus* has a whitish-brown flattened ventral side. The ventral side of the body is scattered with many green to brown podia. The mouth is ventral and the anus is terminal without teeth (Purcell et al. 2012). The average observed length and width of this species were 10.8 cm and 9.0 cm, respectively (Fig. 7.7).

Habitat and distribution: *S. ocellatus* was found inhabiting sandy or muddy-sand bedrocks in a seagrass bed at Kg. Limau-limauan, Kudat (6°49′17″N, 116°51′39″E) and Gaya Island, Kota Kinabalu, Sabah (5°59′43″N, 116°03′48″E).

7.5 Sea Cucumber Species Diversity and Distribution

Initial research in Malaysia about the existence of Holothuroidea started as early as 1985 by Ridzwan and Che Bashah, followed by George and George (1987), Ridzwan (1987), Kaswandi et al. (1990), Ridzwan (1993), Kaswandi et al. (1995), Ridzwan et al. (1995), Ridzwan and Kaswandi (1995), Ridzwan et al. (1996) and Ridzwan et al. (1998). These researches emphasized on Sabah; Semporna, Tuaran and Kota Kinabalu, Peninsular Malaysia; Tioman Island (Pahang), Balik Pulau (Penang), Langkawi Island (Kedah), Besar Island, Aur Island (Johor) and Pangkor Island (Perak), and also in Brunei (Baine and Forbes 1998; Forbes and Ilias 1999; Siti et al. 1999; Zulfigar and Tan 1999; Zulfigar et al. 2000; Zaidnuddin and Forbes 2000; Zaidnuddin 2002; Zulfigar et al. 2007; Zulfigar et al. 2008; Sim et al. 2008; Sim et al. 2009; Kamarudin et al. 2009; Kamarudin et al. 2010). The highest diversity of Holothuroidea was found at ODEC Beach ($S = 5$, $H' = 0.279$, $J' = 0.143$) followed by Gaya Island ($S = 5$, $H' = 0.083$, $J' = 0.043$) and Kg. Limau-limauan ($S = 2$, $H' = 0.083$, $J' = 0.043$). Both the sites, Gaya Island and Kg. Limau-limauan, have the same diversity index but different species

Fig. 7.7 *Stichopus ocellatus* Massin et al., 2002. A = dorsal view, B = ventral view

Table 7.2 Sea cucumber species diversity index for all locations

Location	No. of species (S)	Shannon–Wiener Index (H′)	Shannon evenness (J′)
ODEC Beach	5	0.279	0.143
Gaya Island	5	0.083	0.043
Kg. Limau-limauan	2	0.083	0.043

compositions (Tables 7.2 and 7.3). Of the seven documented species, five are from the genus *Holothuria* and one each from *Stichopus* and *Pearsonothuria.*

Table 7.3 shows the species composition and number of individuals of sea cucumbers at the three selected locations. A total of 32 individuals (69.57%) of *Holothuria (Mertensiothuria) leucospilota* were recorded from ODEC Beach. Single species of (2.17%) of *Holothuria (Metriatyla) scabra* (2.17%) and *Holothuria (Mertensiothuria) hilla* were observed at ODEC Beach and Kg. Limau-limauan, respectively. Three individuals (6.52%) of *Holothuria impatiens*, were spotted at Gaya Island and ODEC Beach. Four individuals (8.70%) of *Holothuria (Lessonothuria) pardalis* were found at Gaya Island and ODEC Beach. Two individuals (4.35%) of *Pearsonothuria graeffei* were observed on Gaya Island. Finally, three individuals (6.52%) of *Stichopus ocellatus* were found at Kg. Limau-limau and Gaya Island. ODEC Beach had the highest percentage (80.46%) of individuals of sea cucumber, but the species composition was similar at Gaya Island, where both locations had five sea cucumber species but different sea cucumber species compositions. Kg. Limau-limauan showed the lowest percentage (6.52%) of individual sea cucumbers and species composition, whereas this location only had two types of sea cucumber species (Table 7.3 and Fig. 7.8).

Holothuria (Mertensiothuria) leucospilota known as *white threads fish* or locally as "bat puntil" was the most dominant species in these three localities in Sabah. This finding was in agreement with the study by Kamarudin et al. (2015) that stated the dominant holothurians species in Malaysia is *H. leucospilota.* It is recognized as "worm" sea cucumbers, where this species is lower-value higher-volume species, usually harvested in artisanal fisheries (Purcell et al. 2012). In addition, the greatest diversity of holothurians was dominated by the order Aspidochirotida which was supported by Ridzwan (1993) and Baine and Forbes (1998) who discovered five genera namely *Actinopyga, Bohadschia, Holothuria, Stichopus* and *Thelenota.* Thus, these findings proved that Aspidochirotida is the most abundant and dominant order within Malaysian coral reef habitats.

Diversity index is crucial to understand the biodiversity of an environment and usually calculated to detect any changes in species composition and/or abundance (Peet 1974). The Shannon's Diversity (H′) is sensitive to the grouping of species because the value is dependent on S (the number of the taxonomic unit). On the contrary, Evenness (J') is relatively insensitive because only the relative abundance of a species is measured. Based on the results obtained from this study, the highest diversity index and species composition of sea cucumber species were recorded at ODEC Beach, particularly in front of the UMS Hatchery. Based on our observations, this area had shallow water with coral rubbles and rocky bottom. Purcell et al. (2012) stated that the habitat preference for *H. leucospilota, H. hilla, H. pardalis, H. impatiens* and *H. scabra* was mostly in shallow water areas with rocky bottoms. The species diversity of sea cucumbers at ODEC Beach is also related to the external morphological characteristics of the holothurian species. The obtained results revealed that *H. leucospilota, H. hilla, H. pardalis, H. impatiens* and *H. scabra* possessed numerous, short and large ventral podia which aid these species in locomotion on the hard substrate as the area at ODEC Beach where these species

Table 7.3 Species composition and distribution of sea cucumbers at various locations in Sabah

No.	Species	Local name	No. of individual			Total	Percentage (%)
			GI	OB	KL		
Order Aspidochirotida Family Holothuriidae							
1	*Holothuria (Mertensiothuria) leucospilota* (Brandt, 1835)	"Bat puntil"/ White threadfish	1	30	1	32	69.57
2	*Holothuria (Metriatyla) scabra* Jaeger, 1833	"Bat putih"/ Sandfish	0	1	0	1	2.17
3	*Holothuria (Mertensiothuria) hilla* Lesson, 1830	"Bat"/Tiger tail sea cucumber	0	1	0	1	2.17
4	*Holothuria impatiens* (Forsskål, 1775)	"Bat"/Brown-spotted sea cucumber	1	2	0	3	6.52
5	*Holothuria (Lessonothuria) pardalis* Selenka, 1867	"Bat"/sea cucumber	1	3	0	4	8.70
Order Aspidochirotida Family Stichopodidae							
6	*Pearsonothuria graeffei* (Semper, 1868)	"Bat"/ Flowerfish	2	0	0	2	4.35
7	*Stichopus ocellatus* Massin, Zulfigar, Hwai & Boss, 2002	"Gamat"/ Curryfish	1	0	2	3	6.52
3	Total	6	37			46	100

GI = Gaya Island, OB=ODEC Beach, KL = Kg. Limau-limauan

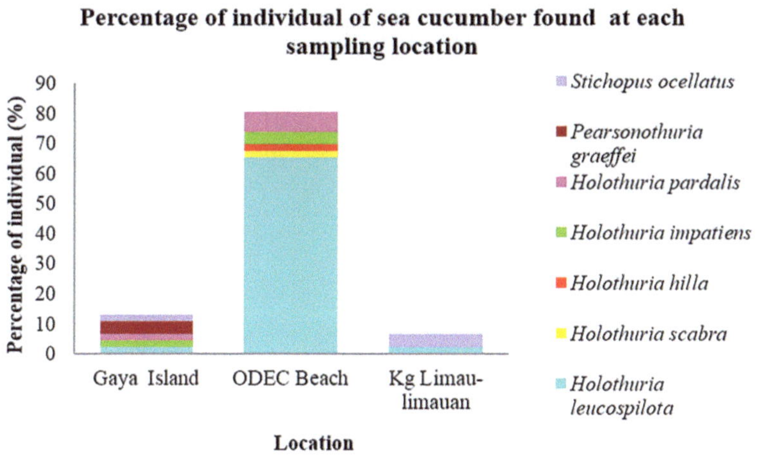

Fig. 7.8 Percentage of sea cucumber individuals found at some locations

found is covered by coral rubbles and rocky bottom. Podia are usually filled with seawater and used for locomotion (Purcell et al. 2012). This is supported by Woodby et al. (2000) who reported that some Holothuroidea prefer harder substrates, such as rocks and coral rubbles associated to their locomotion.

Gaya Island and Kg. Limau-limauan have the same species diversity index but different species composition. Based on the species diversity of Holothuroidea at Gaya Island, the marine conditions of Gaya Island as a protected coral reef area cannot ensure the existence of Holothuroidea from overexploitation due to the many fishermen that periodically enter this area and collect Holothuroidea that have commercial values. *H. leucospilota*, *H. pardalis* and *H. impatiens* are low-value species, and these species were spotted at Gaya Island. *Holothuria scabra* and *H. hilla* were not recorded or sighted at Gaya Island and Kg. Limau-limauan, because these species were reported as fisheries commodities. At Kg. limau-limauan, *H. scabra* is a well-known species as some of the villagers build sea pens for farming and the culture of this species. Hamel et al. (2013) documented that *H. scabra* numbers is decreasing and listed as endangered species under IUCN Red List of Threatened Species.

The same five species of Holothuroidea were found at both Gaya Island and ODEC Beach. Based on the observations, this area consisted of different types of coral reefs with a sandy bottom and marine plants, such as seaweed and seagrass, which explained the reasons why *Stichopus ocellatus* was recorded at Gaya Island. This was supported by Sun et al. (2022), who stated the well-developed coral reef in Sabah provides an appropriate environment for the sea cucumbers. Besides that, in the tropical region, there is a plenteous of organic matter in the coral reef areas such as mucus of the coral, seagrass detritus and remnants of the algae that become food for sea cucumbers along with the mass habitats, and this happens to provide vast and rich feeding areas for the sea cucumbers (Ridzwan 1993). These are the factors that may have led to the miscellaneous Holothuroidea species in the coastal waters of Sabah.

Kg. Limau-limauan has the lowest sea cucumber species composition, with only two species of holothurians (*Stichopus ocellatus* and *Holothuria leucospilota*) recorded at this site. Based on the result of this study, *S. ocellatus* was only recorded at Kg. Limau-limauan and Gaya Island, and this species was found in the shallow water area with sandy and muddy bottoms along with seagrass patches. According to Purcell et al. (2012), the habitat preference for this species is seagrass beds on sandflats or mudflats, which explains why this species was not recorded or sighted at ODEC Beach.

7.6 Conclusion

Sea cucumbers are important ocean dwellers that help to recycle nutrients by feeding on detritus in the marine ecosystem. Some species are overexploited to meet the demand as nutritional delicacies that treat various illnesses in humans. The Holothuroidea species obtained from the West Coast of Sabah were successfully recorded and *Holothuria* (*Mertensiothuria*) *leucospilota* also recognized as "bat puntil," was the most dominant species. ODEC Beach of Kota Kinabalu recorded the highest sea cucumber species diversity due to the nature of the environment that was favourable for Holothuroidea. The molecular phylogeny study of sea cucumbers is imperative to reveal the species diversity, distribution, reconfirmation and detection of new cryptic species in Malaysia especially in Sabah for the sustainability and proper management of sea cucumber fisheries.

Acknowledgements We would like to thank the staff the of Borneo Marine Research Institute, Universiti Malaysia Sabah, for their assistance directly and indirectly during the sampling and laboratory work.

References

Anderson SC, Flemming JM, Watson R, Lotze HK (2011) Serial exploitation of global sea cucumber fisheries. In: Fish and fisheries, vol 12. Wiley, Hoboken, pp 317–339. https://doi.org/10.1111/j.1467-2979.2010.00397.x

Baine M, Forbes B (1998) The taxonomy and exploitation of sea cucumbers in Malaysia. SPC Inf Bull 10:2–7

Choo PS (2008) Population status, fisheries and trade of sea cucumbers in Asia. In: Toral-Granda V, Lovatelli A, Vasconcellos M (eds) Sea cucumbers. A global review of fisheries and trade. FAO fisheries and aquaculture technical paper no. 516. FAO, Rome, pp 81–118

Choo PS (2012) The sea cucumber fishery in Semporna, Sabah, Malaysia. SPC Beche-de-mer Information Bulletin 32

Forbes B, Ilias Z (1999) The taxonomy and ecology of sea cucumbers in Malaysia. SPC Inf Bull 12: 5

George JD, George J (1987) The coral reefs of the Bodgaya Islands (Sabah: Malaysia) and Pulau Sipadan 4. Macroinvertebrates. Malay Nat J 40:225–260

Hamel JF, Mercier A, Conand C, Purcell S, Toral-Granda, TG, Gamboa R (2013) *Holothuria scabra.* The IUCN Red List of Threatened Species 2013: e.180257A1606648

Hasmah I, Woo SP, Zulfigar Y, Azizah MN, Tan SH (2012) Molecular taxonomy of sea cucumber (Holothuroidea) in the upper parts of the straits of Malacca inferred from 16S rRNA gene sequences. Thai Nat Hist Mus J 6(1):11–24

Kamarudin KR, Rehan AM, Lukman AL, Ahmad HF, Anua MH, Nordin NFH, Hashim R, Hussin R, Usup G (2009) Coral reef sea cucumbers in Malaysia. Malay J Sci 28(2):171–186

Kamarudin KR, Ridzwan H, Gires U (2010) Phylogeny of sea cucumber (Echinodermata: Holothuroidea) as inferred from 16S mitochondrial rRNA gene sequences. Sains Malays 39: 209–218

Kamarudin KR, Usup G, Hashim R, Mohamed RM (2015) Sea cucumber (Echinodermata: Holothuroidea) species richness at selected localities in Malaysia. Pertanika J Trop Agric Sci 38(1):7–32

Kaswandi MA, Nor 'Aini AG, Ridzwan BH (1990) Profil protein dinding timun laut. Malays Biochem Soc Conf 15:132–134

Kaswandi MA, Ridzwan B H, Mohd Nor I, Zainab A, Azman S, Rozali MBO (1995) Timun Laut: analisis Semasa. In: Simposium Sumber Alam Kebangsaan Pertama FSSA. UKM Kampus Sabah, Sabah, vol 33, pp 17–32

Massin C (2007) Redescription of *Stichopus naso* Semper, 1868 (Echinodermata, Holothuroidea, Stichopodidae). Biologie 77:123–130

Massin C, Zulfigar Y, Tan SHA, Rizal Boss SZ (2002) The genus *Stichopus* (Echinodermata: Holothuroidea) from the Johore Marine Park (Malaysia) with the description of two new species. Biologie 72:73–99

Minami K, Masuda R, Takahashi K, Sawada H, Shirakawa H, Yamashita Y (2019) Seasonal and interannual variation in the density of visible *Apostichopus japonicus* (Japanese sea cucumber) in relation to sea water temperature. Estuar Coast Shelf Sci 229:106384
Peet RK (1974) The measurement of species diversity. Annu Rev Ecol Syst 5:285–307
Purcell SW, Samyn Y, Conand C (2012). Commercially important sea cucumbers of the world. FAO species catalogue for fishery purposes no 6. FAO, Rome, p 150
Ridzwan BH (1993) Sumber makanan persisiran laut Sabah. Dewan Bahasa dan Pustaka, Kuala Lumpur
Ridzwan BH, Che Bashah CC (1985) Tinjauan awal taburan dan penilaian holothurian di beberapa kawasan perairan Sabah sebagai sumber pemakanan. SUMBER I:143–155
Ridzwan BH (1987) Kajian holothurian di perairan cetek Sabah: nilai pemakanan, kandungan lumen usus dan tumbesaran. In: Biologi Untuk Industri, Noraini M, et al. UKM Publisher, Bangi, pp 339–369
Ridzwan BH, Kaswandi MA (1995) Hidupan marin intertidal: makanan tambahan penduduk daerah Semporna. Malays J Nutr 1:105–114
Ridzwan BH, Kaswandi MA, Dayang Fredalina B, Haniza B (1995) A brief look into current research on sea cucumbers in Malaysia. In: JSPS-VCC joint seminar on marine science. Universiti Putra Malaysia (UPM), 5–8 December 1995
Ridzwan BH, Haniza BH, Dayang Fredalina B, Kaswandi MA, Norlida AL, Suhaina S, Zali BI, Mohd. Nor H (1996) A brief insight into current research on sea cucumber of Malaysia. In: Joint seminar by UKM-Universitas Hassanudin, Dinamika Perkembangan Kebaharian Masyarakat Indonesia dan Malaysia, Ujung Pandang, Sulawesi, 10–12 September 1996
Ridzwan BH, Shukri HI, Fadhil MH, Noor Ibrahim MS, Kaswandi MA, Hawa I, Jamaludin, M, Rozali MBO, Mohd Nor BH, Zali, BI (1998) The species distribution of sea cucumbers in Malaysia. In: Malaysia Science and Technology Congress '98, Kota Kinabalu, Sabah, 23–25 November 1998
Rohayat A, Destiani S, Nurjaman S, Ardiana C, Nawawi I (2021) Sea cucumber ecology Holothuroidea studies in the Sancang marine preserve. IOP Conf Ser Mater Sci Eng 1098: 052040. https://doi.org/10.1088/1757-899X/1098/5/052040
Sim YK, Tan SH, Zulfigar Y, Wong WS, Chai J, Ng BW, Luar L (2008) The diversity and distribution of Holothuroidea in Pulau Aur, Johor. In: Research and information series of Malaysian coast 2. Pusat Penyelidikan Ekosistem Marin (EKOMAR), UKM, pp 13–22
Sim YK, Tan SH, Zulfigar Y (2009) The diversity and abundance of the sea cucumber (Echinodermata: Holothuroidea) from the archipelagos of Beting Patinggi Ali, Beting Raja Jarom and Pulau Layang–Layang, South China Sea. In: 6th IMT-GT UNINET conference 2008: sustaining natural resources towards enhancing the quality of life within the IMT-GT zone. Penang: The Gurney Resort Hotel, 25–30 August 2008, pp 357–360
Siti ZRS, Zulfigar Y, Aileen TSH (1999) Species of sea cucumber found in Pulau Besar, Johor, with special emphasis on the genus *Stichopus*. SPC Inf Bull 12:4
Sun C, Huang D, Xu Q et al (2022) Diverse habitat preferences of two sea cucumber species and the seasonal change in a coral reef area. J Oceanol Limnol 40:1578–1591. https://doi.org/10.1007/s00343-021-1254-z
Sunmugam T, Azhari H, Ng SF, Azmi F (2021) Influence of the oil phase on the wound healing activity of sea cucumber extract-based cream formulations. Sains Malays 50(3):839–847
Tuwo A, Yasir I, Bestari AD, Yanti A, Aprianto R, Tresnati J (2021) High salinity effect on mortality of sandfish *Holothuria scabra*. IOP Conf Ser Earth Environ Sci 763:012019. https://doi.org/10.1088/1755-1315/763/1/012019
Woodby D, Smiley S, Larson R (2000) Depth and habitat distribution of *Parastichopus californicus* near Sitka, Alaska. Alsk Fish Res Bull 7:22–32
Xing L, Sun L, Liu S, Zhang L, Yang H (2021) Comparative metabolomic analysis of the body wall from four varieties of the sea cucumber *Apostichopus japonicus*. Food Chem 352:129339. https://doi.org/10.1016/j.foodchem.2021.129339

Zaidnuddin I (2002) Sea cucumber species diversity and abundance from the south–western waters of Balik Pulau, Penang. In: Choo PS, Ismail I, Chee PE, Chuah TT (eds) Workshop on the impact of development on the coastal fisheries off south–west Penang Island, Malaysia. Fisheries Research Institute, Penang, Malaysia, pp 57–62
Zaidnuddin I, Forbes R (2000) Sea cucumbers of Pulau Payar, Kedah Marine Park Islands. In: Ali A, Arshad MA, Ilias Z, Mohd Sah SA (eds) Proceedings of National Symposium on Pulau Payar Marine Park, pp 174–177
Zulfigar Y, Tan SH (1999) Biodiversity of sea cucumber in the lower part of The South China Sea. In: The joint seminar on marine and fisheries sciences. Melaka, Malaysia, 1–3 December 1999, pp 73–82
Zulfigar Y, Tan SH, Fujita T, Terazaki M (2000) The distribution of sea cucumber associated with seagrass beds found in The South China Sea. In: The 11th JSPS joint seminar on marine science, pp 362–368
Zulfigar Y, Sim YK, Tan SH (2007) The distribution of sea cucumbers in Pulau Aur, Johore, Malaysia. In: Publications of the Seto Marine Biological Laboratory, Special Publication Series Volume 8 (from Selected Paper of NaGISA World Congress 2006). Seto Marine Biological Laboratory, Field Science Education and Research Centre, Kyoto University, pp 73–86
Zulfigar Y, Sim YK, Tan SH, Shirayama Y (2008) Field guide to the echinoderms (sea cucumber and sea stars) of Malaysia. The Nippon Foundation/Kyoto University Press, Japan, p 130

Sea Cucumbers: Source of Nutritional, Medicinal, and Cosmeceutical Products

8

Nurzafirah Binti Mazlan, Nur Nashrah Binti Abd Rahman,
Sarah Syazwani Binti Shukhairi,
and Muhammad Nor Afdall Bin Nazahuddin

Abstract

Sea cucumbers have been known for their medicinal benefits, particularly in Asian countries. Consumption has evolved from mere traditional medicine to various products of a pharmaceutical and cosmeceutical nature. The extracts have also been added to the products as an added value and are being sold over the counter in pharmacies and retail stores. Dried sea cucumbers are being made into delicacies and are known as exotic foods. They are also being sold at high prices, indicating the high value of their nutrition and the fact that they contain lots of medicinal benefits. They are rich in fatty acids, minerals, and vitamins. The nutritional, pharmaceutical, and medicinal profiles also differ based on the species. Studies have shown that the extract of sea cucumbers can be used as an antimicrobial, antifungal, and anticancer agent, and can alleviate the inflammation of osteoarthritis. The extracts are also made into skin products such as lotions, shampoos, and soaps. The most common species in Malaysia are *Holothuria scabra*, *Stichopus horrens*, and *Acaudina molpadioides*. *S. horrens* has been extensively explored for its medicinal potential and is on the verge of extinction. Efforts are being made to restore endangered species and explore the potential of new species.

N. B. Mazlan (✉) · N. N. B. Abd Rahman · S. S. B. Shukhairi · M. N. A. B. Nazahuddin
Borneo Marine Research Institute, Universiti Malaysia Sabah, Kota Kinabalu, Sabah, Malaysia
e-mail: nurzafirah@ums.edu.my

M. D. Shah et al. (eds.), *Marine Biotechnology: Applications in Food, Drugs and Energy*, https://doi.org/10.1007/978-981-99-0624-6_8

8.1 Introduction

Sea cucumber, commonly known as balat, trepang, or gamat in Malay, is a marine invertebrate that belongs to the phylum Echinodermata, a distant relative of starfish and sea urchins (Truongs and Le 2019). To date, there are over 1700 species of sea cucumbers that mainly inhabit the Asia-Pacific region. Out of all species, only 40 are harvested for food and commercial uses. *Holothuria scabra*, *Holothuria fuscogilva*, *Stichopus hermanni*, and *Stichopus horrens* are a few sea cucumber species from Malaysia with the highest commercial values (Harith et al. 2018).

Among Chinese and Malaysians, sea cucumbers have been well recognized as a tonic and traditional remedy for their effectiveness against hypertension, asthma, rheumatism, cuts and burns, impotence, and constipation. Bordbar et al. (2011) found that sea cucumber extracts exert medicinal effects such as wound healing promoters, anticancer, and immunomodulatory properties. The major edible portion of sea cucumbers is the body wall, which is mainly comprised collagen and mucopolysaccharides. Thus, when there is a high concentration of collagen in the body wall, it leads to high value for nutraceutical or pharmaceutical applications. The protein content is higher than the raw material, and its moisture content is lower than the fresh flesh. Interestingly, sea cucumbers are considered a luxury food item alongside abalone, fish maw, and shark fin (Fabinyi et al. 2017).

Sea cucumbers have been made into various products, modern and traditional, such as tonics, massage oils, pills, shampoos, and skincare. Products are sold over the counter and are homemade. There is various potential for sea cucumbers to be exploited for their medicinal, pharmaceutical, and nutritional benefits. However, the population of said species is slowly depleting as they are harvested and traded. Figure 8.1 shows the few commercial species of sea cucumbers in Malaysia.

In Malaysia, there are several methods of sea cucumber processing being practiced traditionally. The most common include the processes of visceral removal, salting, boiling, and drying (Fig. 8.2). The dried sea cucumbers are sold to the supplier, who will then supply them to restaurants and traditional medical practices. In industrial practice, wet sea cucumbers will be freeze-dried and undergo an extraction process to obtain the target bioactive ingredients.

Fig. 8.1 Few commercial sea cucumber species in Malaysia are (**a**) *Acaudina molpadioides*, (**b**) *H. scabra*, and (**c**) *S. horrens*

Fig. 8.2 The traditional method of sea cucumber processing involves the process of visceral removal (**a**), salting, boiling (**b**), and drying (**c**)

8.2 Nutritional Source

Many folklore and mythology, including "The Legend of Iron Crutch Li, one of the Eight Immortals," "The Legend of Liu Gong," and "The Legend of Taishang Laojun," have glorified the sea cucumbers and their health benefits. Dated back to the Ming and Qing dynasties, sea cucumbers are called "haishen," which means "sea ginseng," as it possesses countless nutritional benefits that are incomparable to land-grown superfoods (Pangestuti and Arifin 2018). They have been frequently used in the folk medicine of Asian and Middle Eastern cultures (Halder and Pahari 2020). As documented in the traditional Chinese medicine handbook, sea cucumbers are believed to be able to enhance the urological faculties of men, as in Chinese belief, one can fortify specific body parts by eating foods that are similar in shape. As more health benefits are uncovered in modern medical studies, sea cucumbers are recognized internationally as a superfood (Southey 2020). Figure 8.3 shows the dried sea cucumbers from Sabah, Malaysia, being sold as a luxury food or used as traditional medicine.

Sea cucumbers have gained a reputation as functional ingredients as they carry an impressive amount of valuable nutrients that can cure ailments including hypertension, rheumatism, kidney problems, cuts, and burns, joint and back pain, minor injuries, asthma, constipation, and reproductive disorders (Hossain et al. 2020). This has caused them to be traded either locally or internationally, which makes them susceptible to over-exploitation. However, sea cucumbers have long been used in many Asian cuisines as a valuable food ingredient. They are consumed in various ways: either fresh, dried, or pickled. As of late, the development of sea cucumber-derived food products is on the rise because they provide various nutritional and health benefits. Figure 8.4 shows the local delicacy made from sea cucumbers.

Fig. 8.3 Dried sea cucumbers or bêche-de-mer are sold as a luxury food and used in traditional medicine

Fig. 8.4 Sea cucumbers are used as a food ingredient in different cuisines in local delicacies such as (**a**) kerabu balat and (**b**) sup balat

In terms of nutrition, sea cucumbers contain an array of high-value nutritional components such as amino acids, vitamins, minerals, collagen, and fatty acids (Liu et al. 2021). Vitamins such as vitamin A, B1 (thiamine), B2 (riboflavin), and B3 (niacin) are the main energy suppliers for the body that can boost the immune system and improve heart function (Aminur Rahman et al. 2020; Hossain et al. 2020), while minerals found in the visceral organs of sea cucumbers like zinc, iron, calcium, and magnesium help to improve body metabolism and lower blood sugar levels (Hossain et al. 2020).

Besides that, they are rich in protein (between 43 and 61%) and low in fat (1.01–1.19%), which are perfect for a weight-loss diet (Feng et al. 2021; Kubala 2020; Lalao et al. 2019). In general, a 100 g serving of sea cucumbers contains less than 1 g of fat, sugar, carbohydrates, and fibre, 56 calories, and 13 g of protein (Brennan Dan 2020). The low level of fat in the meat makes it a good source of lean protein, which can be beneficial to the well-being of the heart. Sea cucumbers also contain a high amount of phenol and flavonoid antioxidants to minimize the risk of coronary heart disease and neurodegenerative disorders (Kubala 2020).

Insoluble collagen made up about 70% of the sea cucumbers body wall. This type of collagen is being utilized as a nutrient supplement for haematogenesis (Pangestuti and Arifin 2018). Collagen is commonly used in food products to improve bone integrity, reduce osteoarthritis pain, and increase cell viability (Senadheera et al. 2020). Gelatine from sea cucumbers, on the other hand, is considered more valuable than gelatine from other animals due to the composition of amino acids, particularly essential amino acids (Pangestuti and Arifin 2018). As mentioned in a study by Hossain et al. (2020), the use of sea cucumbers or other marine resources to produce pharmaceutical and nutraceutical compounds, specifically collagen and gelatine, has become an upward trend as they are considered a much safer alternative compared to mammalian sources and void of limitations such as religious restrictions and food safety.

On the other hand, sea cucumbers also carry numerous bioactive compounds that are beneficial to the overall function and health of the body. Fucosylated chondroitin sulphate (FCS) is among the most known bioactive compounds found in the body wall of various sea cucumbers species where the molecular structure of the compound is different than in other invertebrates (Hossain et al. 2020). There is a significant amount of FCS within the body wall of *Cucumaria frondosa* which helps to prevent muscle ageing and facilitate growth, which can be a remedy for osteoarthritis (Hossain et al. 2020). Meanwhile, FCS from *Thelenota ananas* exhibits anticoagulant activity (Pangestuti and Arifin 2018). Researchers also discovered and patented their findings on the antiviral potential of chondroitin sulphates from *T. ananas* that can hinder replication and infection of the human immunodeficiency virus (HIV) (Halder and Pahari 2020). Pangestuti and Arifin (2018) in their study on "Medicinal and Health Benefit Effects of Functional Sea Cucumbers" believe that this compound has the potential as the curative for HIV/AIDS.

Another bioactive compound found in sea cucumbers is glycosaminoglycans (GAGs) There are two types of GAGs which are sulphated and non-sulphated glycosaminoglycan. Both the compounds can be found abundantly in the body wall of *Stichopus hermanni or* commonly known as curry fish, compared to its internal organs and coelomic fluid. Gamat extract from said species contains a high concentration of sulphated glycosaminoglycan causing them to be widely harvested, especially in a few Indo-Pacific countries, namely Malaysia and Indonesia, where the production of traditional medicine from gamat extracts like gamat oil is thriving. Halder and Pahari (2020) concluded that the gamat extract exhibits wound-healing properties that can accelerate wound contraction, and this claim is supported by findings from a study by Pangestuti and Arifin (2018) where over 60% of the wounds in rats healed after daily treatment of sulphated glycosaminoglycan. Their study also mentioned that *S. hermanni* could be used for the treatment of wound healing and neuroprotection. Similarly, fatty acids such as docosahexaenoic acid (DHA) and eicosapentaenoic acid (EPA) in sea cucumber extracts also help to accelerate tissue and wound healing (Achmad et al. 2020).

Commercial values differ between species since the nutritional profile of each species is unique depending on its diet and habitat. According to Feng et al. (2021), sea cucumbers' nutritional composition differs regionally between seasons. In his study, the composition of macro- and micro-elements that are important for enzymatic activities and balancing blood pressure, amino acids for regulating crucial body metabolic pathways, and gut microbes that improve gut health are closely influenced by seasons and origins. Findings as such are fundamental in harvesting the highest quality of sea cucumbers with optimal nutritional benefits. Hence, the overexploitation of sea cucumbers can be reduced and their population in the wild can be restored.

8.3 Medicinal Source

Researchers have found a few properties related to the health and medicinal value of sea cucumbers, such as wound healing, anti-coagulation, tumour prevention/shrinkage, anti-microbial protection, neurological protection, and antioxidant properties. The possibility of sea cucumbers possessing very high value-added compounds in their flesh is due to the presence of bioactive peptides, triterpene glycosides, vitamins, minerals, carotenoids, collagen, fatty acids, amino acids, gelatine, and chondroitin sulphates. These elements contribute to the therapeutic properties of sea cucumbers and eventually act as a medicinal source either in traditional or modern practice. Another finding also showed its potential low in lipids and toxicity, which gave them one of the best values of marine medicinal animals (Liang et al. 2022; Pangestuti and Arifin 2018).

8.3.1 Healing Activity

Traditional therapeutic items made from *Stichopus hermanni* or locally known as "gamat emas" include "gamat" water and "gamat" oil (Fig. 8.5). *S. hermanni* has a high-protein content (47%) and a low-lipid content (0.80%) and sulphated

Fig. 8.5 The "gamat" water and "gamat" oil produced traditionally at Pulau Pangkor, Malaysia

glucosaminoglycan (Pangestuti and Arifin 2018). The sulphated glycosaminoglycan from the integumentary system has been shown to speed up wound healing in rats. After 12 days of daily therapy with sulphated glycosaminoglycan (20 mL of 1 mg/mL), more than 60% of the wound in the rat model was healed (Torres et al. 2018). In wound healing phase I, the activity of sulphated glycosaminoglycan was facilitated by faster wound contraction. Furthermore, 40% of *S. hermanii* extract has been proven to increase the proliferation of lymphocytes cell when being tested on ulcers induced on the oral mucous membrane of Wistar rats, indicating there were healing mechanisms (Arundina et al. 2015). Self-renewing cells called mesenchymal stem cells can differentiate into myocytes, chondrocytes, adipocytes, and osteoblasts. Mesenchymal stem cells were developed into osteoblasts after 4 weeks of treatment with the extract of sea cucumbers *S. hermanni* in the osteogenic generation medium. *S. hermanni* is a type of sea cucumber and is said to have the ability to alleviate wound healing. Furthermore, this sea cucumber species are used to produce various medicinal products such as topical lotion or gel for wound healing (Arundina et al. 2016). Wound healing properties have also been reported for *S. horrens* (Barathi et al. 2013). Although sea cucumbers are best known for their wound-healing action, however, only several species have been studied in detail. Hundreds of species have yet to be discovered to their full potential.

8.3.2 Neurological Protection

Arundina et al. (2015) reported that there was an increase in the number of spinal glial astrocytes cell when being treated with the water extract from *S. hermanni* from Malaysia, indicating good cell growth. Furthermore, GC-MS data revealed that 2-carbamoyl-3-methylquinoaxaline can be detected in the *S. hermanni* extracts. After a spinal cord injury, quinoxaline derivatives have been shown to help with neurological impairments and glia loss (Bordbar et al. 2011; Pangestuti and Arifin 2018). Quinoxaline may also have a role as neuroprotective properties in several species of marine animals. Glycine is abundant in *A. mauritiana* also, which aids in the maintenance of the central nervous system (Suthar et al. 2022).

8.3.3 Anti-Coagulant

Fucosylated chondroitin sulphate is a type of glycosaminoglycan identified in sea cucumbers. It is water soluble and the fucose branch varies by sea cucumber species which determines their characteristics (Torres et al. 2018). Anticoagulant activity in a sea cucumber species, *Thelenota ananas*, follows a logarithmic-like curve that shows a variety of proportions to the molecular weight. The anticoagulant effect of fucosylated chondroitin sulphate from *T. ananas* was recently found to be facilitated by the inhibition of innate tenase (Pangestuti and Arifin 2018). To date, about 25 species of sea cucumbers exhibit anticoagulant activities as reviewed by Li et al. (2021). All species contain the fucosylated glycosaminoglycan extracted from their body walls.

8.3.4 Antioxidant

Enzymatic degradation of sea cucumber *T. ananas* yielded a compound called fucoidan containing a new tetra-fucose repeating unit. With an IC_{50} value reading of 17.46 ± 0.14 mg/mL, fucoidan from *T. ananas* has been shown to have considerable superoxide radical scavenging action (Dhinakaran and Lipton 2014). Fucoidan's radical scavenging ability against superoxide radicals was enhanced as the sulphate level increased. However, different animals raise the radical-scavenging effect with an extra 2-O-sulphation at a specific residue, indicating that the antioxidant activity is dependent on the sulphation pattern rather than the sulphate level alone (Yu et al. 2014).

It has also been shown that the *Holothuria atra* extract demonstrates hepatoprotective and curative benefits in rats against 7,12-dimethylbenz[a]anthracene (DMBA)-induced hepatorenal illnesses (Dakrory et al. 2015) whereby there was an increase in the liver enzymatic reaction with the DMBA consumption. Furthermore, there was evidence of cell and tissue loss and destruction of defence mechanisms (antioxidant activity) with the addition of malondialdehyde at a certain concentration level (Dakrory et al. 2015; Pangestuti and Arifin 2018). However, the addition of *H. atra* extract before or after DMBA toxicity test efficiently repealed the alterations, implying that the effects of *H. atra* extract on liver protection are attributable to its antioxidant function (Dakrory et al. 2015; Esmat et al. 2013). *Holothuria leucospilota* extract and its bioactive components have been shown to exhibit antioxidant action. The antioxidant activity of carotenoids produced from *H. leucospilota* has been investigated using several methods, such as 1,1 diphenyl-2-picrylhydrazyl (DPPH), linoleic acid-free radical killing test, and beta carotene assays (Pangestuti et al. 2016).

8.3.5 Anticancer

Saponins are triterpene glycosides (also known as saponins) that have a sugar part linked to either a triterpene or steroid aglycone. The chemicals can be found in a wide range of plants, marine invertebrates, echinoderms, octocorals, and sponges as secondary metabolites. *T. ananas* and *T. anax* have yielded two triterpene glycosides, which are stichoposide C and stichoposide D. Both differ in terms of the sugar residue with quinovose presence in the stichoposide C and glucose in stichoposide D. Cell culture of HL-60 leukaemia and CT-26 of subcutaneous tumour cells showed a significant reduction with the induction of apoptosis when being tested with Stichoposide C (Pangestuti and Arifin 2018; Yun et al. 2012). The extract of sea cucumber *H. atra* was reported to be able to inhibit cervical cancer cells (HeLa) and human breast cancer cells (MCF-7) (Dhinakaran and Lipton 2014). Furthermore, cytotoxicity detected in the cervical carcinoma cells when being treated with the extract of *H. leucospilota* (Mashjoor and Yousefzadi 2019). It has been found that triterpene glycosides have substantial anticancer effects in the sub-cytotoxic range of concentrations through direct interaction with tumour cells (Aminin et al. 2015). Two types of triterpene glycosides, notably Scabraside A and B isolated from *H. scabra*, were highly toxic to several types of cancer cells such as HL-60, MOLT-4, A549, and BEL-7402 cells (Han et al. 2012). Another type, Scabraside D, a new triterpene glycoside, demonstrated significant inhibitory activity against multiple types of cancer cells such as gastric cancer (MKN-28), colon cancer (HCT116), and breast cancer (MCF-7) cells (Assawasuparerk et al. 2016).

8.3.6 Antifungal and Antiparasitic

A study conducted by Mashjoor and Yousefzadi (2017) discovered that the extract of *H. fuscogilva* exhibited antimicrobial activity against fungi *Candida* sp. and bacteria *Leishmania* sp., and cytotoxicity against human supraclavicular lymph node metastases (LoVo) cells. Meanwhile, *H. fuscogilva* also contains high level of saturated fatty acids (SFA) and monounsaturated fatty acids (MUFA) while low in polyunsaturated fatty acids (PUFA). Moreover, they discovered that the amount of triterpene glycosides in *H. fuscogilva* was also high which may contribute to the cytotoxic activities. Triterpene glycosides are excreted by sea cucumbers as a form of protection, especially in the wild. *H. fuscogilva* using this cytotoxic technique, suggest that it could be used as an anti-infective agent (Pangestuti and Arifin 2018). In addition, *H. atra* was also reported to possess antifungal properties against *Candida albicans*. With a concentration of 0.5%, *H. atra* extract, however, was not toxic to oral mesenchymal stem cells, indicating that the extract is safe and can be developed for further clinical trials (Mashjoor and Yousefzadi 2017; Parisihni and Revianti 2013). Extract of *H. atra* was also tested for its antifungal activities and effectiveness against the *Malassezia furfur* fungus (Alawiyah et al. 2016). Other than that it has also been proven to have antimicrobial activities (Dhinakaran and Lipton 2015).

8.3.7 Antibacterial

Compounds derived from *H. atra* have a wide range of antibacterial activity. The antibacterial activity of a PBS extract of *H. atra*, for example, was able to inhibit the development of both Gram-negative and Gram-positive bacteria (Hossain et al. 2020). Surprisingly, extracts from the inner body wall of *H. atra* had a larger antibacterial effect than extracts from the outer region (Sukmiwati et al. 2019). The presence of microorganisms that are taken in with the dietary ingredients may explain the greater antibacterial activity of the inner component. *H. scabra* extracts have been shown to exhibit antibacterial properties in several experiments (Mashjoor and Yousefzadi 2017). *Aeromonas hydrophila*, *Escherichia coli*, *Enterococcus* sp., *Klebsiella pneumoniae*, *Pseudomonas aeruginosa*, *Salmonella typhi*, *Staphylococcus aureus*, and *Vibrio harveyi*, as well as fish-borne mould, *Aspergillus* sp., were all active against *H. scabra* extracts (Pangestuti and Arifin 2018). Later, it was discovered that the antibacterial action of T-antigen-binding lectin might have a role in *H. scabra*. Lectin is produced by bacterial challenge, and lectin induction is mediated by bacterial cell wall glycoconjugates. Furthermore, lectin demonstrated high-antibacterial action across a broad spectrum (Prompoon et al. 2015; Hossain et al. 2020;).

8.3.8 Anticholesterol

The glycine in *Stichopus vastus* has also been shown to lower total cholesterol levels in the blood. On the other hand, low levels of lysine and arginine were shown to reduce the cholesterol levels in the serum and aorta (Pangestuti and Arifin 2018; Rasyid 2018). In a study conducted by Rasyid (2018), all essential amino acids such as threonine, histidine, valine, tryptophan, isoleucine, phenylalanine, methionine, leucine, cysteine, lysine, and arginine, while non-essential amino acids such as glutamic acid, serine, proline, tyrosine, aspartic acid, alanine, and glycine were found in the sea cucumber *S. vastus*. Arginine (28,651.62 mg/kg) was the most abundant essential amino acid in the *S. vastus* sample. Except for cysteine, which revealed low amounts, all necessary amino acids were present at relatively high levels. The major saturated fatty acid in the present study was palmitic acid (0.07%), followed by myristic acid (0.03%), lauric acid (0.02%), stearic acid (0.02%), caprylic acid (0.01%), arachidonic acid (0.01%), and butyric acid (0.01%). Most of the present amino acids are good for protein to build up and have less cholesterol content, which means palmitic acid shows antioxidant and anti-atherosclerosis activities (Rasyid 2018).

8.3.9 Combined Benefits

The sea cucumber *Thelenota ananas* and *Thelenota anax* contain high protein and low lipid. Eicosapentaenoic acid (EPA) is found to be the major n3-polyunsaturated acid (PUFA) in sea cucumbers (Yahyav 2012; Kareh et al. 2018). EPA consumption has been linked to a lower incidence of cardiovascular disease, tumour, and wound healing (Pangestuti and Arifin 2018; Sroyraya et al. 2017). *H. leucospilota* is a tropical holothurian sea cucumber that has high levels of carotenoids such as b-carotene, ketozeaxanthin, fucoxanthin, b-echinenone, and canthaxanthin that main functions are as energy gatherers as well as antioxidants (Sroyraya et al. 2017). Exposure to light and air produces reactive oxygen species (ROS). Antioxidants may benefit human health by protecting the body from damage caused by ROS, which attack cell structures and organelles and eventually result in a variety of diseases, including diabetes, neurodegenerative diseases, and inflammatory diseases with severe tissue injuries (Taquet et al. 2011). Proteins (63.30% $\pm$ 0.43%) is the most abundant component in *A. mauritiana*, with lipid (1.40% $\pm$ 0.02%) being the least much. Proteins can be broken down into bioactive peptides which produce energy in return. *A. mauritiana* contains a high amount of non-essential amino acids (NEAA), with an essential amino acid, EAA: NEAA ratio of 0.44 $\pm$ 0.01. Glycine, glutamic acid, and proline are the most prevalent amino acids in *Actinopyga* spp., accounting for roughly 11% of amino acids. Glycine is an amino acid that aids in the formation of muscular tissue and the conversion of glucose to energy (Hossain et al. 2020; Nahla 2013; Rahael et al. 2019).

8.4 Cosmeceutical Source

Because of their medicinal and nutritional properties, sea cucumbers may become the next essential item in one's beauty routine. Gelatine and collagen extracted from sea cucumbers have enormous promise as useful components in cosmeceutical products such as creams, gels, makeup, and powders. The use of sea cucumbers as a cosmetic ingredient has been considered for treating skin disorders, particularly skin hydration and wrinkle reduction. Moreover, the bioactive compounds in sea cucumbers provide promising effects that have been proven through various scientific research, and sea cucumbers are also well known for their health benefits. Furthermore, two of the most important compounds in the cosmeceutical industry, such as collagen and hyaluronic acid, have been the focus of much anti-ageing research and development. Both have been found abundant in sea cucumbers.

Collagen is a complicated micro-protein that accounts for 75–80% of the human body's biggest organ, the skin, which is responsible for its overall health, structure, look, and age. Furthermore, protein can help to improve the immune system and create hormones and enzymes that kick-start metabolism. The high-protein content of sea cucumbers can enhance the regeneration of dead cells caused by wounds, allowing them to heal wounds (Kokadir et al. 2021). A major proportion of sea cucumbers' body wall protein comprised collagen. Collagen fibrils and type I collagen, which is proportionally fibre-shaped and short, are the most frequent types of collagens in sea cucumbers. The finding showed that type 1 sea cucumber collagen holds strong potential for use in functional cosmetics due to its exceptional moisture retention and absorption properties (Li et al. 2020; Safira et al. 2022; Senadheera et al. 2020). Collagen, a connective substance found in bones and skin, can be utilized to improve skin appearance and speed up healing by increasing the regeneration of dead cells caused by wounds. As a result, sea cucumbers can be utilized in cosmetics and wound-healing ointments. Furthermore, there is a high demand for gelatine from sea cucumbers as it contains high-essential amino acids compared to other organisms. As mentioned in a study done by Li et al. (2020), glycine (31%) was the most abundant amino acid in the three collagens of *Holothuria cinerascens*, followed by a proline (9–12%) and alanine (10–12%), where these three collagens had higher moisture retention and absorption capacity than glycerol, indicating that the collagen molecules are rich in hydrophilic groups and might be used in cosmetic compositions. It also demonstrated that it had higher moisture retention and absorption ability than collagen isolated from tilapia and pig skin. Because of this distinct property, collagen may interact with water and hence be used in moisturisers. Research on sea cucumber collagen has primarily concentrated on hydrolytic bioactive peptide functions such as damaged tissue healing, antioxidant, and angiotensin-converting enzyme inhibitory action.

Antioxidants are required to protect membrane lipids, DNA, and proteins from ultraviolet-induced reactive oxygen species such as hydroxyl radicals, superoxide anion, and hydrogen peroxide. One of the fundamental causes of wrinkled skin is the oxidation of membrane lipids; consequently, minimizing the generation of reactive oxygen species is critical to preserve wrinkle-free skin. Carotenoids and mycosporine-like amino acids (MAA) have been found as photoprotective chemicals in the epidermis of black sea cucumbers (*H. atra*). MAA has been implicated in photoprotection as it is thought to act as a broad-spectrum UV absorber (Siahaan et al. 2017). Antioxidant activity of carotenoids produced by *H. leucospilota* has been done using several methods, including 1,1-diphenyl-2-picrylhydrazyl (DPPH), linoleic acid-free radical test, and carotene bleaching assays. According to Kim et al. (2016), elastase inhibitory activity from a liquid extract of boiled sea cucumbers that is higher than 50 kDa fractions will protect the skin from ageing by preventing UV or free radical damage to elastic fibres as well as for whitening purposes. The formulated cream with *Holothuria arenicola* extract matched the relevant pharmacological qualities because the antioxidant properties present in the extract help to protect the skin against damage caused by free radicals (Saber Mohamed et al. 2020).

Sea cucumbers contain the bioactive compounds that are responsible for the induction of tissue repair and wound healing processes. The integumentary system, organs, and coelomic fluid of sea cucumbers contain the total O-sulphated glycosaminoglycan, which promotes wound healing. *Stichopus hermannii* hydrogel formulation possesses wound healing properties as it enhanced wound contraction and improved tissue regeneration which stimulates the propagation of fibroblast and enhanced the production of collagen fibre (Masre et al. 2012). Furthermore, fucosylated chondroitin sulphate (FuCS), which is isolated from the body wall of sea cucumber, is a unique glycosaminoglycan in its structure and medical properties (Hossain et al. 2020). Fucoidan is found abundant in the cell wall of sea cucumbers. It can act as an anti-ageing agent in cosmetic manufacturing to prevent photoaging (Siahaan et al. 2017). This polysaccharide contains L-fucose and sulphate groups. Fucoidan from *Thelenota ananas* has been shown to have considerable superoxide radical scavenging action (Yu et al. 2014). Fucoidan's radical scavenging action on superoxide radicals was enhanced as the sulphate level increased. A study showed that fucoidan isolated from *Isostichopus badionotus* demonstrated significant biological activity (Chen et al. 2013). Fucoidan's sulphation concentration and structure are closely related to its biological application.

Arachidonic acid, which is found in sea cucumbers, has been linked to wound healing and growth. The presence of other acids, eicosapentaenoic acid (EPA) and docosahexaenoic acid (DHA), is medicinally significant because these two long-chain fatty acids have been related to a lower risk of cardiovascular disease and several malignancies even though sea cucumbers have a low-fat content (BioMarine 2020; Budzinski et al. 2020).

Vitamins (A, B1, B2, B3), minerals (iron, zinc, copper, calcium, magnesium), and other microelements which are essential to our health can be found in sea cucumber extracts and used in cosmetic formulations (Bordbar et al. 2011). Vitamin and mineral contents are easily absorbed into the skin and provide moisturising and skin-promoting healing and rejuvenation of cells. It could purify, clean, and moisturise the skin.

Stichopus japonicus viscera aqueous extracts are known to have anti-ageing properties and whitening treatments to be applied in cosmeceutical products that activate extracellular-regulated kinase signalling to limit melanin synthesis and enhance collagen synthesis (Kwon et al. 2018). Yoon et al. (2010) reported that *S. japonicus* showed potent cellular anti-melanogenic activity on B16 melanoma cells as it inhibits the expression of tyrosinase and tyrosinase-related proteins (TRP-1 and TRP-2). The report showed that sea cucumbers have the potential to be used as a skin whitening agent for use in skincare cosmetic products or as hypopigmentation agents.

Sea cucumber processing as a medical product has come a long way. Only Stichopus species are traditionally used for medicinal purposes in Southeast Asian nations, and gamat strictly refers to species in the genus Stichopus. In Malaysia, liquid sea cucumber extracts are offered as an essence or in the form of jellies

(Purcell et al. 2014). Gamat water will be prepared from fresh whole *Stichopus* sp. and allowed to simmer under a low flame to get pure liquid. Other than that gamat jelly can be consumed with fruit juice and is believed to improve our health. Sea cucumber emulsion made from *Stichopus* sp. extract is mixed with honey, omega 3 fatty acids, vitamins, and collagen, and can be taken as a tonic to improve wound healing, nourish skin, and reduce joint pains.

In Japan, products derived from sea cucumbers such as soap made from *Stichopus chloronotus* and *H. atra* from Okinawa have acquired appeal among health and nature-conscious consumers (Slater 2015). Renee Alyce, an Australian woman who created her skincare brand (Bescher) made from sea cucumber collagen, claims that her product helps to rejuvenate the appearance of the skin's tone, texture, and firmness in 6 weeks (Rudd 2020). Her ground-breaking moisturiser contains sea cucumbers collagen, squalane, hyaluronic acid, and kakadu plum extract for visible and instant improvement in skin tone, texture, and firmness. The market appears to be amenable to a wide range of value-added sea cucumber goods, with "beauty" and "health" as keywords.

Cosmetics with extracts from marine animals and plants are becoming popular in the cosmeceutical industry, as many people prefer natural-based skincare products. Sea cucumbers have a lot of potential as cosmetic ingredients. However, determining the exact structures of bioactive compounds and their activities in sea cucumbers remain a substantial issue. There are abundant sea cucumber-derived products being made into cosmeceuticals items and some are made as an added-value product from other plants or animal-based products.

8.5 Conclusion

Sea cucumbers are being explored for their medicinal, nutritional, and cosmetic benefits. Their usage as medicine has been proven since hundreds of years ago when they were used as traditional remedies, which eventually were applied to modern medicine. Consumption of sea cucumbers as a healthy food has been practised by Asian people namely the Chinese, Japanese, Koreans, and Malays. They are now being sold as exotic foods at expensive prices and served on special occasions. Their high-nutritional value with protein and fatty acids as the main component has served as a pedestal for their claim as a functional food. Furthermore, the bioactive ingredients in sea cucumbers can also be used as an added value in many cosmeceutical products such as shampoo, face wash, face cream, lotion, and make-ups. Although there are already plenty of studies proving their medicinal, nutritional, and cosmetic benefits, there are still hundreds of species that have yet to be studied. Exploration of the potential of these sea cucumbers should be pursued in order to discover their benefits.

References

Achmad H, Mardiana Adam A, Rezdy Asalui T, Indra Sukmana B, Prahatama Putra A (2020) Use of sea cucumber extract as an alternative treatment of inflammation by chronic periodontitis. Int J Pharm Res 12(4):3567–3575. https://doi.org/10.31838/ijpr/2020.12.04.489

Alawiyah T, Khotimah S, Mulyadi A (2016) Aktivitas antijamur ekstrak teripang darah (*Holothuria atra* Jeager.) terhadap pertumbuhan jamur malassezia furfur penyebab panu. J Protobiont 5(1): 59–67. https://doi.org/10.26418/protobiont.v5i1.14897

Aminin DL, Menchinskaya ES, Pisliagin EA, Silchenko AS, Avilov SA, Kalinin VI (2015) Anticancer activity of sea cucumber triterpene glycosides. Mar Drugs 13(3):1202–1223. https://doi.org/10.3390/md13031202

Aminur Rahman M, Hossain Chowdhury S, Jahidul Hasan M, Habibur Rahman M, Maksuda Yeasmin S, Farjana N, Molla R, Shamim Parvez M (2020) Status, prospects and market potentials of the sea cucumber fisheries with special reference on their proper utilization and trade. Ann Res Rev Biol *35*(7):84–101. https://doi.org/10.9734/ARRB/2020/v35i730250

Arundina I, Suardita K, Setiabudi H, Ariani MD (2016) Golden sea cucumbers (Stichopus Hermanii) as growth factors of stem cells. J Int Dent Med Res *9*(3):242–248

Arundina I, Yuliati Y, Soesilawati P, Damaiyanti DW, Maharani D (2015) The effects of golden sea cucumber extract (Stichopus hermanii) on the number of lymphocytes during the healing process of traumatic ulcer on wistar rat's oral mucous. Dental J (Majalah Kedokteran Gigi) 48(2):100. https://doi.org/10.20473/j.djmkg.v48.i2.p100-103

Assawasuparerk K, Vanichviriyakit R, Chotwiwatthanakun C, Nobsathian S, Rawangchue T, Wittayachumnankul B (2016) Scabraside D extracted from Holothuria scabra induces apoptosis and inhibits growth of human cholangiocarcinoma xenografts in mice. Asian Pac J Cancer Prev 17(2):511–517. https://doi.org/10.7314/APJCP.2016.17.2.511

Barathi SS, Arul A, Pratibha MDA, Arunkumar HD (2013) Efficacy of gamat extract in wound healing in albino wistar rats. Int J Pharm Sci Rev Res 20(1):142–145

BioMarine (2020) Beauty from Northern waters: the sea cucumber has many good reasons to convince you! https://biomarine.org/news/252055. February 3.

Bordbar S, Anwar F, Saari N (2011) High-value components and bioactives from sea cucumbers for functional foods - a review. Mar Drugs *9*(10):1761–1805. https://doi.org/10.3390/MD9101761

Brennan Dan (2020) Sea cucumber: health benefits, nutrients per serving, preparation information, and more. WebMD. https://www.webmd.com/diet/health-benefits-sea-cucumber

Budzinski P, Maimeun M, Mutrakulcharoen P, Wonganu B, Sriariyanun M (2020) Profiling analysis of fatty acids and collagens obtained from sea cucumbers. E3S Web Conf 141: 03006. https://doi.org/10.1051/E3SCONF/202014103006

Chen S, Li G, Wu N, Guo X, Liao N, Ye X, Liu D, Xue C, Chai W (2013) Sulfation pattern of the fucose branch is important for the anticoagulant and antithrombotic activities of fucosylated chondroitin sulfates. Biochim Biophys Acta 1830(4):3054–3066. https://doi.org/10.1016/J.BBAGEN.2013.01.001

Dakrory AI, Fahmy SR, Soliman AM, Mohamed AS, Amer SAM (2015) Protective and curative effects of the sea cucumber Holothuria atra extract against DMBA-induced hepatorenal diseases in rats. BioMed Res Int 2015:563652. https://doi.org/10.1155/2015/563652

Dhinakaran DI, Lipton AP (2014) Bioactive compounds from *Holothuria atra* of Indian ocean. Springerplus 3(1):1–10. https://doi.org/10.1186/2193-1801-3-673

Dhinakaran DI, Lipton AP (2015) Antitumor and antifungal activities of organic extracts of sea cucumber *Holothuria atra* from the southeast coast of India. J Ocean Univ China 14(1): 185–189. https://doi.org/10.1007/s11802-015-2528-x

Esmat AY, Said MM, Soliman AA, El-Masry KSH, Badiea EA (2013) Bioactive compounds, antioxidant potential, and hepatoprotective activity of sea cucumber (Holothuria atra) against thioacetamide intoxication in rats. Nutrition 29(1):258–267. https://doi.org/10.1016/j.nut.2012.06.004

Fabinyi M, Barclay K, Eriksson H (2017) Chinese trader perceptions on sourcing and consumption of endangered seafood. Front Mar Sci 4. https://doi.org/10.3389/FMARS.2017.00181/FULL

Feng J, Zhang L, Tang X, Xia X, Hu W, Zhou P (2021) Season and geography induced variation in sea cucumber (*Stichopus japonicus*) nutritional composition and gut microbiota. J Food Compost Anal 101(103):838. https://doi.org/10.1016/J.JFCA.2021.103838

Halder D, Pahari S (2020) An overview of sea cucumber: chemistry & pharmacology of its metabolites. Indian Res J Pharm Sci 25(2020):2277–2298. https://doi.org/10.21276/irjps.2020.7.2.19

Han H, Li L, Yi Y, Wang X h, Pan M x (2012) Triterpene glycosides from sea cucumber Holothuria scabra with cytotoxic activity. Chinese Herb Med 4(3):183–188. https://doi.org/10.3969/j.issn.1674-6384.2012.03.002

Harith MN, Md Desa MHI, Ilias Z (2018) Holothuria leucospilota population in Satang Besar Island, Sarawak, Malaysia. Int J Zool 2018:8047029. https://doi.org/10.1155/2018/8047029

Hossain A, Dave D, Shahidi F (2020) Northern sea cucumber (*Cucumaria frondosa*): a potential candidate for functional food, nutraceutical, and pharmaceutical sector. Mar Drugs 18(5):274. https://doi.org/10.3390/MD18050274

Kareh M, El Nahas R, Al-Aaraj L, Al-Ghadban S, Al Deen N, Saliba N, El-Sabban M, Talhouk R (2018) Anti-proliferative and anti-inflammatory activities of the sea cucumber *Holothuria pol*i aqueous extract. SAGE Open Med 6:205031211880954. https://doi.org/10.1177/2050312118809541

Kim SJ, Park SY, Hong SM, Kwon EH, Lee TK (2016) Skin whitening and anti-corrugation activities of glycoprotein fractions from liquid extracts of boiled sea cucumber. Asian Pac J Trop Med 9(10):1002–1006. https://doi.org/10.1016/J.APJTM.2016.08.001

Kokadir S, Basilianus E, Swandiny GF, Abdillah S (2021) A review over sea cucumber (*Stichopus variegatus*): a potential candidate for nutraceuticals. J Nutr Food Sci 6:42. www.imedpub.com

Kubala J (2020, February 25) The benefits of eating sea cucumber. Healthline. Retrieved February 27, 2023, from https://www.healthline.com/nutrition/sea-cucumber

Kwon TR, Oh CT, Bak DH, Kim JH, Seok J, Lee JH, Lim SH, Yoo KH, Kim BJ, Kim H (2018) Effects on skin of *Stichopus japonicus* viscera extracts detected with saponin including Holothurin A: down-regulation of melanin synthesis and up-regulation of neocollagenesis mediated by ERK signaling pathway. J Ethnopharmacol *226*:73–81. https://doi.org/10.1016/J.JEP.2018.08.007

Lalao V, Andriamanamisata R, Fidinirina A (2019) The nutritional values of two species of sea cucumbers (Holothuria scabra and *Holothuria lessoni*) from Madagascar. Afr J Food Sci 13(11): 281–286. https://doi.org/10.5897/AJFS2019.1816

Li H, Yuan Q, Lv K, Ma H, Gao C, Liu Y, Zhang S, Zhao L (2021) Low-molecular-weight fucosylated glycosaminoglycan and its oligosaccharides from sea cucumber as novel anticoagulants: a review. Carbohydr Polym 1(251):117034. https://doi.org/10.1016/j.carbpol.2020.117034

Li PH, Lu WC, Chan YJ, Ko WC, Jung CC, le Huynh DT, Ji YX (2020) Extraction and characterization of collagen from sea cucumber (*Holothuria cinerascens*) and its potential application in moisturizing cosmetics. Aquaculture 515(734):590. https://doi.org/10.1016/J.AQUACULTURE.2019.734590

Liang Q, Ahmed F, Zhang M, Sperou N, Franco CMM, Feng Q, Zhang W (2022) In vivo and clinical studies of sea cucumber-derived bioactives for human health and nutrition from 2012–2021. Front Mar Sci 9:917857. https://doi.org/10.3389/fmars.2022.917857

Liu Y, Dave D, Trenholm S, Ramakrishnan VV, Murphy W (2021) Effect of drying on nutritional composition of Atlantic sea cucumber (*Cucumaria frondosa*) viscera derived from newfoundland fisheries. Processes *9*(4):703. https://doi.org/10.3390/PR9040703/S1

Mashjoor S, Yousefzadi M (2017) Holothurians antifungal and antibacterial activity to human pathogens in the Persian Gulf. J Med Mycol 27(1):46–56. https://doi.org/10.1016/j.mycmed.2016.08.008

Mashjoor S, Yousefzadi M (2019) Cytotoxic effects of three Persian Gulf species of Holothurians. Iran J Vet Res 20(1):19–26

Masre SF, Yip GW, Sirajudeen KNS, Ghazali FC (2012) Quantitative analysis of sulphated glycosaminoglycans content of Malaysian sea cucumber *Stichopus hermanni* and *Stichopus vastus*. Nat Prod Res 26(7):684–689. https://doi.org/10.1080/14786419.2010.545354

Nahla ESESO (2013) Nutritional value of some Egyptian sea cucumbers. Afr J Biotechnol 12(35): 5466–5472. https://doi.org/10.5897/ajb2013.13020

Pangestuti R, Arifin Z (2018) Medicinal and health benefit effects of functional sea cucumbers. J Tradit Complement Med 8(3):341–351. https://doi.org/10.1016/J.JTCME.2017.06.007

Pangestuti R, Murniasih T, Putra MY, Rasyid A, Wibowo JT, Ardiansyah A, Untari F (2016) Free radical scavenging activity of selected sea cucumber species from Mataram-Lombok, Indonesia. J Teknol 78(4–2):179–185. https://doi.org/10.11113/jt.v78.8202

Parisihni K, Revianti S (2013) Antifungal effect of *Sticophus hermanii* and Holothuria atra extract and its cytotoxicity on gingiva-derived mesenchymal stem cell. Dental J (Majalah Kedokteran Gigi) 46(4):218. https://doi.org/10.20473/j.djmkg.v46.i4.p218-223

Prompoon Y, Weerachatyanukul W, Withyachumnarnkul B, Vanichviriyakit R, Wongprasert K, Asuvapongpatana S (2015) Lectin-based profiling of coelomocytes in *Holothuria scabra* and expression of superoxide dismutase in purified coelomocytes. Zool Sci 32(4):345–351. https://doi.org/10.2108/zs140285

Purcell S, Akamine J, Fabinyi M, Choo PS (2014) Alternative product forms, consumer packaging and extracted derivatives of tropical sea cucumbers. SPC Inf Bull, p 34. https://www.researchgate.net/publication/262563121

Rahael KP, Rahantoknam SPT, Hamid SK (2019) The amino acid of sandfish sea cucumber (*Holothuria scabra*): dry method with various feeding enzyme. J Phys Conf Ser 1424(1): 012005. https://doi.org/10.1088/1742-6596/1424/1/012005

Rasyid A (2018) Amino acid and fatty acid compositions of sea cucumber Stichopus vastus from Salemo Island waters, Indonesia. Squalen Bull Mar Fish Postharvest Biotechnol 13(1):9. https://doi.org/10.15578/squalen.v13i1.285

Rudd M (2020, July 19) Woman who invented the world's first serum made from sea cucumbers has launched a second product. Daily Mail Online. https://www.dailymail.co.uk/femail/article-8537911/Woman-invented-worlds-serum-sea-cucumbers-launched-second-product.html

Saber Mohamed A, Ramadan Fahmy S, Elsayed AA (2020) Formulation and Evaluation of the sea cucumber, *Holothuria arenicola* extract incorporated skin cream. GSC Biol Pharm Sci 2020(02):232–239. https://doi.org/10.30574/gscbps.2020.13.2.0379

Safira A, Rani CA, Puspitasari RA, Ayuningtyas AK, Mahendra YA, Purnomo A (2022) Amino acid and proximate analysis of type-1 collagen from sea cucumber and tilapia-skin and its potential application as artificial tendon. Pharmacogn J 14(4):358–361. https://doi.org/10.5530/pj.2022.14.107

Senadheera TRL, Dave D, Shahidi F (2020) Sea cucumber derived type I collagen: a comprehensive review. Mar Drugs 18(9). https://doi.org/10.3390/MD18090471

Siahaan EA, Pangestuti R, Munandar H, Kim SK (2017) Cosmeceuticals properties of sea cucumbers: prospects and trends. Cosmetics 4(3):26. https://doi.org/10.3390/COSMETICS4030026

Slater M (2015) Use and exploitation of sea cucumbers. In: Brown NP, Eddy SD (eds) Echinoderm aquaculture. Wiley-Blackwell, Hoboken. https://doi.org/10.1002/9781119005810.ch4

Southey F (2020) Farmed sea cucumbers: the new 'superfood' functional ingredient? FoodNavigator. https://www.foodnavigator.com/Article/2020/03/26/Farmed-sea-cucumbers-The-new-superfood-functional-ingredient

Sroyraya M, Hanna PJ, Siangcham T, Tinikul R, Jattujan P, Poomtong T, Sobhon P (2017) Nutritional components of the sea cucumber *Holothuria scabra*. Funct Food Health Dis 7(3): 168–181. https://doi.org/10.31989/ffhd.v7i3.303

Sukmiwati M, Ilza M, Putri AE, Sidauruk SW (2019) Antibacterial activity of sea cucumber (*Holothuria atra*) against Pseudomonas aeruginosa. IOP Conf Ser Earth Environ Sci 404(1): 012047. https://doi.org/10.1088/1755-1315/404/1/012047

Suthar SK, Chundawat NS, Singh GP, Padrón JM, Jhala YK (2022) Quinoxaline: a comprehension of current pharmacological advancement in medicinal chemistry. Eur J Med Chem Rep 5: 100040. https://doi.org/10.1016/j.ejmcr.2022.100040

Taquet C, Setiawan F, Yasuda N, Suharsono, Nadaoka K (2011) First observation of a large group of *Holothuria leucospilota* sea cucumber juveniles at a nursery in Manado (north Sulawesi, Indonesia). SPC Inf Bull 31:30–34

Torres SMPS, Nader HB, Simões RS, Baracat EC, Simões M d J, Fuchs LFP, Soares JM Jr, Gomes RCT (2018) Concentration of sulfated glycosaminoglycans in the mammary tissue of female rats with the aging and about hormonal influence. Gynecol Endocrinol 34(1):64–68. https://doi.org/10.1080/09513590.2017.1336218

Truongs T, Le T (2019) Characterization of six types of dried sea cucumber product from different countries. Int J Food Sci Agric 3(3):220–231. https://doi.org/10.26855/ijfsa.2019.09.011

Yahyav M (2012) Fatty acid composition in two sea cucumber species, *Holothuria scabra* and *Holothuria leucospilata* from Qeshm Island (Persian Gulf). Afr J Biotechnol 11(12). https://doi.org/10.5897/ajb11.3529

Yoon WJ, Kim MJ, Koh HB, Lee WJ, Lee NH, Hyun CG (2010) Effect of Korean red sea cucumber (*Stichopus japonicus*) on melanogenic protein expression in murine B16 melanoma. Int J Pharmacol 6(1):37–42. https://doi.org/10.3923/IJP.2010.37.42

Yu L, Xue C, Chang Y, Xu X, Ge L, Liu G, Wang Y (2014) Structure elucidation of fucoidan composed of a novel tetrafucose repeating unit from sea cucumber *Thelenota ananas*. Food Chem 146:113–119. https://doi.org/10.1016/J.FOODCHEM.2013.09.033

Yun SH, Park ES, Shin SW, Na YW, Han JY, Jeong JS, Shastina VV, Stonik VA, Park JI, Kwak JY (2012) Stichoposide C induces apoptosis through the generation of ceramide in leukemia and colorectal cancer cells and shows in vivo antitumor activity. Clin Cancer Res 18(21): 5934–5948. https://doi.org/10.1158/1078-0432.CCR-12-0655

Marine Biotechnology and Its Applications in Drug Discovery

9

Vidya Sankarapandian, Bhuvaneshwari Jothirajan, Sumithra Pasumalai Arasu, Sumaliny Subramaniam, and Balu Alagar Venmathi Maran

Abstract

The Ocean is the largest biome on our planet Earth with the greatest extremities of temperature and climatic conditions. High ionic concentrations, high pressures and both low and high temperatures define this particular environment. Marine (Blue) biotechnology is the use of biological resources from the ocean for industrial, medical or environmental purposes. Marine Biotechnology can also be defined as a cutting-edge era of science and technology that majorly focuses on marine resources to support living creatures. Due to the ecologically and environmentally diverse sea environment, the planet's most chemically varied ecosystem, marine biotechnology, has been recently known to produce an increasing number of key medicinal goods, industrial and environmental tools and applications for varied research analysis. This book delves deeper into the prospective use of marine biotechnology for drug discovery, including the

V. Sankarapandian
Faculty of Biomedical Science, Department of Microbiology and Immunology, Kampala International University, Kampala, Uganda

Department of Microbiology, Srimad Andavan Arts and Science College, Bharathidasan University, Trichy, India

B. Jothirajan
Department of Microbiology, Srimad Andavan Arts and Science College, Bharathidasan University, Trichy, India

S. P. Arasu
Department of Biotechnology, Srimad Andavan Arts and Science College, Bharathidasan University, Trichy, India

S. Subramaniam · B. A. Venmathi Maran (✉)
Borneo Marine Research Institute, Universiti Malaysia Sabah, Kota Kinabalu, Sabah, Malaysia
e-mail: bavmaran@ums.edu.my

M. D. Shah et al. (eds.), *Marine Biotechnology: Applications in Food, Drugs and Energy*, https://doi.org/10.1007/978-981-99-0624-6_9

development of enzymes, antibiotics and biopolymers, substances derived from various marine sources. In fact, the potential for natural products offered by marine microbial biodiversity appears to be endless and vast expanding as new methods of measurement are being developed. One of the prerequisite factors for successful innovation in exploiting the enormous diversity of marine life is high-quality research in the field of marine biotechnology.

9.1 Introduction

The field of biotechnology has a long, illustrious history. Biotechnology has been engaged by humans since the dawn of civilization (Roehr 2000). The history of biotechnology can be broadly classified into ancient, classical and modern biotechnology. Before the year 1800, there were discoveries or advancements in the field of biotechnology. The core objective of ancient biotechnology were the domestication of food crops, wild animals and fermented products, such as wine, whiskey, beer and other alcoholic beverages. Classical biotechnology dates from 1800 to the middle of the twentieth century when majority of the significant advances in a fundamental understanding of biotechnology emerged (Verma et al. 2011). Some of the most important achievements of classical biotechnology were the process of pasteurization (Pouyan 2019), Mendel's principle of inheritance (Gayon 2016), Koch's pure culture technique and the process of attenuation (Smith 2012). Modern biotechnology which dates from 1900 to 1953 has been crucial due to its immense innovations and discoveries.

Some of the major accomplishments during this era has been the discovery of antibiotics by Alexander Fleming (Tan and Tatsumura 2015), synthetic antibiotics (Aminov 2010), artificial insemination (Ombelet and Van Robays 2015) and the structure of DNA by Watson and Crick (Schindler 2008) (Fig. 9.1). Although biotechnology has been around for thousands of years, the phrase was first used in 1919 by the Hungarian scientist Károly Ereky (Gupta et al. 2016). Biotechnology was later defined by the European Federation of Biotechnology (1981), as the collective application of biochemistry, microbiology and engineering sciences to attain the technological (industrial) application of microorganisms, cultivated tissue cells and aforesaid components (Aguilar et al. 2013). Key areas of biotechnology comprise industrial biotechnology, medical biotechnology, agricultural biotechnology, marine and freshwater biotechnology and pharmaceutical biotechnology. Furthermore, Kafarski (2012) created a colour code to distinguish the different types of biotechnology: white (industrial), green (agricultural), blue (marine and freshwater) and red (research and development, pharmaceutical) (Barcelos et al. 2018). Among these areas, marine biotechnology has been gaining more attention. Marine biotechnology explores the ocean to develop drugs, enzymes, chemicals and other industrial and agricultural products along with their processes. Since the 1980s, modern marine biotechnology has advanced at a breakneck pace. Promising and interesting developments have been made in the fields of genetics, biochemistry, genomics, aquaculture, bioenergy and other related areas. This chapter covers some fundamental knowledge of marine biotechnology and its potential uses in drug discovery.

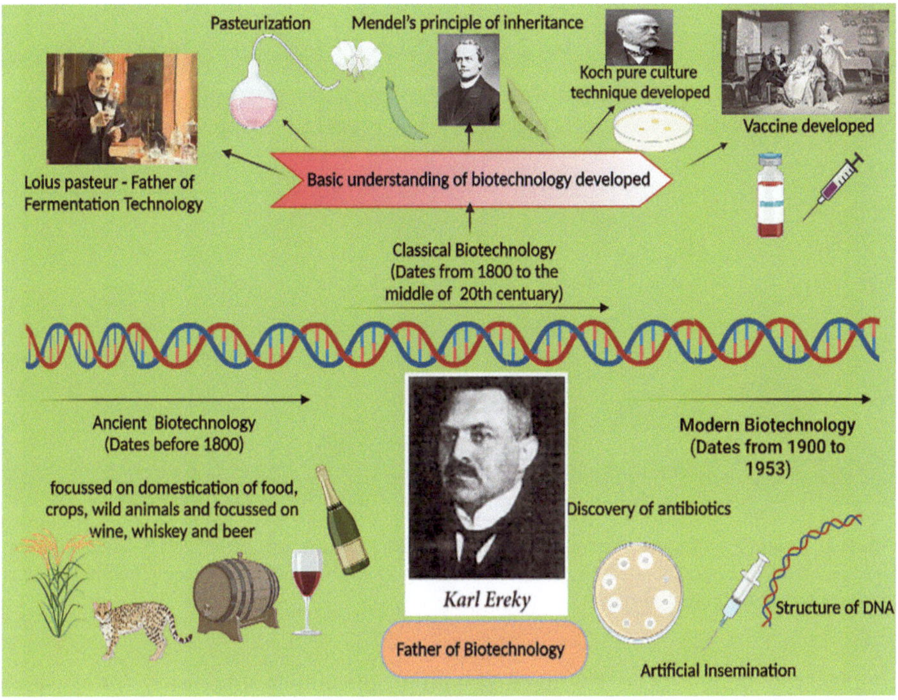

Fig. 9.1 Timeline of the development of biotechnology and important contributions (Created with BioRender.com)

9.2 Marine: Homeland of Diversified Organism

More than 3.8 billion years ago, life first originated in the ocean. Currently, the ocean is home to roughly 235,000 to 250,000 types of marine creatures (Boeuf 2011). Marine life varies in size from microscopic to a blue whale, which may measure up to several metres long (Andersen et al. 2016). Marine life can be classified into different categories: plankton, nekton and benthos (Kankal and Warudkar 2012). Plankton are a vital food source for many sea organisms, both large and tiny creatures. They also play a pivotal role in the ocean's carbon dioxide absorption and oxygen production (Zhang et al. 2017). Plankton are a term that implies 'drifter', which means 'to wander' and is most commonly used to describe the pelagic forms that are transported around by the water's movements rather than by their swimming ability including phytoplankton (plants) and zooplankton (animals) (Tait 1972). Phytoplankton are the primary producers. They photosynthesize in the presence of sunlight and fix carbon dioxide and release oxygen. As a result, they are crucial to preserving both the atmosphere's and the ocean's carbon budget (Irigoien et al. 2004). Zooplankton are animals that live floating in the water. Taxonomically,

Fig. 9.2 Examples of some typical microplankton (20–200 μm, top panel: left to right Diatom, Dinoflagellate, Copepod, Coccolith, Crustacean, Chaetoceros, Cyanobacteria and Chlamydomonas (Created with BioRender.com)

zooplankton ranges from the most primitive unicellular organisms (protists) to vertebrates (fish larvae). Most marine invertebrates and fish that live at the bottom also have a transitory planktonic life, typically the first larval stages of development, and are referred to as 'meroplankton'. This is in contrast to some species that spend their entire lives suspended in water without any contact with solid surfaces (holoplankton) (Ismail and Adnan 2016). Some of the examples of phytoplankton and zooplankton in the marine community are given in Fig. 9.2. The plankton include microbes such as bacteria, cyanobacteria, fungi, algae, protozoa and viruses as well (Trombetta et al. 2020). Microbial ecosystems dominate the oceans. More than 90% of the biomass in the marine environment is made up of microscopic organisms, including most of the primary producers, or photosynthetic organisms, that serve as the foundation of ocean food webs (Dipper 2022).

Nektons (swimmers) are living organisms that can swim and move without being affected by currents. Fish, squid, octopus, sharks and marine animals are examples of nekton, which are heterotrophic and these come in a wide range of sizes. While most nekton are mainly pelagic, or living in the water column, some are demersal or living at the bottom, and they can be found in both oceanic and coastal environments (Brodeur et al. 2019). The benthic zone refers to the organisms that make up the community of benthos. These are often known as the 'seabed' which are made up of entities that dwell on, in, or near the seafloor. The benthos community live in or near marine sedimentary environments, from tidal pools along the foreshore, out to the continental shelf and down to the abyssal depths. Most of the organisms in the benthic zone are scavengers or detritivores (Kroeker et al. 2011).

9.3 Marine: Homeland of Unique Metabolites

To cope with the significant fluctuations in pressure, salinity, temperature and pH prevailing in their environment, marine organisms develop novel metabolites or chemicals, and the compounds produced are unique in terms of diversity, structural and functional properties (Kandasamy et al. 2008). Bioactive compounds are abundantly found in marine creatures such as marine microbes, sponges, tunicates, fishes, soft corals, nudibranchs, sea hares, opisthobranch molluscs, echinoderms, bryozoans, prawns, shells and sea slugs (Donia and Hamann 2003). According to the reports of Carte (1996), many biologically active natural compounds that are found in the marine environment, are not found in terrestrial sources. Bergmann was the first scientist to report the first biologically active marine natural product in the late 1950s (Malve 2016). He disclosed the discovery of the first bioactive compounds that were extracted from the marine Caribbean sponge (Bergmann and Stempien 1957). Since then, several biologically active compounds with varying degrees of activity have been isolated from marine sources. These substances include those with anti-tumour, anti-cancer, anti-microtubule, anti-proliferative, cytotoxic, photoprotective, anti-oxidant, anti-fungal, antiviral, anti-bacterial, anti-parasitic, antibiotic and anti-fouling properties (Bhatnagar and Kim 2010). According to Blunt et al. (2004), sponges (37%), coelenterates (21%), microbes (18%), echinoderms (6%), tunicates (6%), red algae (5%), molluscs (2%), green algae (2%), brown algae (2%) and bryozoans (1%) are the main sources of marine biomedical compounds (Blunt et al. 2004) (Fig. 9.3).

9.4 Biotechnology: A Tool for Exploring Marine Drugs

To find a potential drug candidate, it is crucial to integrate technologies and processes from the beginning to the launch of biotherapeutic pharmaceuticals. A new drug development process can take anywhere from 12 to 15 years to develop (Hughes et al. 2011). Biotechnology helps the pharmaceutical industry to develop new products, new processes, methods and services to improve existing ones. Biotechnological methods have become a significant tool in pharmaceutical drug research and development. Today approximately 15% of drug revenues are derived from biopharmaceuticals (Gaisser and Nusser 2010). By integrating marine and biotechnology, the discovery of a new product from unexplored marine living organisms could be achieved. For discovering new drugs, new extraction protocols to identify bioactive ingredients and integration of different biological streams are imperative. The first biotechnology-based drug was given FDA (The Food and Drug Administration) approval in 1982. Recombinant insulin made from specially engineered bacteria manufactured by Genentech is known as 'Humulin', and it was the first ever formulated drug that was created. It was also one of the first-ever genetically altered products made available to consumers, the first medication that was created using recombinant DNA technology (Stryjewska et al. 2013). The marine biosphere is the world's most important biosphere, and its living conditions

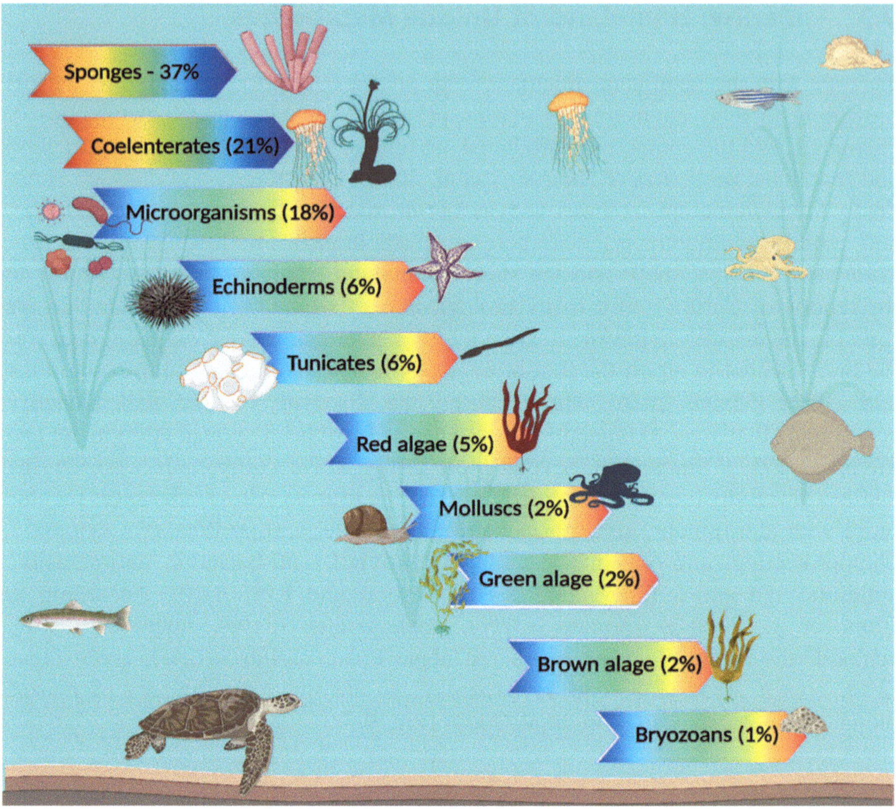

Fig. 9.3 The percentage of marine organisms that contribute to biomedical compounds (Created with BioRender.com)

differ significantly from those on land. One of the most important survival strategies used by marine species is the production of specialized secondary metabolites. These metabolites exhibit biological action, making them ideal for use in human medical treatments. (Lindequist 2016). According to Mayer et al. (2013), 102 marine natural compounds have been shown to exhibit anti-bacterial, anti-fungal, anti-protozoan, anti-tuberculosis and antiviral pharmacological properties (Mayer et al. 2013).

9.5 Marine Bioactive Compounds Used as Drugs

Modern Culture and Lifestyle has increased demand for a diversity of foods, improved health and well-being, novel biomedicines, natural cosmeceuticals, environmental conservation and sustainable energy sources. These socioeconomic needs piqued researchers' interest in marine ecosystems as viable and sustainable sources of biomolecules and biomass, and hence, the ever-growing area of marine (blue) biotechnology was formed to address them (Rotter et al. 2021). Carbohydrates are diversified bioactive compounds that are utilized as drugs in the field of marine biotechnology. This includes fucans, fucanoids, chitin, chitosan, glutathione, alginates, ascophyllan, laminarin, polyuronides, carrageenans and agar-agar (Ghosh et al. 2022). Proteins and peptides: collagen, gelatin and albumin. Fatty acids: omega-3 fatty acids (Ghosh et al. 2022). Phenolic compounds: sinapic acid, catechin, myricetin, kaempferol, protocatechuic acid, vanillic acid, coumaric acid, ferulic acid and rutin (Santana-Casiano et al. 2010). Pigments include β-Cryptoxanthin, α-Carotene, β-Carotene, γ-Carotene (Pereira et al. 2014), chlorophyll a, chlorophyll c, fucoxanthin, xanthophylls, diatoxanthin, diadinoxanthin, violaxanthin, antheraxanthin and zeaxanthin (Kuczynska et al. 2015). Enzymes include protease, lipase, chitinase, chitosanase, Alginate lyases, agarases, carrageenases, cellulose, hemicellulose hydrolase and amylases (Zhang and Kim 2010). These bioactive compounds have potential effects on various diseases and auto-immune conditions. Apart from these bioactive compounds, probiotics from the marine environment have also been used as drugs (Sankarapandian et al. 2022). Marine probiotics are also effective against various diseases such as dermatitis, inflammation, halitosis, diarrhoea, irritable bowel syndrome, hypercholesterolemia, obesity, urogenital infections and cancers (Sankarapandian et al. 2022).

9.6 Marine Biotechnology Drugs Against Various Diseases

Despite great advances in medical science, a handful of diseases continue to pose a threat to the human race. The World Health Organization (WHO), consider the following diseases as the top ten diseases that accounted for 55% of the 55.4 million deaths worldwide. The diseases are heart diseases, stroke, pulmonary diseases, lower respiratory infections, cancer, Alzheimer's diseases, diarrhoeal diseases, diabetes and kidney diseases. Marine population serve as repositories for novel bioactive metabolites with a variety of chemical structures that are useful against a wide range of ailments (Donia and Hamann 2003). The following are the different diseases and their related marine drugs that were developed through biotechnology in the last two decades.

9.6.1 Marine Drugs as Anti-cancer Agents

Cancer is one of the foremost concerns for the public health systems worldwide. As per the American Cancer Society, cancer is the second leading cause of death worldwide, with 1,898,160 new cancer cases and 608,570 cancer deaths expected in the United States in 2021. The total number of cancer fatalities prevented due to the drop in cancer mortality during the early 1990s is large, thanks to advances in diagnostic techniques and the identification of viable treatments (Siegel et al. 2021). Trabectedin (Fig. 9.4) was the first marine anti-cancer treatment approved in the European Union for patients with soft tissue sarcoma in the 1960s. It is a semisynthetic tetrahydroisoquinoline alkaloid produced from the marine tunicate *Ecteinascidia turbinata* (D'Incalci and Galmarini 2010).

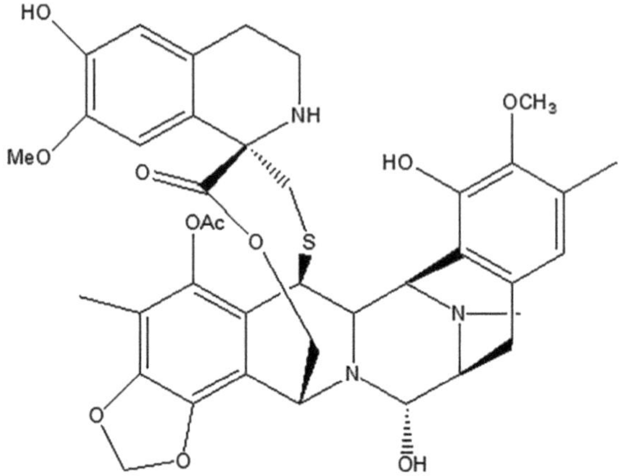

Fig. 9.4 Structure of Trabectedin—the first marine-derived anti-cancer-approved drug

Once after the approval of trabectedin, scientists all over the world focused on marine-derived drugs for their anti-cancer potential. The following is a list of anti-cancer chemicals obtained from marine sources. Some of these recent drugs discovered using marine biotechnology are Kahalalide F, Xinghaiamine A, Fucoxanthin, Bryostatin 1, Halichondrin B, Philinopside A, Philinopside E, Sansalvamide A and Dolastatin (Table 9.1) (Rastogi et al. 2016).

9.6.2 Marine Drugs for Cardiovascular Diseases

According to the World Health Organization, 17.9 million people worldwide die from cardiovascular diseases (CVDs) each year, making them the leading cause of mortality. Coronary heart disease, cerebrovascular illness, rheumatic heart disease and other heart and blood vessel abnormalities are all classified as cardiovascular diseases. Heart attacks and strokes account for more than four out of every five CVD deaths, with one-third of these deaths occurring before the age of 70. Even while the quality of life for people with cardiovascular illnesses is improved by currently existing medications, they tend to experience some side effects that call for the creation of novel lipid-lowering therapies (Chhetry and Jialal 2022). Consequently, marine-extracted compounds are in focus and it has been proved that marine compounds have hypolipidemic activities with less or no side effects (Zhao et al. 2020). Coronary artery disease (CAD) is another disease that causes death worldwide (Liang et al. 2021). Moreover, Liang et al listed the following marine-derived compounds for their anti-coronary activity. Fucoxanthin, saponins, astaxanthin, xyloketal B, DSW, terpenes, benzoic acid derivatives and asperlin are examples of organic small molecules. Fucoidan, alginate, ulvan and chitosan are polysaccharides, while proteins, bioactive peptides and lipid compounds are also included (Liang et al. 2021). The chemicals extracted from the marine environment work predominantly by decreasing cholesterol and lipid levels (Kumar et al. 2017), acting as anti-

Table 9.1 Anti-cancer drugs isolated from the marine source and their applications

Compound name	Compound structure	Source	Type of cancer	Mechanism of action	References
Kahalalide F	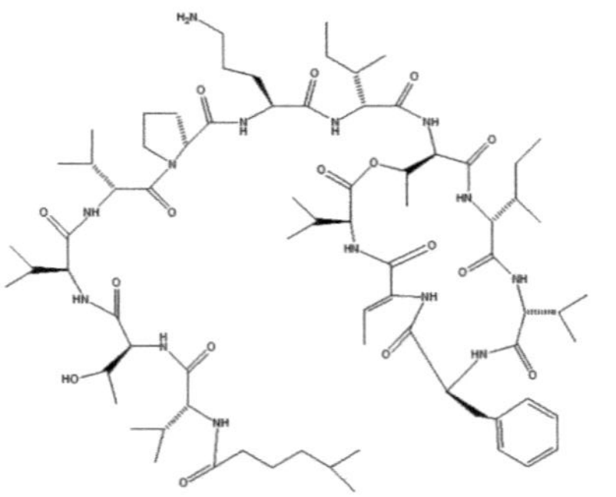	*Elysia rufescens* (sea slug, a marine gastropod mollusc)	Melanoma, prostate, breast and colon cancers	Induces cytotoxicity and blocks the cell cycle in G1 (growth) phase.	Kandasamy et al. (2008) Faircloth and del Carmen Cuevas Marchante (2006)
Xinghaiamine A	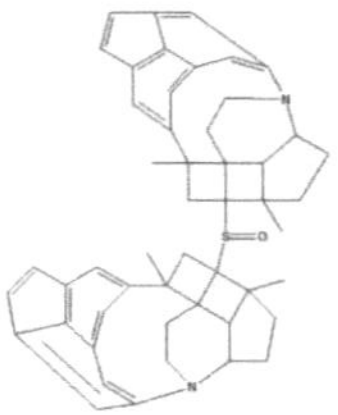	Marine—Actinomycete *Streptomyces xinghaiensis*	Breast cancer, lung cancer	–	Jiao et al. (2013)
Fucoxanthin	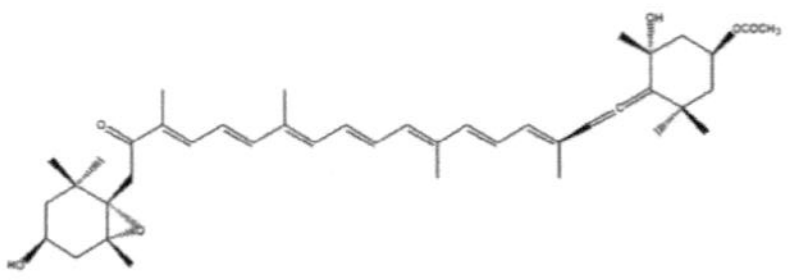	Marine microalgae *Phaeodactylum tricornutum*	Leukaemia, B-cell malignancies	Induces cell growth arrest, apoptosis and/or autophagy	Méresse et al. (2020) Zhang et al. (2020)

(continued)

Table 9.1 (continued)

Compound name	Compound structure	Source	Type of cancer	Mechanism of action	References
Bryostatin 1	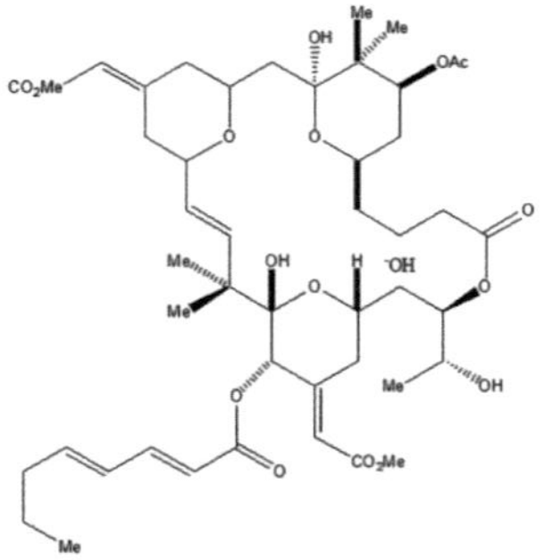	Marine bryozoan—*Bigula neritina*	Renal cancer	Reduction of IL-6 which leads to de-promotion of angiogenesis and cancer progression	Madhusudan et al. (2003)
Halichondrin B	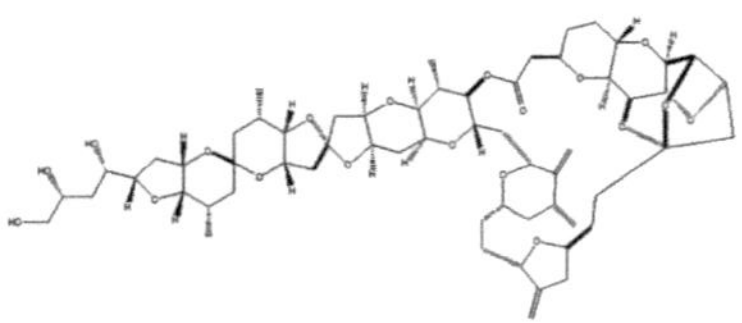	Marine sponge—*Halichondria okadai*	Breast cancer	G2/M cell-cycle block, disruption of mitotic spindles	Nakao and Fusetani (2010)
Philinopside A	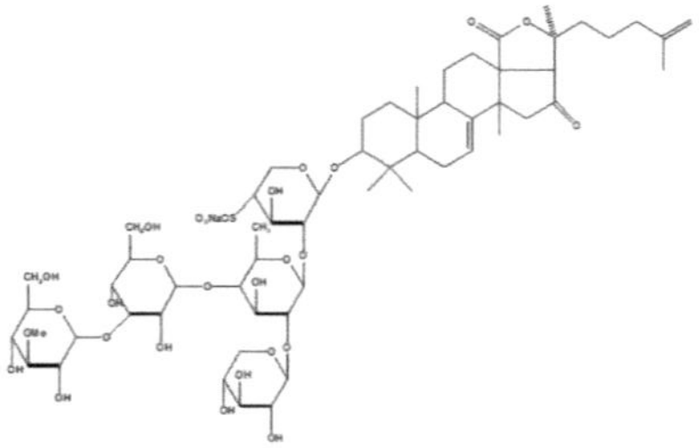	Sea cucumber *Pentacta quadrangulari*	Sarcoma	Prevents angiogenesis	Tong et al. (2005)

Dolastanin 10	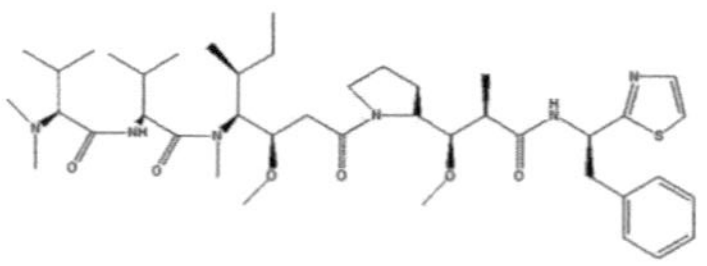	*Dolabella auricularia*	Lymphoma, Lung cancer Human prostate cancer	Inhibition of tubulin polymerization	Turner et al. (1998)
Sansalvamide A	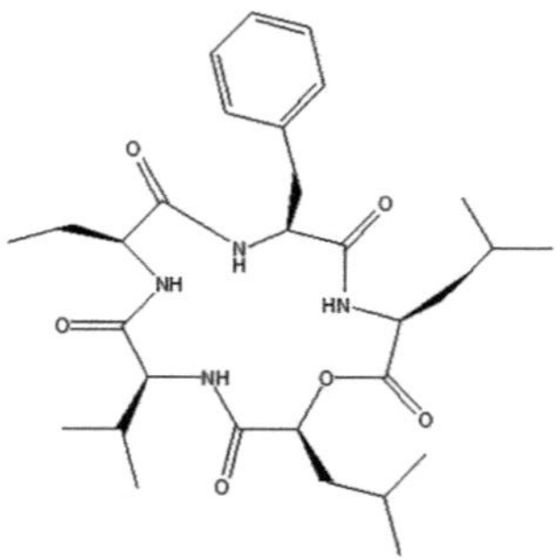	Marine fungus—*Fusarium*	Pancreatic cancer	G1/M cell-cycle block	Heiferman et al. (2010)
Philinopside E	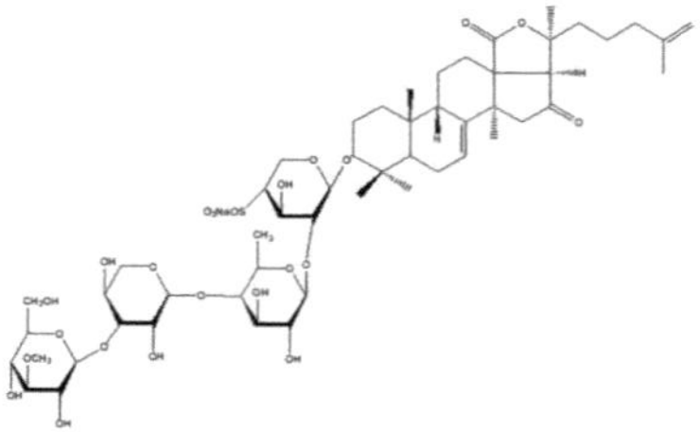	Sea cucumber *Pentacta quadrangulari*	Sarcoma	Induction of apoptosis, inhibition of signalling of vascular endothelial growth factor receptor-2 (VEGFR-2)	Aminin et al. (2015) Li et al. (2013)

platelets (Lindequist 2016), anti-coagulants (Xu et al. 2019), anti-inflammatory and anti-oxidant activities (Qiu et al. 2021).

Furthermore, multiple studies have shown that omega-6 (linoleic acid, -linolenic acid and arachidonic acid) and omega-3 (-linolenic acid, eicosapentaenoic acid and docosahexaenoic acid) fatty acids reduce cardiac risk factors. Marine fishes are the main source of omega-3 (ω3 or n-3), PUFAs (Poly-Unsaturated Fatty Acids) compared to red meat (beef, pork and veal) and vegetable oils (Olgunoglu 2017). Particularly ω3 Fatty acids benefit multiple risk factors including high blood lipids and high blood pressure, and in addition to these, they exhibit anti-inflammatory, anti-thrombotic and anti-oxidative actions (Manson et al. 2019). Most research suggests that omega-3 fatty acids inhibit very low-density lipoprotein (VLDL) particle formation and secretion, thereby increasing TG removal from VLDL and chylomicron particles by upregulating enzymes such as lipoprotein lipase (Bays et al. 2008). The Omega-6 polyunsaturated fatty acids are also considered a broad cholesterol-lowering agent (o Bazinet and Chu 2014).

Probiotics, prebiotics and synbiotics, in addition to the aforementioned chemicals, may act as useful dietary components in the prevention (particularly) and treatment of CVD (Ooi and Liong 2010; Kim et al. 2012). Anti-hypercholesterolemia probiotics could be used as an alternative therapy to minimize the risk of CVD and other disorders (Sivamaruthi et al. 2021). Some of the examples of marine probiotic studies so far include bacterial *Vibrio* spp., *Pseudomonas* spp., *Bacillus* spp. and many other *Lactobacilli* spp. Although there are several probiotics available in the marine environment, marine probiotics vs. cardiac diseases-related research still need attention. Another common heart disease is rheumatic heart disease (RHD), and it is an important determinant of global cardiovascular morbidity and mortality (Auala et al. 2022). The development of a marine-related drug for this condition is still in its infancy (Fig. 9.5).

9.6.3 Marine Drugs for Diabetes

Diabetes is a chronic metabolic disorder marked by high blood glucose (or blood sugar) levels, that can cause debilitating effects on the heart, blood vessels, eyes, kidneys and nerves over a period of time. The most prevalent type is type 2 diabetes, which affects adults and arises when the body becomes insulin resistant or produces insufficient insulin (Mukhtar et al. 2020). Several aquatic creatures, including bacteria, microalgae, macroalgae, seagrasses, sponges, corals, sea anemones and fish, have been evaluated for prospective anti-diabetic effects. Anti-hyperglycemic and anti-diabetic properties of marine species have been tested using both in vitro and in vivo screening (Lauritano and Ianora 2016). Marine algae are the best source of anti-diabetic compounds. Diverse marine algae biosynthetic pathways produce bioactive secondary metabolites, which contribute to a range of chemical and biological properties. Phlorotannins found in marine brown algae has been known to have anti-diabetic properties through a variety of mechanisms, including the inhibition of enzyme targets such as α-amylase, α-glucosidase, angiotensin-

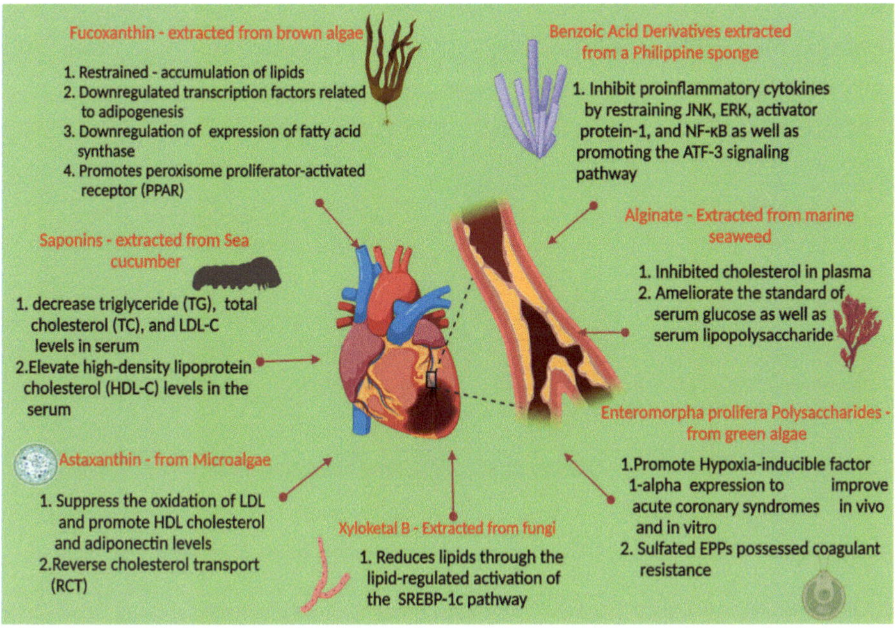

Fig. 9.5 Marine compounds and their mechanism of action in preventing coronary artery disease (Created with BioRender.com)

converting enzymes (ACE), aldose reductase, dipeptidyl peptidase-4 and protein tyrosine phosphatase 1B (PTP 1B) enzyme (Table 9.2) (Lee and Jeon 2013).

9.6.4 Marine Drugs for Malaria

Malaria is an infectious disease caused by Protozoa of the genus *Plasmodium* (*Plasmodium malariae, P. falciparum, P. ovale* and *P. vivax*), with *P. falciparum* causing the most severe and fatal cases. The protozoan infects humans through vector-borne transmission by female mosquitoes of the genus *Anopheles*, which further affects the red blood cells and causes anaemia. The resulting rupture of infected erythrocytes is associated with the release of cell debris into hemozoin into the bloodstream, which causes typical fever spike patterns (Fattorusso and Taglialatela-Scafati 2009). Several marine compounds have been isolated from different marine sources having anti-malarial activity. Higa and co-workers isolated the first malarial drug, 'Manzamines', from a marine source in 1986 from an Okinawan sponge *Haliclona* (Fig. 9.6).

The drug had a high potency and was composed of polycyclic compounds with seven to eight rings or more. Jha and Zi-Rong (2004) extracted more than 300 chemical analogues from the pacific sponge, with approximately 300 chemical analogues being tested in clinical trials as anti-malarial medicines. Manzamines is a highly

Table 9.2 Anti-diabetes drugs isolated from the marine source and their mechanism of action

Compound name	Source	Mechanism of action	References
Phlorotannins (eckol, dieckol, 6,6′-bieckol, phlorofucofuroeckol-A, phloroglucinol and 7-phloroeckol)	*Ecklonia* (brown seaweeds)	Inhibitory action of α-amylase and α-glucosidase due to the presence of different phlorotannins	Lee and Jeon (2013)
Phlorotannins, phloroglucinol, dioxinodehydroeckol, eckol, phlorofurofucoeckol-A, dieckol and 7-phloroeckol	*Eisenia* (brown seaweeds)	Inhibiting α-amylase and α-glucosidase enzymes	Moon et al. (2011)
Phloroglucinol, diphlorethohydroxycarmalol, 6-6-bieckol, octaphlorethol A and ishophloroglucin	*Ishige okamurae* (Brown algae)	Inhibiting α-amylase and α-glucosidase enzymes	Yang et al. (2019)
Phlorotannins (7-phloroeckol and 2-phloroeckol)	*Ecklonia stolonifera*	Inhibiting aldose reductase	Lee and Jeon (2013)
Pheophorbide-A	*Saccharina japonica*	Inhibiting aldose reductase	Jung et al. (2013)
2,3,6-tribromo-4,5-dihydroxybenzyl	*Symphyocladia latiuscula*	Inhibition of tyrosine phosphatase 1B (PTP1B) and α-glucosidase	Paudel et al. (2019)
Bromophenols, 2,4,6-tribromophenol and 2,4-dibromophenol	*Grateloupia elliptica*	High α-glucosidase inhibitory activity	Lauritano and Ianora (2016)

effective drug that inhibits more than 90% of the asexual erythrocytic stages of *P. berghei* after a single intraperitoneal injection. Manzamine has a unique feature, which is its capacity to prolong the survival of extremely parasitemic mice, with a 40% recovery in 60 days after a single injection, which qualifies it as a treatment. Oral administration of a manzamine A oil suspension resulted in a considerable decrease in parasitemia (Fig. 9.7) (Ang et al. 2000).

Due to the emergence of the chloroquine-resistant strain of *P. falciparum*, marine novel drugs are an urgent necessity. Two compounds diterpene and bebrycin A were isolated from octocorals *Bebryce grandis* and *Spongia lamella,* respectively found to be potent against the chloroquine-resistant strain (Wright et al. 2021).

9.7 Future Perspective

Marine biotechnological progressions lead to several achievements in the fields of human health, medicine, drugs, pharmacology, fisheries, bioremediation, food, cosmetics industries and agriculture. These have been major accomplishments, nonetheless, the field of marine biotechnology is still in its infancy. The Ocean,

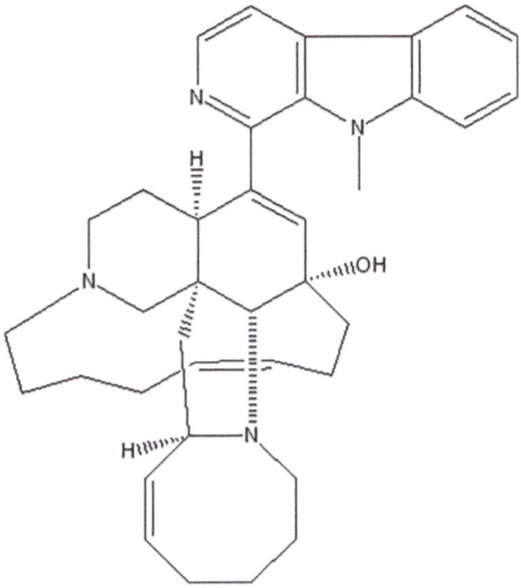

Fig. 9.6 Manzamines—the first malarial drug isolated from an Okinawan *Haliclona*

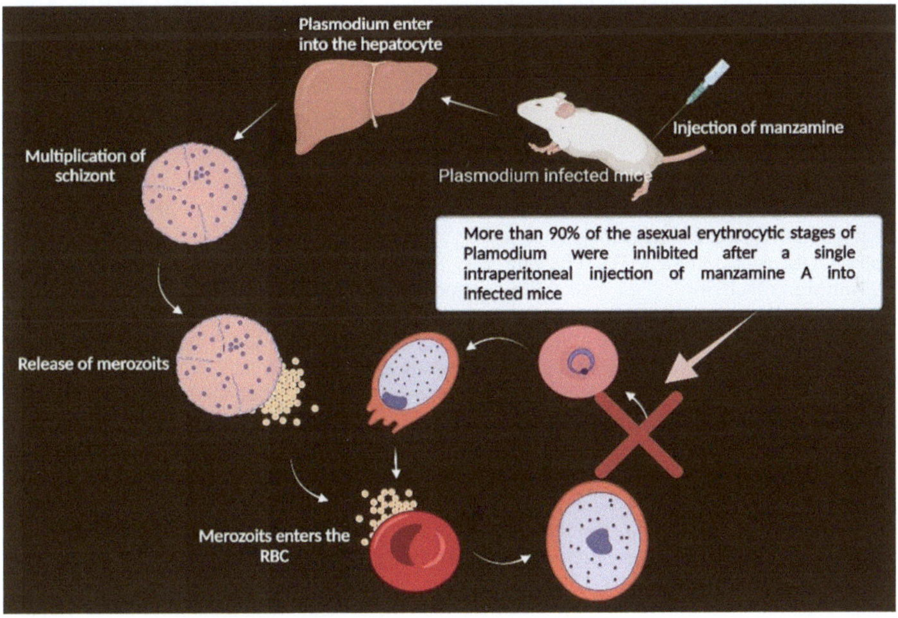

Fig. 9.7 Mechanism of action of manzamine A in controlling a malarial parasite. More than 90% of the asexual erythrocytic stages of Plasmodium were inhibited after a single intraperitoneal injection of manzamine A into the injected mice (Created with BioRender.com)

according to research, looks to be an excellent source of marine bioactive compounds due to enormous marine biodiversity and contains more phyla and classes than land and fresh waters. Thus, the marine habitat is the homeland for an array of biochemical and metabolic pathways. A comprehensive approach is essential to catalyse the expansion of marine biotechnology worldwide and to finally harvest the products of this promising field of research. Besides, interdisciplinary connections and collaborations will be a priority to unleash marine resources.

9.8 Conclusion

Biotechnology enhances the potential of marine biological resources and is a tool to unlock the unleashed potential of marine resources. Although several products like drugs, food, bioenergy, nanomaterials and bioremediation are delivered through marine biotechnology, less than 5% of our vast oceanic environment has been explored. Marine biotechnology is a rapidly growing enterprise that can explore new drugs and medicines every day if it is properly analysed and focused.

Acknowledgement Thanks to the support provided by project no. LPA2007, MABIK, Korea to BAVM.

References

Aguilar A, Magnien E, Thomas D (2013) Thirty years of European biotechnology programmes: from biomolecular engineering to the bioeconomy. New Biotechnol 30(5):410–425. https://doi.org/10.1016/j.nbt.2012.11.014

Aminin DL, Menchinskaya ES, Pisliagin EA, Silchenko AS, Avilov SA, Kalinin VI (2015) Anticancer activity of sea cucumber triterpene glycosides. Mar Drugs 13(3):1202–1223. https://doi.org/10.3390/md13031202

Aminov RI (2010) A brief history of the antibiotic era: lessons learned and challenges for the future. Front Microbiol 1:134. https://doi.org/10.3389/fmicb.2010.00134

Andersen KH, Berge T, Gonçalves RJ, Hartvig M, Heuschele J, Hylander S, Jacobsen NS, Lindemann C, Martens EA, Neuheimer AB, Olsson K, Palacz A, Prowe AEF, Sainmont J, Traving SJ, Visser AW, Wadhwa N, Kiørboe T (2016) Characteristic sizes of life in the oceans, from bacteria to whales. Annu Rev Mar Sci 8:217–241. https://doi.org/10.1146/annurev-marine-122414-034144

Ang KK, Holmes MJ, Higa T, Hamann MT, Kara UA (2000) In vivo antimalarial activity of the beta-carboline alkaloid manzamine A. Antimicrob Agents Chemother 44(6):1645–1649. https://pubmed.ncbi.nlm.nih.gov/10817722/

Auala T, Zavale BG, Mbakwem AÇ, Mocumbi AO (2022) Acute rheumatic fever and rheumatic heart disease: highlighting the role of group a streptococcus in the global burden of cardiovascular disease. Pathogens 11(5):496. https://doi.org/10.3390/pathogens11050496

Barcelos MCS, Lupki FB, Campolina GA, Nelson DL, Molina G (2018) The colors of biotechnology: general overview and developments of white, green and blue areas. FEMS Microbiol Lett 365(21). https://doi.org/10.1093/femsle/fny239

Bays HE, Tighe AP, Sadovsky R, Davidson MH (2008) Prescription omega-3 fatty acids and their lipid effects: physiologic mechanisms of action and clinical implications. Expert Rev Cardiovasc Ther 6(3):391–409. https://doi.org/10.1586/14779072.6.3.391

Bazinet RP, Chu MWA (2014) Omega-6 polyunsaturated fatty acids: is a broad cholesterol-lowering health claim appropriate? Can Med Assoc J 186(6):434–439. https://doi.org/10.1503/cmaj.130253

Bergmann W, Stempien MF (1957) Contributions to the study of marine products. XLIII. The nucleosides of sponges. V. the synthesis of spongosine1. J Org Chem 22(12):1575–1577. https://doi.org/10.1021/jo01363a009

Bhatnagar I, Kim S-K (2010) Immense essence of excellence: marine microbial bioactive compounds. Mar Drugs 8(10):2673–2701. https://doi.org/10.3390/md8102673

Blunt JW, Copp BR, Munro MHG, Northcote PT, Prinsep MR (2004) Marine natural products. Nat Prod Rep 21(1):1–49. https://doi.org/10.1039/B305250H

Boeuf G (2011) Marine biodiversity characteristics. C R Biol 334(5):435–440. https://doi.org/10.1016/j.crvi.2011.02.009

Brodeur RD, Auth TD, Phillips AJ (2019) Major shifts in pelagic micronekton and macrozooplankton community structure in an upwelling ecosystem related to an unprecedented marine heatwave. Front Mar Sci 6:212

Carte BK (1996) Biomedical potential of marine natural products. Bioscience 46(4):271–286. https://www.jstor.org/stable/1312834

Chhetry M, Jialal I (2022) Lipid lowering drug therapy. In: StatPearls. StatPearls Publishing. https://www.ncbi.nlm.nih.gov/books/NBK541128/

D'Incalci M, Galmarini CM (2010) A review of trabectedin (ET-743): a unique mechanism of action. Mol Cancer Ther 9(8):2157–2163. https://doi.org/10.1158/1535-7163.MCT-10-0263

Dipper F (2022) Chapter 4 - Open water lifestyles: marine plankton. In: Dipper F (ed) Elements of marine ecology, 5th edn. Butterworth-Heinemann, pp 193–228. https://doi.org/10.1016/B978-0-08-102826-1.00005-3

Donia M, Hamann MT (2003) Marine natural products and their potential applications as anti-infective agents. Lancet Infect Dis 3(6):338–348. https://doi.org/10.1016/S1473-3099(03)00655-8

Faircloth G, del Carmen Cuevas Marchante M (2006) Kahalalide F and ES285: potent anticancer agents from marine molluscs. In: Cimino G, Gavagnin M (eds) Molluscs: from chemo-ecological study to biotechnological application. Springer, pp 363–379. https://doi.org/10.1007/978-3-540-30880-5_16

Fattorusso E, Taglialatela-Scafati O (2009) Marine antimalarials. Mar Drugs 7(2):130–152. https://doi.org/10.3390/md7020130

Gaisser S, Nusser M (2010) The role of biotechnology in pharmaceutical drug design. Zeitschrift Fur Evidenz, Fortbildung Und Qualitat Im Gesundheitswesen 104(10):732–737. https://doi.org/10.1016/j.zefq.2010.05.001

Gayon J (2016) From Mendel to epigenetics: history of genetics. C R Biol 339:225. https://doi.org/10.1016/j.crvi.2016.05.009

Ghosh S, Sarkar T, Pati S, Kari ZA, Edinur HA, Chakraborty R (2022) Novel bioactive compounds from marine sources as a tool for functional food development. Front Mar Sci 9. https://doi.org/10.3389/fmars.2022.832957

Gupta V, Sengupta M, Prakash J, Tripathy BC (2016) An introduction to biotechnology. Basic Appl Aspects Biotechnol 1–21. https://doi.org/10.1007/978-981-10-0875-7_1

Heiferman MJ, Salabat MR, Ujiki MB, Strouch MJ, Cheon EC, Silverman RB, Bentrem DJ (2010) Sansalvamide induces pancreatic cancer growth arrest through changes in the cell cycle. Anticancer Res 30(1):73–78

Hughes J, Rees S, Kalindjian S, Philpott K (2011) Principles of early drug discovery. Br J Pharmacol 162(6):1239–1249. https://doi.org/10.1111/j.1476-5381.2010.01127.x

Irigoien X, Huisman J, Harris RP (2004) Global biodiversity patterns of marine phytoplankton and zooplankton. Nature 429(6994):863–867. https://doi.org/10.1038/nature02593

Ismail AH, Adnan AAM (2016) Zooplankton composition and abundance as indicators of eutrophication in two small man-made lakes. Trop Life Sci Res 27(Supp 1):31–38. https://doi.org/10.21315/tlsr2016.27.3.5

Jha RK, Zi-Rong X (2004) Biomedical compounds from marine organisms. Mar Drugs 2(3): 123–146. https://www.ncbi.nlm.nih.gov/pmc/articles/PMC3783866/

Jiao W, Zhang F, Zhao X, Hu J, Suh J-W (2013) A novel alkaloid from marine-derived actinomycete streptomyces xinghaiensis with broad-spectrum antibacterial and cytotoxic activities. PLoS One 8(10):e75994. https://doi.org/10.1371/journal.pone.0075994

Jung HA, Islam MN, Lee CM, Oh SH, Lee S, Jung JH, Choi JS (2013) Kinetics and molecular docking studies of an anti-diabetic complication inhibitor fucosterol from edible brown algae Eisenia bicyclis and Ecklonia stolonifera. Chem Biol Interact 206(1):55–62. https://doi.org/10.1016/j.cbi.2013.08.013

Kafarski P (2012) Rainbow code of biotechnology. Chemik 66(8):811–816. https://www.researchgate.net/publication/287253802_Rainbow_code_of_biotechnology

Kandasamy K, Nabeel MA, Manivannan S (2008) Bioprospecting of marine organisms for novel bioactive compounds. J Sci Trans Environ Technol 1:107–120. https://doi.org/10.20894/STET.116.001.003.001

Kankal NC, Warudkar S (2012) Biodiversity of phytoplankton, zooplankton and zoobenthos in east coast, bay of Bengal near Nellore, Andhra Pradesh (INDIA) 1, 14

Kim S-K, Bhatnagar I, Kang K-H (2012) Development of marine probiotics: prospects and approach. Adv Food Nutr Res 65:353–362. https://doi.org/10.1016/B978-0-12-416003-3.00023-8

Kroeker KJ, Micheli F, Gambi MC, Martz TR (2011) Divergent ecosystem responses within a benthic marine community to ocean acidification. Proc Natl Acad Sci U S A 108(35): 14515–14520. https://doi.org/10.1073/pnas.1107789108

Kuczynska P, Jemiola-Rzeminska M, Strzalka K (2015) Photosynthetic pigments in diatoms. Mar Drugs 13(9):5847–5881. https://doi.org/10.3390/md13095847

Kumar R, Salwe KJ, Kumarappan M (2017) Evaluation of antioxidant, hypolipidemic, and antiatherogenic property of lycopene and astaxanthin in atherosclerosis-induced rats. Pharm Res 9(2):161–167. https://doi.org/10.4103/0974-8490.204654

Lauritano C, Ianora A (2016) Marine organisms with anti-diabetes properties. Mar Drugs 14(12): 220. https://doi.org/10.3390/md14120220

Lee S-H, Jeon Y-J (2013) Anti-diabetic effects of brown algae derived phlorotannins, marine polyphenols through diverse mechanisms. Fitoterapia 86:129–136. https://doi.org/10.1016/j.fitote.2013.02.013

Li Y-X, Himaya SWA, Kim S-K (2013) Triterpenoids of marine origin as anti-cancer agents. Molecules (Basel, Switzerland) 18:7886–7909. https://doi.org/10.3390/molecules18077886

Liang B, Cai X-Y, Gu N (2021) Marine natural products and coronary artery disease. Front Cardiovascular Med 8:739932. https://doi.org/10.3389/fcvm.2021.739932

Lindequist U (2016) Marine-derived pharmaceuticals – challenges and opportunities. Biomol Ther 24(6):561–571. https://doi.org/10.4062/biomolther.2016.181

Madhusudan S, Protheroe A, Propper D, Han C, Corrie P, Earl H, Hancock B, Vasey P, Turner A, Balkwill F, Hoare S, Harris AL (2003) A multicentre phase II trial of bryostatin-1 in patients with advanced renal cancer. Br J Cancer 89(8):1418–1422. https://doi.org/10.1038/sj.bjc.6601321

Malve H (2016) Exploring the ocean for new drug developments: marine pharmacology. J Pharm Bioallied Sci 8(2):83–91. https://doi.org/10.4103/0975-7406.171700

Manson JE, Cook NR, Lee I-M, Christen W, Bassuk SS, Mora S, Gibson H, Albert CM, Gordon D, Copeland T, D'Agostino D, Friedenberg G, Ridge C, Bubes V, Giovannucci EL, Willett WC, Buring JE (2019) Marine n−3 fatty acids and prevention of cardiovascular disease and cancer. N Engl J Med 380(1):23–32. https://doi.org/10.1056/NEJMoa1811403

Mayer AMS, Rodríguez AD, Taglialatela-Scafati O, Fusetani N (2013) Marine pharmacology in 2009–2011: marine compounds with antibacterial, antidiabetic, antifungal, anti-inflammatory, antiprotozoal, antituberculosis, and antiviral activities; affecting the immune and nervous systems, and other miscellaneous mechanisms of action. Mar Drugs 11(7):2510–2573. https://doi.org/10.3390/md11072510

Méresse S, Fodil M, Fleury F, Chénais B (2020) Fucoxanthin, a marine-derived carotenoid from brown seaweeds and microalgae: a promising bioactive compound for cancer therapy. Int J Mol Sci 21(23):9273. https://doi.org/10.3390/ijms21239273

Moon HE, Islam MN, Ahn BR, Chowdhury SS, Sohn HS, Jung HA, Choi JS (2011) Protein tyrosine phosphatase 1b and α-glucosidase inhibitory phlorotannins from edible brown algae, *Ecklonia stolonifera* and *Eisenia bicyclis*. Biosci Biotechnol Biochem 75(8):1472–1480. https://doi.org/10.1271/bbb.110137

Mukhtar Y, Yunusa U, Galalain AM (2020) A modern overview on diabetes mellitus: a chronic endocrine disorder. Eur J Biol 5:1–14. https://doi.org/10.47672/ejb.409

Nakao Y, Fusetani N (2010) 2.10 - marine invertebrates: sponges. In: Liu H-WB, Mander L (eds) Comprehensive natural products II. Elsevier, pp 327–362. https://doi.org/10.1016/B978-008045382-8.00043-5

Olgunoglu İ (2017) Review on Omega-3 (n-3) fatty acids in fish and seafood. J Biol Agric Healthc 7:37–45

Ombelet W, Van Robays J (2015) Artificial insemination history: hurdles and milestones. Facts Views Vis ObGyn 7(2):137–143

Ooi L-G, Liong M-T (2010) Cholesterol-lowering effects of probiotics and prebiotics: a review of in vivo and in vitro findings. Int J Mol Sci 11(6):2499–2522. https://doi.org/10.3390/ijms11062499

Paudel P, Seong SH, Park HJ, Jung HA, Choi JS (2019) Anti-diabetic activity of 2,3,6-tribromo-4,5-dihydroxybenzyl derivatives from *Symphyocladia latiuscula* through PTP1B downregulation and α-glucosidase inhibition. Mar Drugs 17(3):166. https://doi.org/10.3390/md17030166

Pereira DM, Valentão P, Andrade PB (2014) Marine natural pigments: chemistry, distribution and analysis. Dyes Pigments 111:124–134. https://doi.org/10.1016/j.dyepig.2014.06.011

Pouyan N (2019) Pasteur, A chemistry graduate who revoulotionized the medicine 148–151. https://doi.org/10.5923/j.microbiology.20140403.04

Qiu Y, Jiang H, Fu L, Ci F, Mao X (2021) Porphyran and oligo-porphyran originating from red algae *Porphyra*: preparation, biological activities, and potential applications. Food Chem 349: 129209. https://doi.org/10.1016/j.foodchem.2021.129209

Rastogi A, Imadi S, Gul A (2016) Marine biotechnology: focus on anticancer drugs. In: Applied molecular biotechnology: the next generation of genetic engineering, pp 555–568. https://doi.org/10.1201/b19543-28

Roehr M (2000) History of biotechnology in Austria. Adv Biochem Eng Biotechnol 69:125–149. https://doi.org/10.1007/3-540-44964-7_5

Rotter A, Barbier M, Bertoni F, Bones AM, Cancela ML, Carlsson J, Carvalho MF, Cegłowska M, Chirivella-Martorell J, Conk Dalay M, Cueto M, Dailianis T, Deniz I, Díaz-Marrero AR, Drakulovic D, Dubnika A, Edwards C, Einarsson H, Erdoğan A, Vasquez MI (2021) The essentials of marine biotechnology. Front Mar Science 8. https://doi.org/10.3389/fmars.2021.629629

Sankarapandian V, Venmathi Maran BA, Rajendran RL, Jogalekar MP, Gurunagarajan S, Krishnamoorthy R, Gangadaran P, Ahn B-C (2022) An update on the effectiveness of probiotics in the prevention and treatment of cancer. Life 12(1):59. https://doi.org/10.3390/life12010059

Santana-Casiano JM, González-Dávila M, González AG, Millero FJ (2010) Fe(III) reduction in the presence of catechol in seawater. Aquat Geochem 16(3):467–482. https://doi.org/10.1007/s10498-009-9088-x

Schindler S (2008) Model, theory, and evidence in the discovery of the DNA structure. Br J Philos Sci 59(4):619–658. https://doi.org/10.1093/bjps/axn030

Siegel RL, Miller KD, Fuchs HE, Jemal A (2021) Cancer statistics, 2021. CA Cancer J Clin 71(1): 7–33. https://doi.org/10.3322/caac.21654

Sivamaruthi BS, Bharathi M, Kesika P, Suganthy N, Chaiyasut C (2021) The administration of probiotics against hypercholesterolemia: a systematic review. Appl Sci 11(15):6913. https://doi.org/10.3390/app11156913

Smith K (2012) Louis Pasteur, the father of immunology? Front Immunol 3. https://doi.org/10.3389/fimmu.2012.00068

Stryjewska A, Kiepura K, Librowski T, Lochyński S (2013) Biotechnology and genetic engineering in the new drug development. Part I. DNA technology and recombinant proteins. Pharmacol Rep 65(5):1075–1085. https://doi.org/10.1016/s1734-1140(13)71466-x

Tait RV (1972) Marine plankton. In: Tait RV (ed) Elements of marine ecology: an introductory course. Springer, pp 18–33. https://doi.org/10.1007/978-1-4613-8297-3_2

Tan S, Tatsumura Y (2015) Alexander Fleming (1881–1955): discoverer of penicillin. Singap Med J 56(07):366–367. https://doi.org/10.11622/smedj.2015105

Tong Y, Zhang X, Tian F, Yi Y, Xu Q, Li L, Tong L, Lin L, Ding J (2005) Philinopside A, a novel marine-derived compound possessing dual anti-angiogenic and anti-tumor effects. Int J Cancer 114(6):843–853. https://doi.org/10.1002/ijc.20804

Trombetta T, Vidussi F, Roques C, Scotti M, Mostajir B (2020) Marine microbial food web networks during phytoplankton bloom and non-bloom periods: warming favors smaller organism interactions and intensifies trophic cascade. Front Microbiol 11. https://doi.org/10.3389/fmicb.2020.502336

Turner T, Jackson WH, Pettit GR, Wells A, Kraft AS (1998) Treatment of human prostate cancer cells with dolastatin 10, a peptide isolated from a marine shell-less mollusc. Prostate 34(3):175–181. https://doi.org/10.1002/(SICI)1097-0045(19980215)34:3<175::AID-PROS4>3.0.CO;2-H

Verma AS, Agrahari S, Rastogi S, Singh A (2011) Biotechnology in the realm of history. J Pharmacy Bioallied Sci 3(3):321–323. https://doi.org/10.4103/0975-7406.84430

Wright AE, Collins JE, Roberts B, Roberts JC, Winder PL, Reed JK, Diaz MC, Pomponi SA, Chakrabarti D (2021) Antiplasmodial compounds from deep-water marine invertebrates. Mar Drugs 19(4):179. https://doi.org/10.3390/md19040179

Xu Y, Zhu W, Wang T, Jin L, Liu T, Li X, Guan Z, Jiang Z, Meng X, Wang J, Guo Y (2019) Low molecule weight fucoidan mitigates atherosclerosis in ApoE (−/−) mouse model through activating multiple signal pathway. Carbohydr Polym 206:110–120. https://doi.org/10.1016/j.carbpol.2018.10.097

Yang H-W, Fernando KHN, Oh J-Y, Li X, Jeon Y-J, Ryu B (2019) Anti-obesity and anti-diabetic effects of *Ishige okamurae*. Mar Drugs 17(4):202. https://doi.org/10.3390/md17040202

Zhang C, Kim S-K (2010) Research and application of marine microbial enzymes: status and prospects. Mar Drugs 8(6):1920–1934. https://doi.org/10.3390/md8061920

Zhang Y, Zhang J, Liang Y, Li H, Li G, Chen X, Liu J (2017) Carbon sequestration processes and mechanisms in coastal mariculture environments in China. Sci China Earth Sci 60(12):2097–2107

Zhang W, Tang X, Yang Y, Zhang X, Zhang X (2020) Elevated pCO2 level affects the extracellular polymer metabolism of *Phaeodactylum tricornutum*. Front Microbiol 11. https://doi.org/10.3389/fmicb.2020.00339

Zhao J, Cao Q, Xing M, Xiao H, Cheng Z, Song S, Ji A (2020) Advances in the study of marine products with lipid-lowering properties. Mar Drugs 18(8):390. https://doi.org/10.3390/md18080390

Bacteriophage as Therapeutic Strategy Against Pathogenic Vibrio

10

Mohammad Tamrin Bin Mohamad Lal and Julian Ransangan

Abstract

Disease outbreaks frequently impede aquaculture's expanded expansion. Bacterial infections are one of the main issues among them. Antibiotics are frequently used in the treatment of bacterial diseases in aquaculture. Bacteriologists must create alternative control agents due to the emergence of bacteria that are resistant to standard antibiotics and bactericides as well as their possible adverse effects on the environment and human health. Therefore, new bacterial disease control methods are required. Bacteriophage therapy is thus one of the tactics. Bacteriophages, viruses that can only infect and kill highly particular types of bacteria, are potential agents with no known harmful impacts on the environment or human health. Numerous bacteriophages have been discovered to combat various fish pathogenic bacteria, and numerous studies have demonstrated how effectively they may control the spread of disease in both closed and open environments. This chapter contains details on potential bacteriophages that can fight off illnesses brought on by fish pathogenic bacteria. Bacteriophages must be bactericidal, highly specific to their host, accurately identified, free of virulence factors and stable in a variety of environmental conditions for bacteriophage therapy to be successful. With these qualities, the phage may be useful for treating vibriosis in aquaculture.

M. T. B. Mohamad Lal (✉) · J. Ransangan
Borneo Marine Research Institute, Universiti Malaysia Sabah, Kota Kinabalu, Sabah, Malaysia
e-mail: mdtamrin@ums.edu.my

M. D. Shah et al. (eds.), *Marine Biotechnology: Applications in Food, Drugs and Energy*, https://doi.org/10.1007/978-981-99-0624-6_10

10.1 Introduction

Globally, fish is regularly one of the most popular and affordable dietary sources of animal protein (Maulu et al. 2021). It is an excellent source of critical nutrients, including high-quality protein and lipids, vitamins and minerals (micronutrients), which are crucial for global food and nutrition security (FAO 2020). Aquaculture helps produce fish globally, and it is anticipated that this production will keep rising every year. As a result, aquaculture has experienced growth in recent years (Rocha et al. 2022). Marine aquaculture in Malaysia is currently growing thanks to the introduction of new fish species. With the increased development of aquaculture, diseases, especially bacterial diseases, have become major hurdles to successful fish production.

Both in their natural habitat and captivity, fish are susceptible to a variety of bacterial illnesses, including vibriosis, streptococcosis and bacterial kidney disease (BKD) (Toranzo et al. 2005). Among the *Vibrios* that might cause aquaculture infections are *V. anguillarum*, *V. harveyi*, *V. alginolyticus* and *V. parahaemolyticus* (Austin and Austin 2007; Won and Park 2008; Toranzo et al. 2005; Ina-Salwany et al. 2019). Numerous marine organisms have been implicated in the alleged global outbreak of vibriosis (Austin and Austin 2007). The typical symptoms of this disease in fish include skin lesions, exophthalmia, ocular opacity and haemorrhage at the base of the fins (Mohamad et al. 2019). In the meantime, the dead fish will suffer from severe anaemia, which will show in pale gills (Toranzo et al. 2005). In Sabah, *V. harveyi* is a causative agent for disease in Asian seabass (*Lates calcarifer*) (Ransangan and Mustafa 2009). This opportunistic disease may lead to fish production delays, declines in production and occasionally fish fatalities in farmed fish.

Antibiotic treatments are frequently used in aquaculture to treat bacterial infections. However, most nations forbid using antibiotics to control fish infections due to health and environmental concerns (Musa et al. 2008). In addition, using antibiotics might encourage the growth of microorganisms resistant to them (Tendencia and de la Pena 2001). Therefore, a new strategy is required to enhance the number of alternative techniques for bacterial disease prevention in aquaculture. The use of bacteriophages in aquaculture has increased as a result (Nakai and Park 2002). Bacteriophage may therefore present advantageous chances for the development of disease-controlling strategies as an alternative to antibiotics.

The bacteriophage was virulent to the bacteria (Guenther et al. 2009; Maura and Debarbieux 2012; Park et al. 1997; Yang et al. 2010). Because of its virulence, the entire phage has been used as a therapy for bacterial infections (Elbreki et al. 2014; Inal 2003; Hargreaves and Clokie 2014). However, a few requirements must be met for the phage to be beneficial as a treatment method (Oliveira et al. 2012).

10.2 Bacteriophages

The viruses known as bacteriophages feed on bacteria (Gillis and Mahillon 2014). They are obligate parasites of bacteria, like other viruses (Kutter and Sulakvelidze 2005). Federick Twort and Felix d'Herelle discovered this virus in 1915 and 1917 (Duckworth 1976). Felix d'Herelle independently identified this virus and gave it the name bacteriophage, which means 'bacterial eater' (d'Herelle 1917). The nature of the bacteriophage is still under research in the following decades. The bacteriophage is utilized as a model organism to research different virological topics, including the structure, genetics and mechanism of viral replication (Keen 2015). According to Hershey and Chase (1952), bacteriophages carry their genetic information in their DNA. Okazaki et al. (1968) employed the T4 bacteriophage to explore the discontinuous replication of DNA. The bacteriophage lambda has been widely employed for a variety of research projects, including the study of gene regulation and as a vector for gene analysis (Chauthaiwale et al. 1992; Ptashne 2004). The thorough analysis of the bacteriophage genome has also shed light on the discovery of novel biochemical pathways (Miller et al. 2003; Harada et al. 2018).

In 1920, a thorough investigation into bacteriophage therapy was launched (Carlton 1999). The investigation of bacteriophage's therapeutic potential was given up after the creation of the first antibiotic, Penicillin, in 1928, in favour of the more widespread application of antibiotics (Alharbi et al. 2014; Gil and Hyman 2010). Bacteriophage research was only continued in Eastern Europe and the former Soviet Union (Sulakvelidze et al. 2001). This work was not readily available to the globe due to a lack of international peer review and a lack of English-language articles. Since the emergence of bacteria resistant to antibiotics, interest in bacteriophage therapy has increased (Keary et al. 2013; Pirnay 2020).

10.3 Bacteriophage Therapy

Pathogenic bacterial infections have been treated with bacteriophage (Chan et al. 2013; Kutateladze and Adamia 2010; Mahony et al. 2013; Matsuzaki et al. 2005). It has evolved into a tool for the treatment of infectious bacterial diseases in aquaculture due to its advantages over anti-microbials. However, bacteriophage selection and the identification of disease-causing organisms are necessary in order to construct phage therapy for bacterial infections (Oliveira et al. 2012). Oliveira and associates described several sequential steps to do this in aquaculture.

There are seven steps in this process, which are:

1. isolation of the lytic bacteriophage;
2. phage propagation;
3. phage characterization;
4. phage typing;
5. selection of the lytic bacteriophage;
6. evaluation of therapeutic efficacy; and

7. identification of the presence of virulence genes or other toxic factors in the phage.

Finally, the phage needs to be ready for large-scale treatment and long-term preservation methods to be used for commercial reasons (Nakai 2010; Rosner and Clark 2021). To date, not many reports are available on the virulent bacteriophage against fish pathogenic bacteria. The reports were summarized in Table 10.1.

10.3.1 Bacteriophage Therapy in Aquaculture

The research on bacteriophage therapy in aquaculture began in 1982, and the Japanese eel, *Anguila japonica*, was used to test the therapeutic potential of bacteriophages (Wu and Chao 1982). The results indicated that the bacteriophage may inactivate the pathogen in the water system and that *Edwerdsiella tarda* was the target bacterium. Then, in 1997, the second effort was made and *Lactococcus garvieae* examined the protective effect of bacteriophage on experimentally infected yellowtail (*Seriola quinqueradiata*) (Nakai et al. 1999). The results showed low fish mortality after bacteriophage treatment, which was a good sign. By applying bacteriophages to aquaculture, Park et al. (2000) and Park and Nakai (2003) reported a substantial outcome. They can steer the proposal for the use of bacteriophages for preventative and therapeutic purposes in fish by a few intriguing findings, such as the fact that food with bacteriophages protects fish against infection, the bacteriophage reduces disease transmission, the fish do not produce phage-resistant bacteria or bacteriophage-neutralizing antibodies, and the bacteriophage reduces mortality and infection outbreaks.

The results of Park and Nakai's (2003) study served as a guide for bacteriophage-based biocontrol strategies. Treatment of *Aeromonas salmonicida* infection was the next effort in bacteriophage therapy (Imbeault et al. 2006). However, the therapy failed when it was unable to treat furunculosis in farmed Atlantic salmon and brook trout (Imbeault et al. 2006; Verner-Jeffreys et al. 2007). Both parties contend that using bacteriophages to treat the *Aeromonas salmonicida* infection was not immediately successful. However, the catfish *Clarias bartrachusm* illness columnaris was successfully treated using bacteriophage therapy (Prasad et al. 2011). The infected *C. batrachus* survived, showing that the bacteriophage has a wide host range.

Data on bacteriophage therapy against infection in shrimp is scarce compared to fish. According to Vinod et al. (2006), bacteriophages were used to control luminous vibriosis. The marine bacterium *V. harveyi* was reported to be controlled by the bacteriophage PW2 in a controlled laboratory setting (Phumkhachorn and Rattanachaikunsopon 2010). However, the application of bacteriophage therapy in aquaculture is still relatively new. The influence of biocontrol technology on the aquaculture system and products must be considered even though the therapy has excellent results. As a result, bacteriophages are widely used in aquaculture, and the possibility of using naturally occurring lytic bacteriophages for therapy seems encouraging.

Table 10.1 Research on bacteriophage isolation infecting fish pathogenic bacteria

Animal source	Causative agent	Disease	Phage source	References
Rainbow trout *Salmo gairdnerii*	*Yersinia ruckeri*	Enteric red mouth disease	Sewage	Stevenson and Airdrie (1984)
Japanese eel *Anguilla japonica*	*Edwardsiella tarda*	Edwardsiellosis	Pond water	Yamamoto and Maegawa (2008)
Ayu fish *Plecoglossus altivelis*	*Pseudomonas plecoglossicida*	Bacterial haemorrhagic ascites	Diseased ayu and water	Park et al. (2000) Park and Nakai (2003)
Yellowtail *Seriola quimqueradiata*	*Lactococcus garvieae*	Lactococcosis	Culture of *L. garvieae*	Park et al. (1997)
			Seawater	Nakai et al. (1999)
Japanese eel *Anguilla japonica*	*Aeromonas hydrophila*	Haemorrhagic septicaemia	Pond	Hsu et al. (2000)
Rainbow trout *Oncorhynchus mykiss*	*Flavobacterium psychrophilum*	Enteric red mouth disease	Water and sediment sample	Stenholm et al. (2008)
Brook trout *Onchorhynchus fontinalis*	*Aeromonas salmonicida*	Furunculosis	Unknown	Imbeault et al. (2006)
Atlantic salmon *Salmo salar*	*Aeromonas salmonicida* subsp. *salmonicida*	Furunculosis	Unknown	Verner-Jeffreys et al. (2007)
Japanese flounder *Paralichthys olivaceus*	*Streptococcus iniae*	Streptococcosis	Fish culture environment	Matsuoka et al. (2007)
Shrimp *Penaeus monodon*	*Vibrio harveyi*	Luminous vibriosis	Seawater	Vinod et al. (2006) Okano et al. (2007) Chrisolite et al. (2008) Phumkhachorn and Rattanachaikunsopon (2010)
Catfish	*Edwardsiella ictaluri*	Enteric septicaemia	Unknown	Walakira et al. (2008)
Catfish	*Edwardsiella ictaluri*	Enteric septicaemia	Water of aquaculture pond	Carrias et al. (2011)

(continued)

Table 10.1 (continued)

Animal source	Causative agent	Disease	Phage source	References
Catfish *Clarias batrachus*	*Flavobacterium columnare*	Columnaris disease	Water and sediment	Prasad et al. (2011)
Rock lobster *Panulirus ornatus*	*Vibrio harveyi*	Luminous vibriosis	Water	Crothers-Stomps et al. (2010)
Kuruma prawn	*Vibrio campbellii*	Vibriosis	Seawater	Li et al. (2021)
Salmon	*Vibrio anguillarum*	Vibriosis	Unknown	Tan et al. (2014)
Oyster	*Vibrio alginolyticus*	Unknown	Unknown	Le et al. (2020)
Unknown	*Vibrio parahaemolyticus*	Unknown	Sewage	Yang et al. (2020)
Seafood	*Vibrio parahaemolyticus*	Unknown	Seafood	Tan et al. (2021)
Clinical isolate	*Vibrio mimicus*	Unknown	Sewage	Gao et al. (2022)
Whiteleg Shrimp *Litopenaeus vannamei*	*Vibrio alginolyticus*	Vibriosis	Seawater	Rezaee et al. (2020)
Gilthead Seabream *Sparus aurata*	*Vibrio harveyi*	Vibriosis	Seawater	Droubogiannis and Katharios (2022)

10.3.2 Bacteriophage's Benefits and Drawbacks

Both benefits and drawbacks to bacteriophage therapy are present. The development of bacteriophage stocks, its impact in field settings, its mode of distribution, its impact on the bacterial community as a whole, horizontal gene transfer of virulence genes, bacteriophage resistance and issues with the law and public opinion are a few of the topics covered. Table 10.2 provided a breakdown of the benefits and drawbacks.

10.4 Selection of Bacteriophage Therapy Candidate

10.4.1 Lytic to Host

The entire clearing zone that the phage caused on the bacterial lawn indicated that the virus is lytic to its host. This matched the potential phage therapy's initial quality. A small number of studies on putative phage therapies started with phage isolation with clear plaque (Sillankorva et al. 2008; Synnott et al. 2009; Mirzaei and Nilsson 2015). The lysis on the bacterial lawn shows how virulent the phage is to its host (Abedon and Yin 2009). This virulent trait is advantageous since it prevented the phage from becoming a stable entity, which finally caused a lysogenic event to develop in the bacterial host (Broxmeyer et al. 2002). The preference is not to use lysogenic-capable phage as a candidate for phage therapy (Owens et al. 2013). Turbid lysis on the bacterial lawn revealed the presence of the lysogenic (Refardt 2011). This lysogenic experiment demonstrated that the phage stabilized in the bacterial cell after assimilating into the host DNA (Lood and Collin 2011). This is unfavourable because this lysogenic phage has reported an occurrence of pathogenicity increase (Stevens et al. 2013; Kraushaar et al. 2017). However, the most recent research has demonstrated that lysogenic phages can be used to cure bacterial infections (Lu and Collins 2009). Therefore, to employ the entire phage for treating bacterial infections, the lytic phage has to be retrieved during the screening of possible phage isolates. The study on the use of this phage is still in progress; thus, it is not necessary to remove the lysogenic phage that has been collected.

10.4.2 Host Specificity

The phage's specificity has a role in the selection process as well. The phage should be injected into several bacterial species for this part. Most phage isolates can infect their host, indicating that they have strong species/strain specificity. Other factors, such as the infection state (Garbe et al. 2010) and the effectiveness of plating (Kutter 2009), may also play a role in the phage's inability to infect closely related species of bacteria, in addition to the slight variations in the antigenic structure on the bacterial surface (Mahony et al. 2013). There are instances where a single phage can infect different bacterial species, though (Lal et al. 2017). Indeed, the quality of candidate

Table 10.2 Bacteriophage's benefits and drawbacks

Issue	Benefits	Drawbacks	References
Specificity	1. Narrow host range 2. Reduce the possibility of secondary infection	1. Correct identification of target host 2. Strain specific 3. Difficult preparation of bacteriophage with diverse host variants	Barrow and Soothill (1997) Carlton (1999) Nakai and Park (2002) Mathur et al. (2003)
Bacterial debris	1. Complete removal can be achieved	1. Might cause therapy to fail	Carlton (1999) Inal (2003) Morrison and Rainnie (2004)
Abundance	1. Ubiquitous	1. Lytic bacteriophage must be selected	Morrison and Rainnie (2004)
Dose	1. Determination of precise initial dose	1. Limited data available for an effective dose	Inal (2003) Mathur et al. (2003) Nakai (2010)
Multiplication	1. Rapid exponential replication	1. Difficult to confirm from in vitro to in vivo	Inal (2003) Weld et al. (2004)
Multiple infections	1. Bacteriophage cocktails can be applied	1. Target bacteria must be recognized	Carlton (1999)
Immunology	1. High dose can compensate for those that have been cleared by neutralizing antibodies	1. Antibodies might prevent the adsorption of bacteriophage	Morrison and Rainnie (2004)
Treatment	1. The bacteriophage activity is faster compared to the production of bacteriophage-neutralizing antibodies	1. Bacteriophage-neutralizing antibodies might prevent the treatment	Sulakvelidze et al. (2001) Inal (2003)
Bacteriophage resistance bacteria	1. High abundance of bacteriophage in the environment and rapid mutation of bacteriophage	1. Bacteriophage-resistant mutant appearance is rapid	Levin and Bull (2004) Merril et al. (2006) Sandeep (2006) Nakai (2010)

phage therapy is its capacity for selective infection (Chan et al. 2013). The narrowness of the phage, which is a drawback of phage therapy due to its specificity (Loc-Carrillo and Abedon 2011). The application of multiple types of phages in one mixture will be required for the treatment of the entire phage (Chan et al. 2013). This was necessary due to the target host's several strains' phage enrichment (Switt et al. 2013). On the other hand, the high specificity is beneficial because it indicates that the phage will not disrupt other bacterial flora (Rea et al. 2011), which is helpful for the development of phage therapy.

10.4.3 Identification of Bacteriophage

Whole genome sequencing and morphological characterization can be used to identify phage isolates. Typically, Caudovirales are chosen for phage therapy. *V. alginolyticus*' bacteriophage has previously been recognized as a member of the *Podoviridae* family (Liu et al. 2014; Lin et al. 2012; Zhang et al. 2014). The *V. harveyi* phage is also identified as a member of the *Myoviridae* (Crothers-Stomps et al. 2010; Surekhamol et al. 2014) and *Siphoviridae* (Crothers-Stomps et al. 2010; Surekhamol et al. 2014; Vinod et al. 2006) families in the meantime. The filamentous phage (Nasu et al. 2000) is found to infect *V. paraahemolyticus* in addition to *Myoviridae* (Alanis Villa et al. 2012), *Siphoviridae* (Kim et al. 2012) and *Podoviridae* (Hardies et al. 2003). More research revealed that *V. parahaemolyticus* phages belonged to the *Siphoviridae* family (Lal et al. 2016b), while *V. alginolyticus* and *V. harveyi* phages belonged to the *Myoviridae* family (Lal et al. 2016a, 2017). The properties of phage therapy have previously been described as the tailed phage (Letarov et al. 2010). Figure 10.1 indicates the tail phage under an electron microscope.

10.4.4 Absence of Virulence Factor

According to Oliveira et al. (2012), the evaluation of the virulence factor is required to be completed before the application of the medication. It is possible to sequence the genomes of all phage isolates. Utilizing bioinformatics software, the sequence

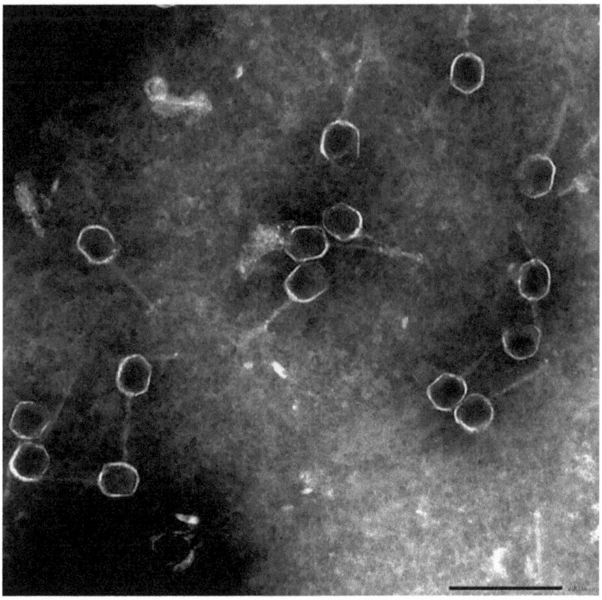

Fig. 10.1 Electron micrograph of negative-stained phage. Bars = 100 nm (Bar = 200 nm)

was examined to determine whether the genome included any potential virulence factors (Jothi et al. 2008). Currently, two virulence databases can be used which are VFDB: http://www.mgc.ac.cn/VFs/search_VFs.htm (Chen et al. 2012) and MvirDB: http://mvirdb.llnl.gov/ (Zhou et al. 2007). The phage's potential as a treatment could be aided by the absence of virulence factors.

10.4.5 Environmental Stability

The development of phage therapy also relied heavily on the phage's resistance to various environmental factors. Temperature, pH and bile salt concentration were reported as variables that can indicate whether the phage is effective when administered to fish as a treatment (Silva et al. 2013; Ly-Chatain 2014). Phages that can resist high temperatures may be helpful for treatment since they can withstand difficult circumstances, including abrupt temperature changes (Jończyk et al. 2011). However, the fish gut component might potentially have an impact on the phage's stability (Krasowska et al. 2015). A relatively high pH level was found in the digestive tracts of marine fish (Yu et al. 2007). The phage's stability under high pH may guarantee its survival in fish guts. In addition to pH, the fish gut contains bile, which may have an impact on the stability of phages like bile (Ma et al. 2008).

10.5 Conclusion

A solid scientific basis is crucial for the creation of therapeutic phages for aquaculture. A recently identified phage needs to be scrutinized. The bacteriophage must have notable lytic activity and be identified as a DNA phage by molecular studies. The family of phage, preferably the Caudovirales, might be identified via morphological examination (tailed phage). The phage's potential as a therapeutic candidate is increased by the lack of genes necessary for lysogenic conversion and pathogenicity. Additional data on the phage's capacity for bactericidal action and the non-harmful nature of its effects on the target species must be provided to establish its candidacy for disease treatment. Additional model application studies are required to increase the usage of phage isolates and assess the phage activities in the experimentally infected fish.

References

Abedon ST, Yin J (2009) Bacteriophage plaques: theory and analysis. Methods Mol Biol 501:161–174

Alanis Villa A, Kropinski AM, Abbasifar R, Griffiths MW (2012) Complete genome sequence of *Vibrio parahaemolyticus* bacteriophage vB_VpaM_MAR. J Virol 83(23):13138–13139

Alharbi SA, Wainwright M, Alahmadi TA, Salleeh HB, Faden AA, Chinnathambi A (2014) What if Fleming had not discovered penicillin? Saudi J Biol Sci 21:289–293

Austin B, Austin DA (2007) Bacterial fish pathogen: diseases of farmed and wild fish, 4th edn. Springer, Edinburgh

Barrow PA, Soothill JS (1997) Bacteriophage therapy and prophylaxis: rediscovery and renewed assessment of potential. Trends Microbiol 5(7):268–271

Broxmeyer L, Sosnowska D, Miltner E, Chacón O, Wagner D, McGarvey J, Barletta RG, Bermudez LE (2002) Killing of *Mycobacterium avium* and *Mycobacterium tuberculosis* by a mycobacteriophage delivered by a nonvirulent mycobacterium: a model for phage therapy of intercellular bacterial pathogens. J Infect Dis 186:1155–1160

Carlton RM (1999) Phage therapy: past history and future prospects. Arch Immunol Ther Exp 47: 267–274

Carrias A, Welch TJ, Waldbieser GC, Mead DA, Terhune JS, Liles MR (2011) Comparative genomic analysis of bacteriophages specific to the channel catfish pathogen *Edwardsiella ictaluri*. Virol J 8(1):6

Chan BK, Abedon ST, Loc-Carrillo C (2013) Phage cocktails and the future of phage therapy. Future Microbiol 8(6):769–783

Chauthaiwale VM, Therwath A, Deshpande VV (1992) Bacteriophage lambda as a cloning vector. Microbiol Rev 56(4):577–591

Chen L, Xiong Z, Sun L, Yang J, Jin Q (2012) VFDB 2012 update: toward the genetic diversity and molecular evolution of bacterial virulence factors. Nucleic Acids Res 40(Database Issue):D641–D645

Chrisolite B, Thiyagarajan S, Alavandi SV, Abhilash EC, Kalamani N, Vijayan KK, Santiago TC (2008) Distribution of luminescent *Vibrio harveyi* and their bacteriophages in a commercial shrimp hatchery in South India. Aquaculture 275:13–19

Crothers-Stomps C, Høj L, Bourne DG, Hall MR, Owens L (2010) Isolation of lytic bacteriophage against *Vibrio harveyi*. J Appl Microbiol 108(5):1744–1750

d'Herelle F (1917) An invisible microbe that is antagonistic to the dysentery bacillus. C R Acad Sci 165:373–375

Droubogiannis S, Katharios P (2022) Genomic and biological profile of a novel bacteriophage, *Vibrio* phage Virtus, which improves survival of *Sparus aurata* larvae challenged with *Vibrio harveyi*. Pathogens 11:630

Duckworth DH (1976) Who discovered bacteriophage. Bacteriol Rev 40(4):793–802

Elbreki M, Ross RP, Hill C, O'Mahony J, McAuliffe O, Coffey A (2014) Bacteriophages and their derivatives as biotherapeutic agents in disease prevention and treatment. J Viruses 382539, 20 p. https://doi.org/10.1155/2014/382539

FAO (2020) The State of World Fisheries and Aquaculture 2020. Sustainability in action. FAO, Rome

Gao L, Ouyang M, Li Y, Zhang H, Zheng XF, Li HX, Rao SQ, Yang ZQ, Gao S (2022) Isolation and characterization of a lytic Vibriophage OY1 and its biocontrol effects against *Vibrio* spp. Front Microbiol 13:830692

Garbe J, Wesche A, Bunk B, Kazmierczak M, Selezska K, Rohde C, Sikorski J, Rohde M, Jahn D, Schobert M (2010) Characterization of JG024, a *Pseudomonas aeruginosa* PB1-like broad host range phage under simulated infection conditions. BMC Microbiol 10:301

Gil JJ, Hyman P (2010) Phage choice, isolation and preparation for phage therapy. Curr Pharm Biotechnol 11:2–14. https://doi.org/10.2174/138920110790725311

Gillis A, Mahillon J (2014) Phages preying on *Bacillus anthracis*, *Bacillus cereus* and *Bacillus thuringiensis*: past, present and future. Viruses 6:2623–2672

Guenther S, Huwyler D, Richard S, Loessner MJ (2009) Virulent bacteriophage for efficient biocontrol of *Listeria monocytogenes* in ready-to-eat foods. Appl Environ Microbiol 75(1): 93–100

Harada LK, Silva EC, Campos WF, Del Fiol FS, Vila M, Dabrowska K, Krylov VN, Balcão VM (2018) Biotechnological applications of bacteriophages: state of the art. Microbiol Res 212–213:38–58

Hardies SC, Comeau AM, Serwer P, Suttle CA (2003) The complete sequence of marine bacteriophage VpV262 infecting *Vibrio parahaemolyticus* indicates that an ancestral component of a T7 viral supergroup is wide spread in the marine environment. Virology 310:359–371

Hargreaves KR, Clokie MRJ (2014) Clostridium difficile phages: still difficult? Front Microbiol 5:184

Hershey AD, Chase M (1952) Independent functions of viral protein and nucleic acid in growth of bacteriophage. J Gen Physiol 36(1):39–56

Hsu CH, Lo CY, Liu JK, Lin CS (2000) Control of the eel. (*Anguilla japonica*) pathogens, *Aeromonas hydrophilla* and *Edwardsiella tarda*, by bacteriophages. J Fish Soc Taiwan 27(1):21–31

Imbeault S, Parent S, Lagacé M, Uhland CF, Blais JF (2006) Using bacteriophages to prevent furunculosis caused by *Aeromonas salmonicida* in farmed Brook trout. J Aquat Anim Health 18(3):203–214

Inal JM (2003) Phage therapy: a reappraisal of bacteriophages as antibiotics. Arch Immunol Ther Exp 51:237–244

Ina-Salwany MY, Al-Saari N, Mohamad A, Mursidi FA, Mohd-Aris A, Amal MNA, Kasai H, Mino S, Sawabe T, Zamri-Saad M (2019) Vibriosis in fish: a review on disease development and prevention. J Aquat Anim Health 31(1):3–22

Jończyk E, Kłak M, Międzybrodzki R, Górski A (2011) The influence of external factors on bacteriophages-review. Folia Microbiol 56(3):191–200

Jothi R, Parthasarathy S, Ganesan K (2008) Comparison of the virulence factors and analysis of hypothetical sequences of the strains TIGR4, D39, G54 and R6 of *Streptococcus pneumonia*. J Comput Sci Syst Biol 1:103–118

Keary R, McAuliffe O, Ross RP, Hill C, O'Mahony J (2013) Bacteriophages and their endolysins for control of pathogenic bacteria. In: Méndez-Vilas A (ed) Microbial pathogens and strategies for combating them: science, technology and education. Formatex Research Center, Badajoz, pp 1028–1040

Keen EC (2015) A century of phage research: bacteriophages and the shaping of modern biology. BioEssays 37(1):6–9

Kim JH, Jun JW, Choresca CH, Shin SP, Han JE, Park SC (2012) Complete genome sequence of a novel marine siphovirus, pVp-1, infecting *Vibrio parahaemolyticus*. J Virol 86(12):7013–7014

Krasowska A, Biegalska A, Augustyniak D, Łoś M, Richert M, Łukaszewicz M (2015) Isolation and characterization of phages infecting *Bacillus subtilis*. Biomed Res Int Article ID 179597. https://doi.org/10.1155/2015/179597

Kraushaar B, Hammerl J, Kienöl M, Heinig ML, Sperling N, Thanh MD, Reetz J, Jäckel C, Fetsch A, Hertwig S (2017) Acquisition of virulence factors in livestock-associated MRSA: lysogenic conversion of CC398 strains by virulence gene-containing phages. Sci Rep 7:2004

Kutateladze M, Adamia R (2010) Bacteriophages as potential new therapies to replace or supplement antibiotics. Trends Biotechnol 28(12):591–595

Kutter E (2009) Phage host range and efficiency of plating. Methods Mol Biol 501:141–149

Kutter E, Sulakvelidze A (2005) Bacteriophages: biology and applications. CRC Press, Florida

Lal TM, Sano M, Hatai K, Ransangan J (2016a) Complete genome sequence of a giant Vibrio phage ValKK3 infecting *Vibrio alginolyticus*. Genom Data 8:837–838

Lal TM, Sano M, Ransangan J (2016b) Genome characterization of a novel vibriophage VpKK5 (*Siphoviridae*) specific to fish pathogenic strain of *Vibrio parahaemolyticus*. J Basic Microbiol 56(8):872–888

Lal MT, Sano M, Ransangan J (2017) Isolation and characterization of large marine bacteriophage (*Myoviridae*) VhKM4 infecting *Vibrio harveyi*. J Aquat Anim Health 29(1):26–30. https://doi.org/10.1080/08997659.2016.1249578

Le TS, Southgate PC, Connor WO, Vu SV, Kurtböke DI (2020) Application of bacteriophages to control *Vibrio alginolyticus* contamination in oyster (*Saccostrea glomerata*) larvae. Antibiotics 9:415

Letarov AV, Golomidova AK, Tarasyan KK (2010) Ecological basis for rational phage therapy. Acta Nat 2(1):60–71

Levin BR, Bull JJ (2004) Population and evolutionary dynamics of phage therapy. Nat Rev Microbiol 2(2):166–173

Li C, Wang Z, Zhao J, Wang L, Xie G, Huang J, Zhang Y (2021) A novel vibriophage vB_VcaS_HC containing lysogeny-related gene has strong lytic ability against pathogenic bacteria. Virol Sin 36:281–290

Lin YR, Chiu CW, Chang FY, Lin CS (2012) Characterization of a new phage, termed ϕA318, which is specific for *Vibrio alginolyticus*. Arch Virol 157(5):917–926

Liu W, Lin YR, Lu MW, Sung PJ, Wang WH, Lin CS (2014) Genome sequences characterizing five mutations in RNA polymerase and major capsid of phages ϕA318 and ϕAs51 of *Vibrio alginolyticus* with different burst efficiencies. BMC Genomics 15:505

Loc-Carrillo C, Abedon ST (2011) Pros and cons of phage therapy. Bacteriophage 1(2):111–114

Lood R, Collin M (2011) Characterization and genome sequencing of two *Propionibacterium acnes* phages displaying pseudolysogeny. BMC Genomics 12:198

Lu TK, Collins JJ (2009) Engineered bacteriophage targeting gene networks as adjuvants for antibiotic therapy. Proc Natl Acad Sci U S A 106(12):4629–4634

Ly-Chatain MH (2014) The factors affecting effectiveness of treatment in phages therapy. Front Microbiol 5:51

Ma Y, Pacan JC, Wang Q, Xu Y, Huang X, Korenevsky A, Sabour PM (2008) Microencapsulation of bacteriophage Felix O1 into chitosan-alginate microspheres for oral delivery. Appl Environ Microbiol 74(15):4799–4805

Mahony J, Kot W, Murphy J, Ainsworth S, Neve H, Hansen LH, Heller KJ, Sørensen SJ, Hammer K, Cambillau C, Vogensen FK, van Sinderen D (2013) Investigation of the relationship between lactococcal host cell wall polysaccharide genotype and 936 phage receptor binding protein phylogeny. Appl Environ Microbiol 79(14):4385–4392

Mathur MD, Vidhani S, Mehndiratta PL (2003) Bacteriophage therapy: an alternative to conventional antibiotics. J Assoc Physicians India 51:593–596

Matsuoka S, Hashizume T, Kanzaki H, Iwamoto E, Chang PS, Yoshida T, Nakai T (2007) Phage therapy against beta-hemolytic streptococcicosis of Japanese flounder *Paralichthys olivaceus*. Fish Pathol 42(4):181–189

Matsuzaki S, Rashel M, Uchiyama J, Sakurai S, Ujihara T, Kuroda M, Ikeuchi M, Tani T, Fujieda M, Wakiguchi H, Imai S (2005) Bacteriophage therapy: a revitalized therapy against bacterial infectious diseases. J Infect Chemother 11:211–219

Maulu S, Nawanzi K, Abdel-Tawwab M, Khalil HS (2021) Fish nutritional value as an approach to children's nutrition. Front Nutr 8:780844

Maura D, Debarbieux L (2012) On the interactions between virulent bacteriophages and bacteria in the gut. Bacteriophage 2(4):229–233

Merril CR, Scholl D, Adhya S (2006) Phage therapy. In: Calendar R (ed) The bacteriophage. Oxford University Press, New York, pp 725–741

Miller ES, Kutter E, Mosig G, Arisaka F, Kunisawa T, Rüger W (2003) Bacteriophage T4 genome. Microbiol Mol Biol Rev 67(1):86–156

Mirzaei MK, Nilsson AS (2015) Isolation of phages for phage therapy: a comparison of spot tests and efficiency of plating analyses for determination of host range and efficacy. PLoS One 10(3): e0118557

Mohamad N, Amal MNA, Yasin ISM, Saad MZ, Nasruddin NS, Al-Saari N, Mino S, Sawabe T (2019) Vibriosis in cultured marine fishes: a review. Aquaculture 512(3):1–17

Morrison S, Rainnie DJ (2004) Bacteriophage therapy: an alternative to antibiotic therapy in aquaculture? Can Tech Rep Fish Aquati Sci 2532:23

Musa N, Wei LS, Wee W (2008) Phenotypic and genotypic characteristics of *Vibrio harveyi* isolated from Black Tiger Shrimp (*Penaeus monodon*). World Appl Sci J 3(6):885–902

Nakai T (2010) Application of bacteriophages for control of infectious diseases in aquaculture. In: Sabour PM, Griffiths MW (eds) Bacteriophages in the control of food- and waterborne pathogens. American Society for Microbiology Press, Washington, pp 257–272

Nakai T, Park SC (2002) Bacteriophage therapy of infectious diseases in aquaculture. Res Microbiol 153:13–18

Nakai T, Sugimoto R, Park KH, Matsuoka S, Mori K, Nishioka T, Maruyama K (1999) Protective effects of bacteriophage on experimental *Lactococcus garvieae* infection in yellowtail. Dis Aquat Org 37:33–41

Nasu H, Iida T, Sugahara T, Yamaichi Y, Park KS, Yokoyama K, Makino K, Shinagawa H, Honda T (2000) A filamentous phage associated with recent pandemic *Vibrio parahaemolyticus* O3:K6 strains. J Clin Microbiol 38(6):2156–2161

Okano S, Yoshikawa T, de la Cruz AA, Sakata T (2007) Characterization of *Vibrio harveyi* bacteriophages isolated from aquaculture tanks. Mem Fac Fish Kagoshima Univ 56:55–62

Okazaki R, Okazaki T, Sakabe K, Sugimoto K, Sugino A (1968) Mechanism of DNA chain growth. I. Possible discontinuity and unusual secondary structure of newly synthesized chains. Proc Natl Acad Sci U S A 59(2):598–605

Oliveira J, Castilho F, Cunha A, Pereira MJ (2012) Bacteriophage therapy as a bacterial control strategy in aquaculture. Aquac Int 20:879–910

Owens J, Barton MD, Hauzenroeder MW (2013) The isolation and characterization of *Campylobacter jejuni* bacteriophages from free range and indoor poultry. Vet Microbiol 162:144–150

Park SC, Nakai T (2003) Bacteriophage control of *Pseudomonas plecoglossicida* infection in ayu *Plecoglossus altivelis*. Dis Aquat Org 53:33–39

Park KH, Matsuoka S, Nakai T, Muroga K (1997) A virulent bacteriophage of *Lactococcus garvieae* (formerly *Enterococcus seriolicida*) isolated from yellowtail *Seriola quinqueradiata*. Dis Aquat Org 29:145–149

Park SC, Shimamura I, Fukunaga M, Mori KI, Nakai T (2000) Isolation of bacteriophages specific to a fish pathogen, *Pseudomonas plecoglossicida*, as a candidate for disease control. Appl Environ Microbiol 66(4):1416–1422

Phumkhachorn P, Rattanachaikunsopon P (2010) Isolation and partial characterization of a bacteriophage infecting the shrimp pathogen *Vibrio harveyi*. Afr J Microbiol Res 4(16):1794–1800

Pirnay JP (2020) Phage therapy in the year 2035. Front Microbiol 11:1171

Prasad Y, Arpana KD, Sharma AK (2011) Lytic bacteriophages specific to *Flavobacterium columnare* rescue catfish, *Clarias batrachus* (Linn.) from columnaris disease. J Environ Biol 32:161–168

Ptashne M (2004) Two "what if" experiments. Cell 116(Supplement 2):S71–S72

Ransangan J, Mustafa S (2009) Identification of *Vibrio harveyi* isolated from diseased Asian seabass *Lates calcarifer* by use of 16S ribosomal DNA sequencing. J Aquat Anim Health 21: 150–155

Rea MC, Dobson A, O'Sullivan O, Crispie F, Fouhy F, Cotter PD, Shananhan F, Kiely B, Hill C, Ross RP (2011) Effect of broad- and narrow- spectrum antimicrobials on *Clostridium difficile* and microbial diversity in a model of the distal colon. Proc Natl Acad Sci U S A 108(Suppl. 1): 4639–4644

Refardt D (2011) Within-host competition determines reproductive success of temperate bacteriophages. ISME J 5:1451–1460

Rezaee S, Rajabzadeh GE, Farahmand H, Zamani I (2020) Characterization of Vibrio alginolyticus bacteriophage recovered from shrimp ponds in south west of Iran. Iran J Fish Sci 19(6): 3183–3200

Rocha CP, Cabral HN, Marques JC, Gonçalves AMM (2022) A global overview of aquaculture food production with a focus on the activity's development in transitional systems—the case study of a south European country (Portugal). J Mar Sci Eng 10:417

Rosner D, Clark J (2021) Formulations for bacteriophage therapy and the potential uses of immobilization. Pharmaceuticals 14:359

Sandeep K (2006) Bacteriophage precision drug against bacterial infections. Curr Sci 90(5): 631–633

Sillankorva S, Neubauer P, Azeredo J (2008) Isolation and characterization of a T7-like lytic phage for *Pseudomonas fluorescens*. BMC Biotechnol 8:80

Silva YJ, Costa L, Pereira C, Cunha Å, Calado R, Gomes NCM, Almeida A (2013) Influence of environmental variables in the efficiency of phage therapy. Microb Biotechnol 7:401–413

Stenholm AR, Dalsgaard I, Middelboe M (2008) Isolation and characterization of bacteriophages infecting the fish pathogen *Flavobacterium psychrophilum*. Appl Environ Microbiol 74(13): 4070–4078

Stevens RH, de Moura Martins Lobo Dos Santos C, Zuanazzi D, de Accioly Mattos MB, Ferreira DF, Kachlany SC, Tinoco EM (2013) Prophage induction in lysogenic *Aggregatibacter actinomycetemcomitans* cells co-cultures with human gingival fibroblasts, and its effect on leukotoxin release. Microb Pathog 54:54–59

Stevenson RMW, Airdrie DW (1984) Isolation of *Yersinia ruckeri* bacteriophages. Appl Environ Microbiol 74(13):4070–4078

Sulakvelidze A, Alvidze Z, Morris JGJ (2001) Bacteriophage therapy. Antimicrob Agents Chemother 45:649–659

Surekhamol IS, Deepa GD, Somnath Pai S, Sreelakshmi B, Varghese S, Bright Singh IS (2014) Isolation and characterization of broad spectrum bacteriophages lytic to *Vibrio harveyi* from shrimp farms of Kerala. India Lett Appl Microbiol 58(3):197–204

Switt AIM, den Bakker HC, Vongkamjan K, Hoelzer K, Warnick LD, Cummings KJ, Wiedmann M (2013) *Salmonella* bacteriophage diversity reflects host diversity on dairy farms. Food Microbiol 36(2):275–285

Synnott AJ, Kuang Y, Kurimoto M, Yamamichi K, Iwano H, Tanji Y (2009) Isolation from sewage influent and characterization of novel *Staphylococcus aureus* bacteriophages with wide host range and potent lytic capability. Appl Environ Microbiol 75(13):4483–4490

Tan D, Gram L, Middleboe M (2014) Vibriophages and their interactions with the fish pathogen *Vibrio anguillarum*. Appl Environ Microbiol 80(10):3128–3140

Tan CW, Rukayadi Y, Hasan H, Abdul-Mutalib NA, Jambari NN, Hara H, Thung TY, Lee E, Radu S (2021) Isolation and characterization of six *Vibrio parahaemolyticus* lytic bacteriophages from seafood samples. Front Microbiol 12:616548

Tendencia EA, de la Pena LD (2001) Antibiotic resistance of bacteria from shrimp ponds. Aquaculture 195:193–204

Toranzo AE, Magariños B, Romalde JL (2005) A review of the main bacterial fish diseases in mariculture systems. Aquaculture 246:37–61

Verner-Jeffreys DW, Algeot M, Pond MJ, Virdee HK, Bagwell NJ, Roberts EG (2007) Furunculosis in Atlantic salmon (*Salmo salar* L.) is not readily controllable by bacteriophage therapy. Aquaculture 270(1–4):475–484

Vinod MG, Shivu MM, Umesha KR, Rajeeva BC, Krohne G, Karunasagar I, Karunasagar I (2006) Isolation of *Vibrio harveyi* bacteriophage with a potential for biocontrol of luminous vibriosis in hatchery environments. Aquaculture 255:117–124

Walakira JK, Carrias AA, Hossain MJ, Jones E, Terhune JS, Liles MR (2008) Identification and characterization of bacteriophages specific to the catfish pathogen, *Edwardsiella ictaluri*. J Appl Microbiol 105(6):2133–2142

Weld RJ, Butts C, Heinemann JA (2004) Models of phage growth and their applicability to phage therapy. J Theor Biol 227(1):1–11

Won KM, Park SI (2008) Pathogenicity of *Vibrio harveyi* to cultured marine fishes in Korea. Aquaculture 285:8–13

Wu JL, Chao WJ (1982) Isolation and application of a new bacteriophage, ET-1, which infect *Edwardsiella tarda*, the pathogen of edwardsiellosis. CAPD Fisheries Series 8:8–17

Yamamoto A, Maegawa T (2008) Phage typing of *Edwardsiella tarda* from eel farm and diseased eel. Suisan Zoshoku 56:611–612

Yang H, Liang L, Lin S, Jia S (2010) Isolation and characterization of a virulent bacteriophage AB1 of *Acinetobacter baumannii*. BMC Microbiol 10:131

Yang M, Liang Y, Huang S, Zhang J, Wang J, Chen H, Ye Y, Gao X, Wu Q, Tan Z (2020) Isolation and characterization of the novel phages vB_VpS_BA3 and vB_VpS_CA8 for lysing *Vibrio parahaemolyticus*. Front Microbiol 11:259

Yu D, Wang G, Xie J, Guan S, Hu Z, Wu L (2007) Activity change of protease and amylase in digestive organs of grouper, *Epinephelus coioides*. J Zhejiang Univ Sci 26:246–251

Zhang J, Cao Z, Xu Y, Li X, Li H, Wu F, Wang L, Cao F, Li Z, Li S, Jin L (2014) Complete genomic sequence of the *Vibrio alginolyticus* lytic bacteriophage PVA1. Arch Virol 159(12): 3447–3451

Zhou CE, Smith J, Lam M, Zemla A, Dyer MD, Slezak T (2007) MvirDB-a microbial database of protein toxins, virulence factors and antibiotic resistance genes for bio-defence application. Nucleic Acids Res 35(Database Issue):D391–D394

Bacterial Diversity in the Marine Environment and the Cutting-Edge Tools for Isolation, Identification and Characterization of the Marine Microbiome

11

Imran Khan, Riaz Ullah, Sajid Ali, and Muhammad Dawood Shah

Abstract

The marine environment is a major source of biodiversity, food, energy and therapeutics. The vastness and significance of marine could be measured by the fact that about 90% of the marine environment is comprised of microbes and covers about 70% of the surface of planet Earth. Conditions like the wide range of temperature, pH, pressure and salinity further facilitate the diversification of marine organisms. Since the marine environment carries a wide range of organisms, therefore, this chapter is limited to marine bacteria only. We explained the therapeutic applications of different bacteria. Furthermore, to discover and harness the potential of marine microorganisms, we have proposed several methods and sophisticated tools that need to be developed for isolation, culturing and identification.

11.1 Introduction

The marine environment is one of the extremely valuable ecosystems of planet Earth that holds an immense wealth of supplies for the sustenance of this world and its population. Marine environments are comprised of a multitude of organisms ranging

I. Khan (✉) · S. Ali
Department of Biotechnology, Abdul Wali Khan University Mardan, Khyber, Pakhtunkhwa, Pakistan
e-mail: imrankhan31@awkum.edu.pk

R. Ullah
Department of Food Science and Technology, The University of Haripur, Haripur, Pakistan

M. D. Shah
Borneo Marine Research Institute, Universiti Malaysia Sabah, Kota Kinabalu, Sabah, Malaysia

M. D. Shah et al. (eds.), *Marine Biotechnology: Applications in Food, Drugs and Energy*, https://doi.org/10.1007/978-981-99-0624-6_11

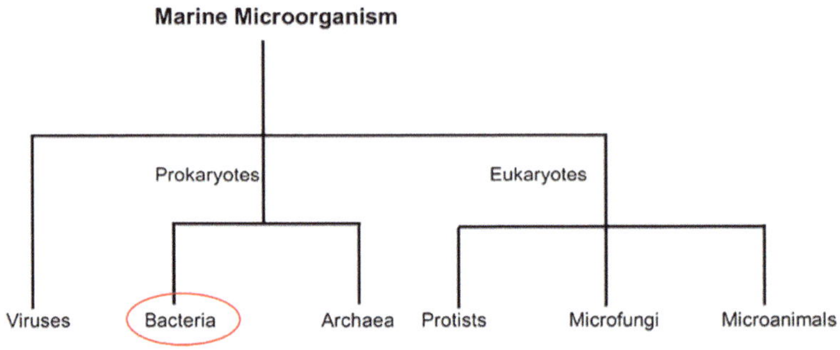

Fig. 11.1 Types of marine microorganisms. This chapter focuses on the marine bacteria

from whales to microorganisms. It is estimated that about 90% of marine biomass is microbial. Marine microorganisms are characterized by their habitation. They cover about 70% of Earth's surface and range from the saltwater of a sea or ocean or the brackish water of a coastal estuary (Jørgensen and Boetius 2007). These microorganisms have an abundance of 10^5 cells per millilitre, and it is calculated that the oceans altogether contain 3.6×10^{29} microbial cells. Most marine microorganisms are composed of archaea, bacteria, protists and unicellular fungi (Baharum et al. 2010). Marine microorganisms are well-known for maintaining and regulating the biological and chemical cycles in the marine environment. They are fixing carbon and nitrogen and playing a key role in the re-mineralization of organic matter. These microbes are also fundamental to the ocean food webs (Azam and Malfatti 2007). Marine environments are promising ecosystems for biodiscovery. The most noticeable marine microorganisms include archaea, algae, bacteria, fungi, protozoa and viruses (Fig. 11.1) (Brock et al. 2003).

The marine microorganism is so diverse and holds so many industrial applications that a book chapter is not enough to grasp such immense information. For this reason, we limited this chapter to marine bacteria only. This chapter describes the diversity and composition of marine bacteria, as well as tools for isolating living marine bacteria, innovative tools for bacterial identification and their application in various industries.

Marine bacteria, unicellular organisms, are structured like small orbits, rods and spirals. These organisms are about 1/100th the width of a human hair and can perform all sorts of chemical activities in the sea and ocean. For instance, *Trichodesimium* (a genus of Cyanobacteria) and *Crocosphaera* fix nitrogen through the nitrogenase enzyme from nitrogen gas into biologically useful nitrogen compounds. Since nitrogenase is sensitive to the presence of oxygen, therefore, these photosynthetic bacteria have adopted a special mechanism through which they can separate photosynthesis and nitrogen fixation separately (**a**) either spatially by using diverse types of cells for each procedure or (**b**) temporally, by carrying out photosynthesis in the daytime and nitrogen-fixation at night. In addition, some marine bacteria are involved in ammonium oxidation and convert ammonium to nitrite and nitrate. Betaproteobacteria and gamma proteobacteria are well-known classes of bacteria for nitrification. Bacteria related to these classes encode the ammonia monooxygenase enzyme (*amoA*). There are other bacteria such as *Nitrobacter*, *Nitrospina* and *Nitrospira* that can subsequently convert nitrite into nitrate—the second step in the nitrification process (Pajares and Ramos 2019). Bacterial-converted nitrates are produced by other bacteria as a source of nutrients, e.g., *Prochlorococcus* and *Synechococcus*. Although *Prochlorococcus* are photosynthetic, they require nitrogen as a nutrient (Partensky et al. 1999).

Marine environments present a diverse and extreme environment for living organisms as it provides temperature that ranges from −2 °C to more than 100 °C. Hydrothermal vents (<200 m in depth) temperature varies with a fluid temperature range of 10 to 119 °C (Tarasov et al. 2005). Such conditions are hostile for most organisms; however, specialized microbes could thrive in these conditions despite the tough environments (Jannasch and Taylor 1984). Bacteria have built an exceptional approach to live in extreme ecological environments through the production of an extracellular polymeric matrix, giving protection to the cells in the hostile microenvironment. The major structural elements of this composite network constitute high-molecular weight hydrophilic macromolecules, known as exopolysaccharides. The deep-sea hydrothermal vent bacteria's exopolysaccharides are enriched with hexosamines and uronic acids. *Alteromonas*, *Pseuodoalteromonas* and *Vibrio* are a few of the exopolysaccharide-secreting bacteria that dwell in hydrothermal vents and have been cultured in lab conditions (Zykwinska et al. 2019). Contrary to bacteria adapted to hot temperatures, the psychrophile bacteria can thrive and reproduce at low temperatures varying from 10 °C to −20 °C. The psychrophilic enzymes are distinctive in catalytic action and have low activation energy that yields high efficiency at lower temperatures (Podar and Reysenbach 2006).

Besides temperature variation, marine environments are also characterized by different spatial-temporal and other properties—such as salinity, pH, variation in nutrients, etc.—so different microbes inhabit differently according to their nutrient and growth condition requirements. Even, the water depth is favouring the growth of different bacteria. For instance, a study compared bacterial richness and diversity between the sea surface and subseafloor and found marked variation in the bacterial operational taxonomic units (OTUs). Abundant OTUs were spotted in the subseafloor sediments (Walsh et al. 2016).

Nonetheless, the bacterial diversity changes with changes in season. For instance, the South China Sea holds higher bacterial diversity during the summer compared to other seasons. In particular, Alpha proteobacteria is the most abundant taxon in summer and spring samples followed by Cyanobacteria and Gamma proteobacteria. On the contrary, during the winter season, the South China Sea is enriched with cyanobacteria. Even the coastal and offshore waters hold distinct bacterial communities (Du et al. 2013). Most of the bacteria that inhabit the South China Sea are related to phyla Firmicutes, Planctomycetes, Actinobacteria and Chloroflexi Gamma proteobacteria (Zhu et al. 2013).

Since the marine environment is diversified, these environments inhabit different bacteria. Therefore, the following literature is categorized to describe environmental-bacteria associations and their potential industrial application.

11.2 Co-inhabiting Niches for Marine Bacteria

11.2.1 Mangroves

Mangrove forests are a hitherto undiscovered source of microbial diversity that are found at the meeting point of marine and terrestrial ecosystems (Nedwell et al. 1994; Andreote et al. 2012). More than a hundred nations are home to mangroves, which encompass an area of around 150,000 km^2. The American continent has the most mangroves, followed by Brazil, which accounts for 8.5% of the world's mangroves (Colares and Melo 2013). Mangrove jungles are well-known for their great quantities of organic matter (OM) and nutrients, making them very productive habitats (Nedwell et al. 1994). Fish may reproduce and develop in the sediments of mangroves, and they can also find food and safety there. They serve as a transitional area between the terrestrial and marine habitats and are crucial for preserving sea levels (Duke et al. 2007). Additionally, mangrove forests serve as a terminal sink for local waste and nutrient-rich sediments.

Complex mangrove ecosystems offer a dynamic microbial environment. Because of elevated relative moisture and recurring tidal floods, the microbial colonies are exceptionally diverse and adaptable. The key determinants of the development of mangrove ecosystems are environmental elements like nutrient availability, light, salinity and temperature (Holguin et al. 2006). The general bacteria of the mangroves are N2-fixing (*Klebsiella* sp., *Azospirillum*, *Rhizobium*, *Azotobacter*, *Clostridium*, etc.), sulphate-reducing (*Desulfosarcina*, *Desulfovibrio*, *Desulfococcus* sp., *Desulfotomaculum*, etc.), phosphate-solubilizing (*Vibrio proteolyticus*, *Enterobacter*, *Kluyvera*, *Bacillus*, *Paenibacillus*, *Chryseomonas*, *Pseudomonas* sp., etc.), methanogenic (*Methanoccoides methylutens* sp., etc.) and photosynthetic anoxygenic (*Beggiatoa*, *Chloronema*, *Thiopedia*, *Leucothiobacteria* sp., etc.) bacteria (Kathiresan and Bingham 2001). Despite the extensive studies conducted on mangrove bacteria—it is estimated that more than 95% of the species remain uncharacterized. In a study carried out in the pristine and anthropogenically influenced mangrove ecosystems on the Red Sea, a research group recorded 32 bacterial phyla in the mangrove rhizospheres which was dominated by bacteria form phylum Proteobacteria (Ullah et al. 2017).

Mangrove-associated bacteria participate in nitrogen fixation (combining N_2 to ammonia or organic nitrogen), phosphate-solubilization, sulphur-oxidization and cellulose degradation. Mangrove environments dwelling bacteria are a valuable source of industrially valuable agents. These bacteria produce a wide range of vital therapeutic compounds that include insecticides, enzymes, anti-tumour agents, immunosuppressants, vitamins and immune modulators. These bacteria encode industrially useful enzymes that are used in the production of biosurfactants, bioplastics and natural bioproducts (Hong et al. 2009).

11.2.2 Marine Sponges: A Niche for Marine Bacteria

In certain species of sponges, up to 60% of the tissue volume is comprised of microbes—that said—a sponge tissue inhabitant more than 10^9 microbial cells per mL (Webster and Hill 2001). Sponges inhabit mutualistic microbes and develop mutualistic relationships. Sponges provide shelter and food to these microbes and in turn, microbial by products protect sponges against various infectious diseases (Taylor et al. 2007).

So far, at least 10 bacterial phyla that are associated with sponges have been reported, including Bacteroidetes, Actinobacteria, Proteobacteria and Verrucomicrobia (Schloss and Handelsman 2003). *Pseudoalteromonas piscicida* (a bacterium) is associated with *Hymeniacidon perleve* (a sponge) to produce β-carboline alkaloids that possess anti-microbial activities, particularly against *Staphylococcus aureus*, *Bacillus subtilis* and *Agrobacterium tumefaciens* (Zheng et al. 2005). Similarly, metabolites extracted from a strain of *Suberites domuncula* have shown cytotoxic and anti-microbial activities. These metabolites have limited the growth of *Staphylococcus aureus* and *Staphylococcus epidermidis* (Thakur et al. 2005). In addition to their anti-microbial action, most of these compounds have also shown anti-cancer properties (Aoyagi et al. 1992). These mutualistic microbes are diverse and produce Sponge-dwelling microbes including bacteria producing alkaloids, peptides, polyketides, fatty acids and terpenes that have exhibited a wide array of biological functions of biotechnological and medical significance such as anti-fungal, immunosuppressive, anti-tumour and anti-protozoal (Imhoff et al. 2011).

Moreover, sponges are in direct contact with pollutants. The hydrolases-producing microbes that inhabit sponges help in the breakdown of organic pollutants into nutrients. For instance, Cytophaga bacteria dwell on *Halichondria panacea* (a sponge) and encode the agarase enzyme. Other sponge-associated bacteria, such as *Arthrobacter*, produce another important enzyme called acetylcholinesterase (Imhoff and Stöhr 2003).

11.2.3 Algae–Bacteria Association

(a) Macroalgae (Seaweeds)

Macroalgae, also known as seaweeds, refer to thousands of species of macroscopic and multicellular marine algae. Several studies have been conducted on the interaction of macroalgae-bacteria and bacterial role in macroalgae physiology. The interlinked evolution of macroalgae and bacteria permitted the development of a wide range of relationships defined by symbiotic assistance with growth factors, nutrient interchange and quorum sensing facilitation. Bacteria could inhabit the seaweeds and chemical boundary layer of the seaweed in the microenvironment (Mieszkin et al. 2013; Wichard 2015). Some of the known seaweeds-associated bacteria include *Leucobacter* sp., *Micrococcus* and *Kocuria palustri* (Villarreal-Gómez et al. 2010; Zhuang et al. 2003).

Seaweed-associating bacteria secrete certain molecules, known as algal development and morphogenesis-promoting factors, that are important for the seaweed's development and facilitate macroalgal adaptation to environmental changes (Wichard 2015). For instance, the transition of *Ectocarpus* (macroalgae) from high-saline water to lower-saline water is dependent on intrinsic microbial communities. In a study Dittami et al. (2016) treated *Ectocarpus* with antibiotics to remove the associated bacteria and noted that *Ectocarpus* could not adapt to a new environment without associated bacteria. Similarly, macroalgae-bacteria share several strategies for metal detoxification. For example, bacteria (such as *Bacillus* spp. and *Pseudomonas aeruginosa*) eliminate cadmium and lead metals through their setup in extracellular polymeric substances (De et al. 2008). Seaweed-associated bacteria produce secondary metabolites that have the potential to inhibit colorectal cancer cells. For instance, *Bacillus* sp. which is associated with seaweed produces metabolites that have inhibited the development of HCT-116 cancer cells (Villarreal-Gómez et al. 2010).

Algal–bacterial interaction is historically well-known and has been explored for wastewater treatment as early as the 1950s (Oswald and Gotaas 1957). This interaction holds great promise in the metal bioremediation and degradation of organic pollutants (Boivin et al. 2007; Tang et al. 2010).

(b) Microalgae (Diatoms)

Microalgae are photosynthetic and unicellular microorganisms—that regularly live in aqueous surroundings such as marine systems, blackish and freshwater. The most important types of microalgae include Bacillariophyta (diatoms), golden algae (chrysophyta), green algae (chlorophyta) and blue–green algae (cyanophyta) (Chen et al. 2010).

Diatoms contribute to one-fifth of the photosynthesis on Earth, whereas bacteria re-mineralize a big part of the fixed carbon in the water. Diatoms and bacteria have lived in common habitats for millions of years and have fostered connections with world-wide biogeochemical significances. The diatoms-bacterial interaction occurs in the microenvironment surrounding diatoms, known as the phycosphere (Amin et al. 2012). Bacteria inhabit the diatoms' phycosphere either through random encounters, chemotaxis or vertical transmission (Seymour et al. 2017). Several organic nutrients that are essential for the growth of microalgae are provided by bacteria in marine environments. For instance, microalgae, including diatoms, need ammonia but these organisms cannot obtain ammonia directly by fixing N_2. Instead, nitrogen-fixing bacteria and cyanobacteria provide ammonia to algae in the marine setting (Foster et al. 2011).

Some of the most commonly diatom-associated bacteria include Gammaproteobacteria (*Marinobacter*), Alphaproteobacteria (*Sulfitobacter*), Bacteroidetes (*Croceibacter*) and Betaproteobacteria (*Limnobacter*) (Schäfer et al. 2002). However, different diatoms harbour different bacterial species. A study that investigated the bacterial communities of six diatoms genera (namely *Ditylum*, *Asterionella*, *Thalassiosira*, *Chaetoceros*, *Coscinodiscus* and *Leptocylindrus*) found distinctive bacteria associated with each genus (Grossart et al. 2005; Sapp et al. 2007a, b; Schäfer et al. 2002). Moreover, the evaluation of diatom-bacteria interaction is important to decipher nutrients and biochemical rounds in the marine setting. Their mutualistic existence cycles nutrients between reduced and oxidized states that facilitate the bioavailability of nutrients to higher trophic levels.

11.2.4 Fish

Fish-linked symbiotic gut bacteria contribute to nutritional provisioning, metabolic equilibrium and immunological defence, just like humans and other mammals (Sullam et al. 2012). Fish gut microbe research has been going on since the first part of the twentieth century, but lately, as the aquaculture sector has grown, interest in this area has accelerated. Some early studies that investigated bacteria related to fish eggs suggest that the dominating species at this point are related to the bacterial phylum Cytophaga, and the genera *Flavobacterium* and *Pseudomonas* (Austin 1982). The first colonizing bacteria are now recognized as being species-specific, with distinctions being governed by variations in binding glycoproteins on the egg exterior (Larsen 2014). Fascinatingly, as fish grow, so does the diversity of bacteria, just like in humans. In the various fish GIT parts, there are variations in microbiota density, composition and function (Clements et al. 2014).

The gut microbiota (GM), gut-dwelling microbes, has been assessed in a few model fish species, such as rainbow trout, guppy and zebrafish. In addition, the gut microbes of a few other valuable marine animals, such as Atlantic salmon, Atlantic cod and sturgeon, have also been catalogued (Kim et al. 2021). It is observed that the GM of fish is strongly influenced by the host habitat.

Much focus has been given to the nutritional exploitation and alteration of GM to meet the demands of fish farming whereas maintaining the quality and welfare of fish. Various prebiotic and probiotic methodologies are being developed to replace the use of chemicals and antibiotics in fish farming (Abelli et al. 2009). Some of the commonly examined probiotics include lactic acid producing *Lactococcus*, *Shewanella* and *Aeromonas* (Merrifield and Carnevali 2014).

11.3 Tools for the Identification of Marine Bacteria

The discovery and extraction of bioactive compounds from the marine environment demands the identification and isolation of marine bacteria in in-vitro conditions. Since the microbial world is so diverse, their growth conditions are difficult to replicate in the lab conditions. Even though many new methods have brought previously uncultured bacteria into laboratory culture, there are still numerous most wanted bacteria that need to be cultured from marine environments. In addition, the screening of marine bacteria is also needed as they could transfer antibiotic resistance genes from marine resources to humans (Khan et al. 2019). Following, we will discuss the innovative tools that are in best practice to identify and isolate bacterial colonies from the marine environment.

11.3.1 Culturomics

Culturomics is a culturing approach that uses multiple culture conditions and identification technologies (such as MALDI-TOF mass spectrometry and 16S rRNA sequencing) to bring the bacterial species into the lab conditions. A cultivation-based method is a gold standard for isolating and characterizing bacteria. This method is essential for being able to isolate pure colonies. The bacterial community belong to widespread groups, isolation and characterization of bacteria offer insight into their role and functions in natural ecosystems. In addition, culture-based techniques enable us to understand metabolic interactions between microorganisms and their surroundings (Kaeberlein et al. 2002). Culturomics is an indispensable technique used to explore cultivable bacterial diversity.

There are several reasons for our inability to cultivate marine microorganisms in the laboratory, including a lack of adequate growth conditions, low growth rates, poor development of colonies, requirements for metabolites generated by other microbes and the presence of dormant cells (Gutleben et al. 2018). Numerous attempts have been made to unveil the hidden secrets of bacteria through culture-based techniques. Cultivation-based and molecular-based techniques have been used for decades to investigate untapped environments.

Many functions and applications of marine bacteria are unknown as growing these microbes in a laboratory is challenging. Therefore, the field of Culturomics came into being to design and formulate growth media and conditions that could support maximum bacterial diversity. Besides, the rigorous cultivation methods, the number of bacteria gained through Culturomics is lesser than that seen via sequencing.

(a) Identification Through Matrix-Assisted Laser Desorption Ionization-Time of Flight Mass Spectrometry (MALDI-TOF MS)

This is the fastest technique for purified bacterial identification. Samples are obtained from the marine setup and then pass-through homogenization and serial dilution. The dilution is then evenly spread on the growth medium to obtain the master culture. After the predetermined culture time, the master plate is sub-plated for isolation of the pure colonies. These pure colonies are then processed through the preparation phase for identification through the MALDI-TOF MS. Briefly, each bacterial isolate is spread on a MALDI-TOF target plate and then poured with a 1 μL matrix solution. Matrix solution is then obtained by adding 475 μL HPLC grade water in 500 μL acetonitrile and 25 μL trifluoroacetic acid—in the end, 5 mg of α-cyano-4-hydroxycinnamic acid is dissolved. The solution is vortexed until it becomes clear. Each spot on the plate is targeted with a laser and spectra are obtained through flex control software and analysed by MALDI-Biotyper software. If the spectra aren't clear, then the colony is speared again until the spectrum is measured with a score $\geq$ 1.9). If a colony remained unidentified, then 16S RNA sequencing is employed for the amplification and sequencing of the 16S rRNA gene. This technique is graphically illustrated in Fig. 11.2.

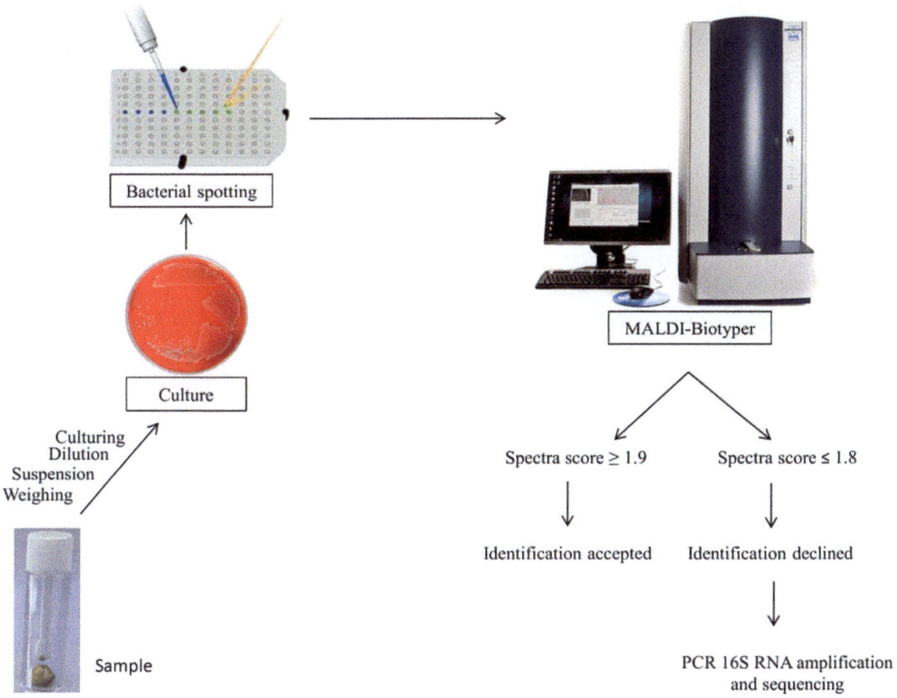

Fig. 11.2 Graphical illustration of the culturomics mechanism for isolation and identification of marine bacteria

(b) Colony PCR and Sanger Sequencer-Based Identification

This is the second fastest method of bacterial identification. Purified bacterial colonies on the media plates are selected and processed for DNA isolation. First, a 5% Chelex is prepared in sterile distilled water (1 g of chelex/20 mL). A volume of 150 μL of 5% Chelex is pipetted into 1.5 mL Eppendorf tubes. Approximately, five to six pure colonies of sample bacteria are carefully mixed with this 150 μL CHELEX and the contents were thoroughly mixed. This step is conducted inside a laminar flow hood setup to restrict contamination. All the remaining samples are processed in the same manner. The samples are warmed in a water bath set at 95 °C for 25 to 30 min. At this temperature, bacterial cells lyse. The main function of Chelex is to protect DNA from DNases. After heating, the sample tubes are centrifuged at 5000 rpm for 1 min. This centrifugation pellets out the Chelex and cell debris and DNA being obtained in the supernatant. Finally, the supernatant is collected and transferred to new, sterile and properly labelled Eppendorf tubes for further processing. The obtained DNA is stored at 4 °C until used for PCR.

Generally, PCR is performed with the universal set of primers 27F 5′-AGAGTTTGATCMTGGCTCAG-3′ and 1492R 5′-GGTTACCTTGTTACGACTT-3′ to amplify the highly variable region of the 16S rRNA gene. The master mix is prepared by mixing 12.5 μL of master mix with

1 μL of each reverse and forward primer and 9 μL of de-ionized water. The mixture is slightly vortexed for 5 s and then 1.5 μL of the template DNA is added and again vortexed. The tubes are then centrifuged for 5 s. After the PCR step, the amplified region of the 16S rRNA gene is then targeted for sequencing using Sanger sequencer by following the manufacturer protocol. This technique is graphically illustrated in Fig. 11.3.

11.3.2 Culture-Independent Method (Metagenomics)

Metagenomics is a beautiful combination of biological sciences and bioinformatics. Arguably, the invention of metagenomics is one of the greatest events in the field of microbiology in the past two decades. Metagenomics describes the genetic analysis of the genomes of an environment. This field originally began with the cloning of environmental DNA, followed by functional expression screening. As has been discussed below, metagenomics can be classified into two categories:

(a) 16S Amplicon Sequencing

The study of the interaction between organisms and their habitat is fundamental for the understanding of ecology and evolution. This mutuality led to a novel view of complex lifeforms, known as metaorganisms. A metaorganism is a grouping of all the organisms (mostly microorganisms) that live in a given habitat. But it is extremely tough and time-taking to catalogue the metaorganism of a habitat timely and efficiently. Currently, metagenomics is the gold standard to identify and characterize the metaorganism. Metagenomic was first utilized to describe the diversity and composition of dense microbial communities (Venter et al. 2004). Metagenomics has been broadly applied to recognize microorganisms that are related to an ecological and physiological state.

Unlike bacterial Culturomics, metagenomics suffers to determine the ratio of dead or alive microorganisms in a sample. Metagenomics sequences a part of the total DNA pool. At the start, metagenomics was struggling with bacterial identification at the species level due to the shorter read length, which hindered the wider application and precision of metagenomics in various applied fields (Scher et al. 2013). However, recent advancements in sequencing have supported bacterial detection beyond the strain level (Truong et al. 2017). The big development in bioinformatics has aided the investigation of complete sequence data sets by utilizing homogeneity algorithms or clustering tools without taxonomic annotation (Truong et al. 2017). Before the metagenomic experiment, the isolated DNA must be quantified using nanodrop equipment and subsequently visualize through gel electrophoresis to ensure that the DNA is not denatured. To this day, several sequencing platforms are available for 16S rRNA gene sequencing such as (Illumina, PacBio and Oxford Nanopore Sequencing). And these sequencing platforms have adopted different chemistry. Designing and enriching cultivation media and conditions based on information obtained from metagenomic datasets and growth condition is a promising strategy in development for cataloguing the most wanted marine bacteria.

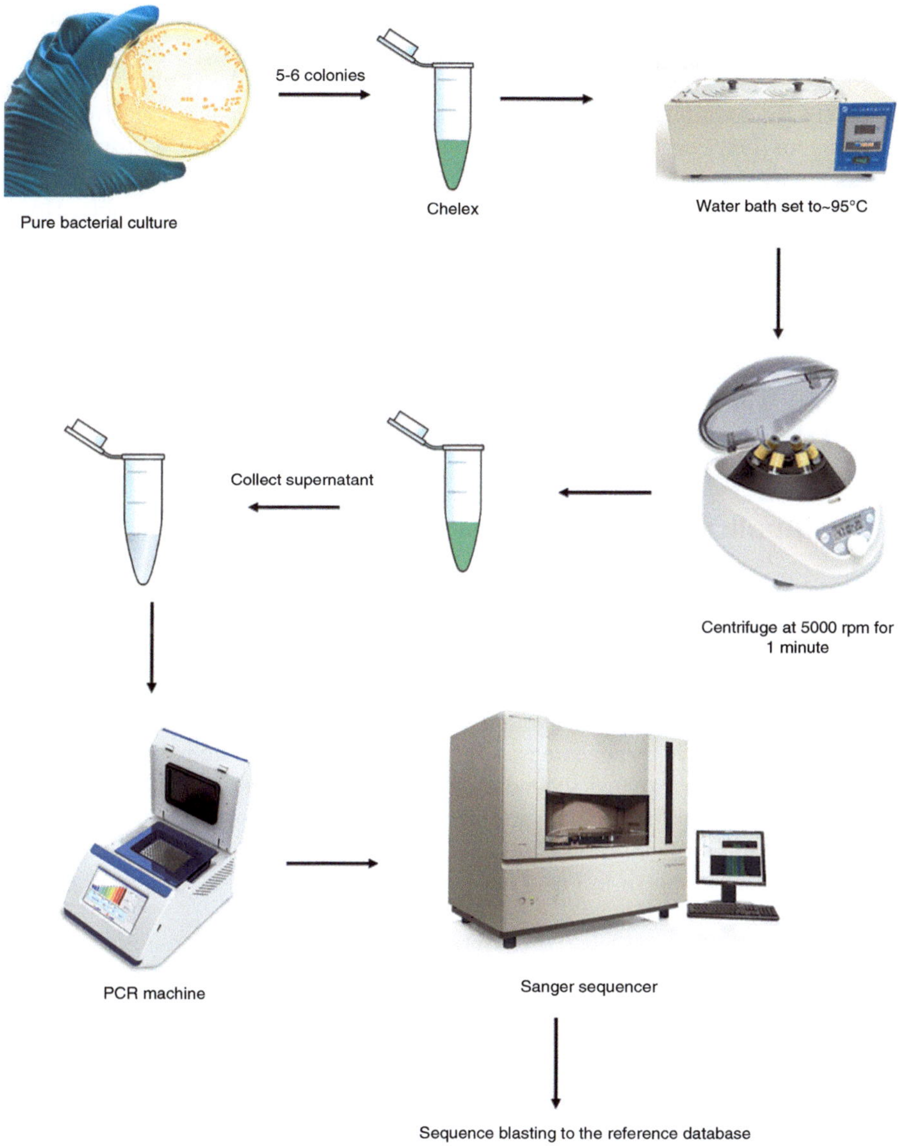

Fig. 11.3 Graphical illustration of the culturomics mechanism for isolation and identification of the marine bacteria

(b) Metagenomics Shotgun Sequencing

Beyond taxonomic identification that relies on the sequencing of the 16S rRNA gene, shotgun sequencing empowers the functional assignment of genes and provides the details about the functional potential of a community (Glass and Meyer 2011). In other words, shotgun sequencing enables scientists to comprehensively catalogue all genes in the meta-organism present in each complex environment.

Shotgun sequencing holds the lead in species and strain-level classification of bacteria. In addition to that, it enables researchers to assess the functional associations between hosts/habitats and metaorganisms (bacteria, viruses, fungi) by identifying the functional content. Beyond that, shotgun sequencing enables scientists to explore the unknown microbial life that remains unclassifiable. However, its high cost is limiting the shotgun sequencing application from a wider perspective (Jovel et al. 2016). Hopefully, the cost of shotgun metagenomics will go down in the same manner as 16S rRNA gene sequencing methods. I expect that this technique will become a standard approach for researchers around the world to comprehensively analyse and understand the function of microbial colonies and their association with the host and surrounding. A workflow of the metagenomic experiment is shown in Fig. 11.4.

11.4 Application of Marine Bacteria in the Medicinal Industry

Despite the massive interest in isolating and storing marine bacteria, metabolites obtained from these organisms are increasingly captivating to researchers due to their wide-ranging application, especially those with unique colour pigments. Great interests are being developed to explore the application of marine bacteria in various industries such as food, medicine, animal feed and colouring agents. Particularly, marine bacteria have been extensively investigated for medicinal purposes and several studies have isolated secondary metabolites from marine bacteria that possess anti-cancer, anti-microbial, anti-parasitic and immunosuppressive activities (Ramesh et al. 2019). Some of the studied metabolites are discussed below and their structures are shown in Fig. 11.5.

1. Phenazine (C6H4)2 N2, is a small nitrogen-containing organic compound that is produced by several marine bacteria such as *Actinomycetes*, *Pelagibacter*, *Vibrio* and *Pseudomonas*. Phenazine holds a broad spectrum of anti-microbial, insecticidal, anti-protozoal and anti-tumour activities (Pierson and Pierson 2010; Soliev et al. 2011).
2. Carotene, C40 Polyunsaturated hydrocarbon, is a fat-soluble precursor of vitamin A that is produced plants, fungi, yeast and bacteria. The carotenes include β-carotene, α-carotene and γ-carotene. Marine bacteria that produce carotene include *Arthrobacter* sp., *Rhodotorula rubra* and *Rhodotorula brasiliensis*. They are well-known for their strong antioxidant activities. In addition, carotene is an important product for cosmetics and a good source of natural food colourants (Asker 2017; Ligia et al. 2017).
3. Marine bacteria can also produce melanin pigment, which has exhibited anti-viral, anti-cancer and anti-bacterial activities. Most melanin-producing bacteria are sponge symbionts. Bacteria that produce melanin include *Vibrio* sp., *Bacillus* sp., *Providencia* sp., *S. algae, S. roseus, S. sciuri, P. maritimus* and

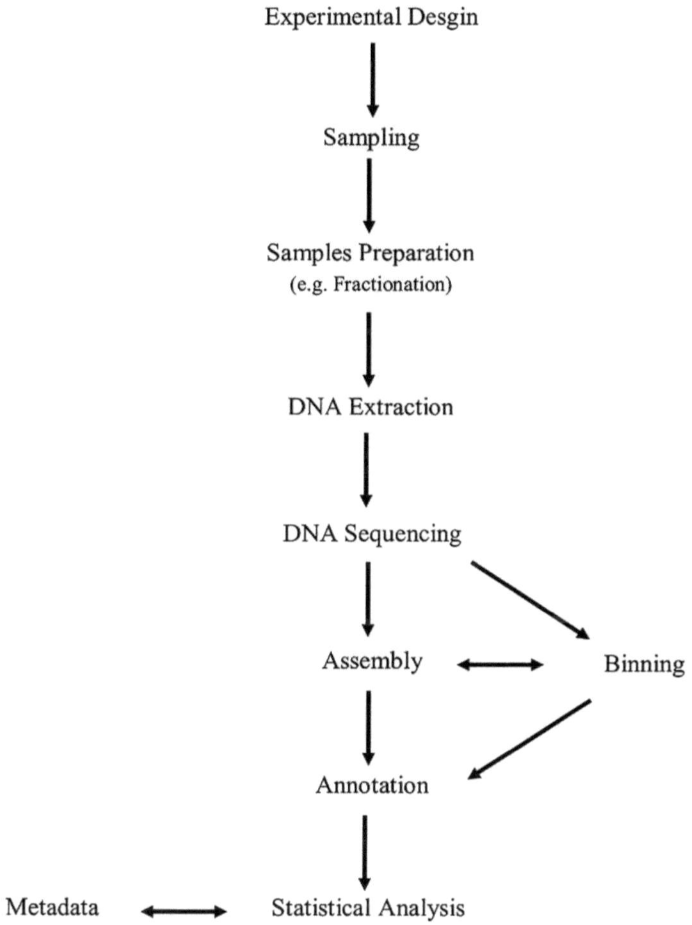

Fig. 11.4 Graphical illustration of the workflow of the metagenomic experiment

G. creatinolyticus (Vijayan et al. 2017). In addition, the bacterial produce melanin also possesses heavy metal treatment and preventing fouling (Manirethan et al. 2018, 2020).

4. Prodigiosin is a red-coloured secondary metabolite compound that is produced by bacteria such as *Serratia marcescens*, *Zooshikella* sp. S2.1 and *Streptomyces* sp. Prodigiosin has shown solid anti-microbial, anti-fungal, cytotoxic and immune suppressive activities (Huryn and Wipf 2008; Ramesh et al. 2020). This compound has also shown anti-bacterial activity against *S. aureus* (Danevčič et al. 2016).
5. Violacein is another compound which is water-insoluble in nature and purple-coloured produced by *Janthinobacterium lividum*, *Pseudoalteromonas tunicata* D2, *Chromobacterium violaceum* and *Janthinobacterium lividum* (Masuelli et al. 2016). The formation of violacein originates from the oxidation of tryptophan.

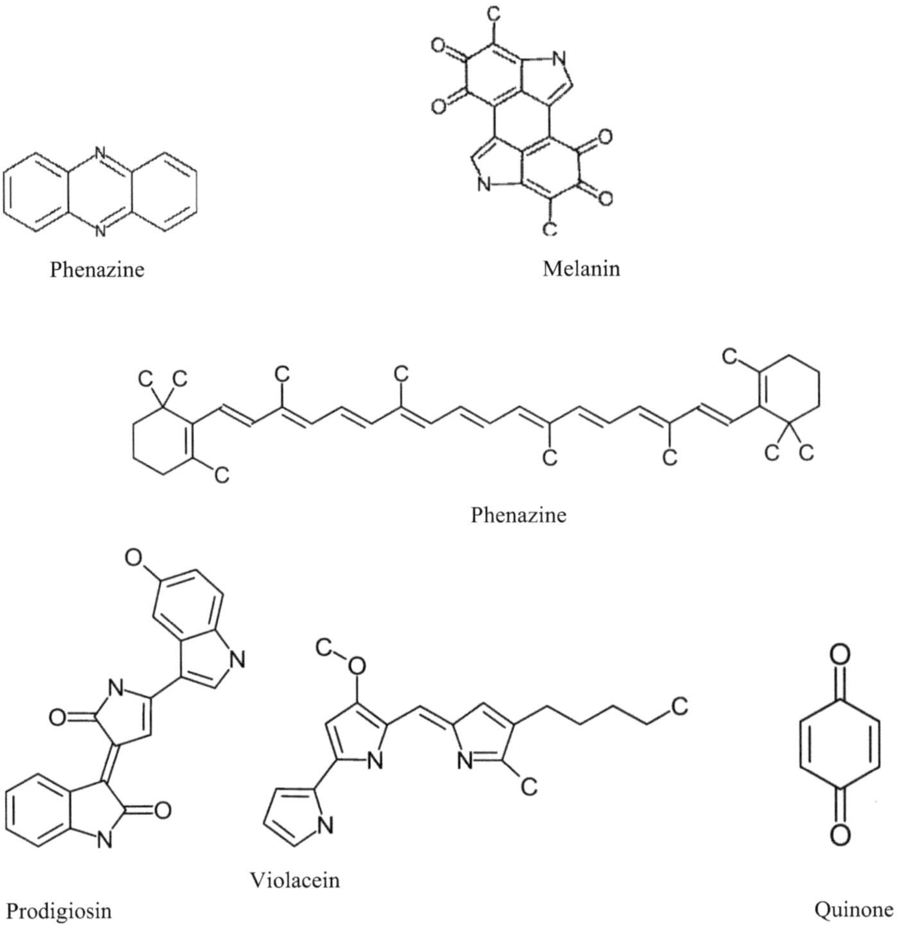

Fig. 11.5 Some of the medicinally and industrially important compounds that marine bacteria produce

This compound has been comprehensively investigated and reported to be an effective anti-bacterial, anti-viral, anti-tumour, anti-protozoal and anti-parasitic agent (Masuelli et al. 2016). This compound exhibits anti-cancer properties by enhancing mitochondrial membrane hyperpolarization (Masuelli et al. 2016).

6. Quinones are abundant biological pigments that have a yellow crystalline appearance and a distinctive irritating odour like that of chlorine. Quinones are produced by bacteria, fungi, plants and a few animals. Quinone displays anti-viral, anti-cancer, anti-microbial and insecticidal activities where the colour ranges from yellow to red (Soliev et al. 2011). *Streptomyces* sp. is a well-known bacterium for quinone production (Liang et al. 2016). This compound is usually prepared by the oxidation of aromatic amines, polyhydric phenols and polynuclear hydrocarbons.

References

Abelli L, Randelli E, Carnevali O, Picchietti S (2009) Stimulation of gut immune system by early administration of probiotic strains in Dicentrarchus labrax and Sparus aurata. Ann N Y Acad Sci 1163(1):340–342

Amin SA, Parker MS, Armbrust EV (2012) Interactions between diatoms and bacteria. Microbiol Mol Biol Rev 76(3):667–684

Andreote FD, Jiménez DJ, Chaves D, Dias ACF, Luvizotto DM, Dini-Andreote F, Fasanella CC, Lopez MV, Baena S, Taketani RG, de Melo IS (2012) The microbiome of Brazilian mangrove sediments as revealed by metagenomics. PLoS One 7(6):e38600. https://doi.org/10.1371/journal.pone.0038600

Aoyagi T, Hatsu M, Kojima F, Hayashi C, Hamada M, Takeuchi T (1992) Benarthin: a new inhibitor of pyroglutamyl peptidase. I. Taxonomy, fermentation, isolation and biological activities. J Antibiot 45(7):1079–1083. https://doi.org/10.7164/antibiotics.45.1079

Asker D (2017) Isolation and characterization of a novel, highly selective Astaxanthin-producing marine bacterium. J Agric Food Chem 65(41):9101–9109. https://doi.org/10.1021/acs.jafc.7b03556

Austin B (1982) Taxonomy of bacteria isolated from a coastal, marine fish-rearing unit. J Appl Bacteriol 53(2):253–268

Azam F, Malfatti F (2007) Microbial structuring of marine ecosystems. Nat Rev Microbiol 5(10): 782–791. https://doi.org/10.1038/nrmicro1747

Baharum SN, Beng EK, Mokhtar MAA (2010) Marine microorganisms: potential application and challenges

Boivin MEY, Greve GD, Garcia-Meza JV, Massieux B, Sprenger W, Kraak MHS, Breure AM, Rutgers M, Admiraal W (2007) Algal–bacterial interactions in metal contaminated floodplain sediments. Environ Pollut 145(3):884–894

Brock TD, Madigan MT, Martinko JM, Parker J (2003) Brock biology of microorganisms. Prentice-Hall, Upper Saddle River (NJ), p 2003

Chen P, Min M, Chen Y, Wang L, Li Y, Chen Q, Wang C, Wan Y, Wang X, Cheng Y (2010) Review of biological and engineering aspects of algae to fuels approach. Int J Agric Biol Eng 2(4):1–30

Clements KD, Angert ER, Montgomery WL, Choat JH (2014) Intestinal microbiota in fishes: what's known and what's not. Wiley Online Library

Colares GB, Melo VMM (2013) Relating microbial community structure and environmental variables in mangrove sediments inside Rhizophora mangle L. habitats. Appl Soil Ecol 64:171–177

Danevčič T, Borić Vezjak M, Tabor M, Zorec M, Stopar D (2016) Prodigiosin induces autolysins in actively grown Bacillus subtilis cells. Front Microbiol 7:27

De J, Ramaiah N, Vardanyan L (2008) Detoxification of toxic heavy metals by marine bacteria highly resistant to mercury. Mar Biotechnol (New York, NY) 10(4):471–477. https://doi.org/10.1007/s10126-008-9083-z

Dittami SM, Duboscq-Bidot L, Perennou M, Gobet A, Corre E, Boyen C, Tonon T (2016) Host-microbe interactions as a driver of acclimation to salinity gradients in brown algal cultures. ISME J 10(1):51–63. https://doi.org/10.1038/ismej.2015.104

Du J, Xiao K, Li L, Ding X, Liu H, Lu Y, Zhou S (2013) Temporal and spatial diversity of bacterial communities in coastal waters of the South China Sea. PLoS One 8(6):e66968. https://doi.org/10.1371/journal.pone.0066968

Duke NC, Meynecke J-O, Dittmann S, Ellison AM, Anger K, Berger U, Cannicci S, Diele K, Ewel KC, Field CD, Koedam N, Lee SY, Marchand C, Nordhaus I, Dahdouh-Guebas F (2007) A world without mangroves? Science 317(5834):41–42. https://doi.org/10.1126/science.317.5834.41b

Foster RA, Kuypers MMM, Vagner T, Paerl RW, Musat N, Zehr JP (2011) Nitrogen fixation and transfer in open ocean diatom-cyanobacterial symbioses. ISME J 5(9):1484–1493. https://doi.org/10.1038/ismej.2011.26

Glass EM, Meyer F (2011) The metagenomics RAST server: a public resource for the automatic phylogenetic and functional analysis of metagenomes. In: Handbook of molecular microbial ecology I: metagenomics and complementary approaches. https://doi.org/10.1002/9781118010518.ch37

Grossart H-P, Levold F, Allgaier M, Simon M, Brinkhoff T (2005) Marine diatom species harbour distinct bacterial communities. Environ Microbiol 7(6):860–873. https://doi.org/10.1111/j.1462-2920.2005.00759.x

Gutleben J, Chaib De Mares M, van Elsas JD, Smidt H, Overmann J, Sipkema D (2018) The multi-omics promise in context: from sequence to microbial isolate. Crit Rev Microbiol 44(2): 212–229. https://doi.org/10.1080/1040841X.2017.1332003

Holguin G, Gonzalez-Zamorano P, de Bashan LE, Mendoza R, Amador E, Bashan Y (2006) Mangrove health in an arid environment encroached by urban development--a case study. Sci Total Environ 363(1–3):260–274. https://doi.org/10.1016/j.scitotenv.2005.05.026

Hong K, Gao A-H, Xie Q-Y, Gao H, Zhuang L, Lin H-P, Yu H-P, Li J, Yao X-S, Goodfellow M, Ruan J-S (2009) Actinomycetes for marine drug discovery isolated from mangrove soils and plants in China. Mar Drugs 7(1):24–44. https://doi.org/10.3390/md7010024

Huryn DM, Wipf P (2008) Natural product chemistry and anticancer drug discovery. Cancer Drug Design Discov:91–120

Imhoff JF, Stöhr R (2003) In: Müller WEG (ed) Sponge-associated bacteria: general overview and special aspects of bacteria associated with Halichondria panicea BT - sponges (Porifera). Springer, Berlin, pp 35–57. https://doi.org/10.1007/978-3-642-55519-0_2

Imhoff JF, Labes A, Wiese J (2011) Bio-mining the microbial treasures of the ocean: new natural products. Biotechnol Adv 29(5):468–482. https://doi.org/10.1016/j.biotechadv.2011.03.001

Jannasch HW, Taylor CD (1984) Deep-sea microbiology. Annu Rev Microbiol 38:487–514. https://doi.org/10.1146/annurev.mi.38.100184.002415

Jørgensen BB, Boetius A (2007) Feast and famine—microbial life in the deep-sea bed. Nat Rev Microbiol 5(10):770–781. https://doi.org/10.1038/nrmicro1745

Jovel J, Patterson J, Wang W, Hotte N, O'Keefe S, Mitchel T, Perry T, Kao D, Mason AL, Madsen KL, Wong GK-S (2016) Characterization of the gut microbiome using 16S or shotgun metagenomics. Front Microbiol 7. https://doi.org/10.3389/fmicb.2016.00459

Kaeberlein T, Lewis K, Epstein SS (2002) Isolating "uncultivable" microorganisms in pure culture in a simulated natural environment. Science (New York, NY) 296(5570):1127–1129. https://doi.org/10.1126/science.1070633

Kathiresan K, Bingham BL (2001) Biology of mangroves and mangrove ecosystems

Khan I, Yasir M, Farman M, Kumosani T, Albasri SF, Bajouh OS, Azhar EI (2019) Evaluation of gut bacterial community composition and antimicrobial resistome in pregnant and non-pregnant women from Saudi population. Infection Drug Resistance 12:1749–1761. https://doi.org/10.2147/IDR.S200213

Kim PS, Shin N-R, Lee J-B, Kim M-S, Whon TW, Hyun D-W, Yun J-H, Jung M-J, Kim JY, Bae J-W (2021) Host habitat is the major determinant of the gut microbiome of fish. Microbiome 9(1):166. https://doi.org/10.1186/s40168-021-01113-x

Larsen A (2014) Studies on the microbiota of fishes and the factors influencing their composition

Liang Y, Xie X, Chen L, Yan S, Ye X, Anjum K, Huang H, Lian X, Zhang Z (2016) Bioactive polycyclic Quinones from marine Streptomyces sp. 182SMLY. Mar Drugs 14(1):10. https://doi.org/10.3390/md14010010

Ligia A d CC, Karen YFK, Susan GK (2017) Microbial production of carotenoids A review. Afr J Biotechnol 16(4):139–146

Manirethan V, Raval K, Rajan R, Thaira H, Balakrishnan RM (2018) Data on the removal of heavy metals from aqueous solution by adsorption using melanin nanopigment obtained from marine source: Pseudomonas stutzeri. Data Brief 20:178–189. https://doi.org/10.1016/j.dib.2018.07.065

Manirethan V, Raval K, Balakrishnan RM (2020) Adsorptive removal of trivalent and pentavalent arsenic from aqueous solutions using iron and copper impregnated melanin extracted from the marine bacterium Pseudomonas stutzeri. Environ Pollut (Barking, Essex: 1987) 257:113576. https://doi.org/10.1016/j.envpol.2019.113576

Masuelli L, Pantanella F, La Regina G, Benvenuto M, Fantini M, Mattera R, Di Stefano E, Mattei M, Silvestri R, Schippa S, Manzari V, Modesti A, Bei R (2016) Violacein, an indole-derived purple-colored natural pigment produced by Janthinobacterium lividum, inhibits the growth of head and neck carcinoma cell lines both in vitro and in vivo. Tumour Biol 37(3): 3705–3717. https://doi.org/10.1007/s13277-015-4207-3

Merrifield DL, Carnevali O (2014) Probiotic modulation of the gut microbiota of fish. Aquaculture nutrition: gut health, probiotics and prebiotics, 185–222

Mieszkin S, Callow ME, Callow JA (2013) Interactions between microbial biofilms and marine fouling algae: a mini review. Biofouling 29(9):1097–1113. https://doi.org/10.1080/08927014.2013.828712

Nedwell D, Blackburn T, Wiebe W (1994) Dynamic nature of the turnover of organic carbon, nitrogen and Sulphur in the sediments of a Jamaican mangrove forest. Mar Ecol Prog Ser 110: 223–231

Oswald WJ, Gotaas HB (1957) Photosynthesis in sewage treatment. Trans Am Soc Civ Eng 122(1): 73–105

Pajares S, Ramos R (2019) Processes and microorganisms involved in the marine nitrogen cycle: knowledge and gaps. Front Mar Sci 6. https://doi.org/10.3389/fmars.2019.00739

Partensky F, Hess WR, Vaulot D (1999) Prochlorococcus, a marine photosynthetic prokaryote of global significance. Microbiol Mol Biol Rev 63(1):106–127. https://doi.org/10.1128/MMBR.63.1.106-127.1999

Pierson LS, Pierson EA (2010) Metabolism and function of phenazines in bacteria: impacts on the behavior of bacteria in the environment and biotechnological processes. Appl Microbiol Biotechnol 86(6):1659–1670. https://doi.org/10.1007/s00253-010-2509-3

Podar M, Reysenbach A-L (2006) New opportunities revealed by biotechnological explorations of extremophiles. Curr Opin Biotechnol 17(3):250–255. https://doi.org/10.1016/j.copbio.2006.05.002

Ramesh CH, Vinithkumar NV, Kirubagaran R (2019) Marine pigmented bacteria: a prospective source of antibacterial compounds. J Nat Sci Biol Med 10:104

Ramesh C, Vinithkumar NV, Kirubagaran R, Venil CK, Dufossé L (2020) Applications of prodigiosin extracted from marine red pigmented bacteria Zooshikella sp. and actinomycete streptomyces sp. Microorganisms 8(4). https://doi.org/10.3390/microorganisms8040556

Sapp M, Schwaderer AS, Wiltshire KH, Hoppe H-G, Gerdts G, Wichels A (2007a) Species-specific bacterial communities in the phycosphere of microalgae? Microb Ecol 53(4):683–699. https://doi.org/10.1007/s00248-006-9162-5

Sapp M, Wichels A, Wiltshire KH, Gerdts G (2007b) Bacterial community dynamics during the winter-spring transition in the North Sea. FEMS Microbiol Ecol 59(3):622–637. https://doi.org/10.1111/j.1574-6941.2006.00238.x

Schäfer H, Abbas B, Witte H, Muyzer G (2002) Genetic diversity of "satellite" bacteria present in cultures of marine diatoms. FEMS Microbiol Ecol 42(1):25–35. https://doi.org/10.1111/j.1574-6941.2002.tb00992.x

Scher JU, Sczesnak A, Longman RS, Segata N, Ubeda C, Bielski C, Rostron T, Cerundolo V, Pamer EG, Abramson SB, Huttenhower C, Littman DR (2013) Expansion of intestinal Prevotella copri correlates with enhanced susceptibility to arthritis. elife 2:e01202. https://doi.org/10.7554/eLife.01202

Schloss PD, Handelsman J (2003) Biotechnological prospects from metagenomics. Curr Opin Biotechnol 14(3):303–310. https://doi.org/10.1016/s0958-1669(03)00067-3

Seymour JR, Amin SA, Raina J-B, Stocker R (2017) Zooming in on the phycosphere: the ecological interface for phytoplankton-bacteria relationships. Nat Microbiol 2:17065. https://doi.org/10.1038/nmicrobiol.2017.65

Soliev AB, Hosokawa K, Enomoto K (2011) Bioactive pigments from marine bacteria: applications and physiological roles. Evid Based Complement Alternat Med 2011:670349. https://doi.org/10.1155/2011/670349

Sullam KE, Essinger SD, Lozupone CA, O'Connor MP, Rosen GL, Knight R, Kilham SS, Russell JA (2012) Environmental and ecological factors that shape the gut bacterial communities of fish: a meta-analysis. Mol Ecol 21(13):3363–3378. https://doi.org/10.1111/j.1365-294X.2012.05552.x

Tang X, He LY, Tao XQ, Dang Z, Guo CL, Lu GN, Yi XY (2010) Construction of an artificial microalgal-bacterial consortium that efficiently degrades crude oil. J Hazard Mater 181(1–3):1158–1162

Tarasov VG, Gebruk AV, Mironov AN, Moskalev LI (2005) Deep-sea and shallow-water hydrothermal vent communities: two different phenomena? Chem Geol 224(1–3):5–39

Taylor MW, Radax R, Steger D, Wagner M (2007) Sponge-associated microorganisms: evolution, ecology, and biotechnological potential. Microbiol Mol Biol Rev 71(2):295–347. https://doi.org/10.1128/MMBR.00040-06

Thakur AN, Thakur NL, Indap MM, Pandit RA, Datar VV, Müller WEG (2005) Antiangiogenic, antimicrobial, and cytotoxic potential of sponge-associated bacteria. Marine Biotechnol (New York, NY) 7(3):245–252. https://doi.org/10.1007/s10126-004-4085-y

Truong DT, Tett A, Pasolli E, Huttenhower C, Segata N (2017) Microbial strain-level population structure and genetic diversity from metagenomes. Genome Res 27(4):626–638. https://doi.org/10.1101/gr.216242.116

Ullah R, Yasir M, Khan I, Bibi F, Sohrab SS, Al-Ansari A, Al-Abbasi F, Al-Sofyani AA, Daur I, Lee S-W, Azhar EI (2017) Comparative bacterial community analysis in relatively pristine and anthropogenically influenced mangrove ecosystems on the Red Sea. Can J Microbiol 63(8):649. https://doi.org/10.1139/cjm-2016-0587

Venter JC, Remington K, Heidelberg JF, Halpern AL, Rusch D, Eisen JA, Wu D, Paulsen I, Nelson KE, Nelson W, Fouts DE, Levy S, Knap AH, Lomas MW, Nealson K, White O, Peterson J, Hoffman J, Parsons R, Smith HO et al (2004) Environmental genome shotgun sequencing of the Sargasso Sea. Science (New York, NY) 304(5667):66–74. https://doi.org/10.1126/science.1093857

Vijayan V, Jasmin C, Anas A, Parakkaparambil Kuttan S, Vinothkumar S, Perunninakulath Subrayan P, Nair S (2017) Sponge-associated bacteria produce non-cytotoxic melanin which protects animal cells from photo-toxicity. Appl Biochem Biotechnol 183(1):396–411. https://doi.org/10.1007/s12010-017-2453-0

Villarreal-Gómez LJ, Soria-Mercado IE, Guerra-Rivas G, Ayala-Sánchez NE (2010) Antibacterial and anticancer activity of seaweeds and bacteria associated with their surface. Rev Biol Mar Oceanogr 45(2):267–275

Walsh EA, Kirkpatrick JB, Rutherford SD, Smith DC, Sogin M, D'Hondt S (2016) Bacterial diversity and community composition from seasurface to subseafloor. ISME J 10(4):979–989. https://doi.org/10.1038/ismej.2015.175

Webster NS, Hill RT (2001) The culturable microbial community of the great barrier reef sponge Rhopaloeides odorabile is dominated by an α-Proteobacterium. Marine Biol 138(4):843–851. https://doi.org/10.1007/S002270000503

Wichard T (2015) Exploring bacteria-induced growth and morphogenesis in the green macroalga order Ulvales (Chlorophyta). Front Plant Sci 6:86

Zheng L, Chen H, Han X, Lin W, Yan X (2005) Antimicrobial screening and active compound isolation from marine bacterium NJ6-3-1 associated with the sponge Hymeniacidon perleve. World J Microbiol Biotechnol 21(2):201–206. https://doi.org/10.1007/s11274-004-3318-6

Zhu D, Tanabe S-H, Yang C, Zhang W, Sun J (2013) Bacterial community composition of South China Sea sediments through pyrosequencing-based analysis of 16S rRNA genes. PLoS One 8(10):e78501. https://doi.org/10.1371/journal.pone.0078501

Zhuang W-Q, Tay J-H, Maszenan AM, Krumholz LR, Tay ST-L (2003) Importance of gram-positive naphthalene-degrading bacteria in oil-contaminated tropical marine sediments. Lett Appl Microbiol 36(4):251–257. https://doi.org/10.1046/j.1472-765x.2003.01297.x

Zykwinska A, Marchand L, Bonnetot S, Sinquin C, Colliec-Jouault S, Delbarre-Ladrat C (2019) Deep-sea hydrothermal vent bacteria as a source of glycosaminoglycan-mimetic exopolysaccharides. Molecules (Basel, Switzerland) 24(9):1703. https://doi.org/10.3390/molecules24091703

Biotechnological Applications of Jellyfish-Derived Products

12

B. A. Venmathi Maran, Aurelia Jackson, Thyviaah Ananthan, and Mogana Kumar

Abstract

Jellyfish have been deemed as a viable food source for centuries, however little is known regarding its collagen and its bioactive qualities. With the rise of demand in marine resources, the potential of jellyfish collagen to be applied in the pharmaceutical, medicinal, cosmeceutical and nutraceutical sectors cannot be ignored. In this chapter, we collate the prevalent methods to extract the jellyfish collagen, including acid-solubilized and pepsin-soluble-solubilized method, and physical assisted acid-extraction method, each yielding various percentage of jellyfish collagen extract. Aside from that, this chapter also discusses the applications of jellyfish collagen in numerous industries, thanks to studies investigating their benefits to mankind when consumed. More studies should be done to further characterize jellyfish collagen from different species; however, preventive measures should be taken note of so that exploitation of jellyfish does not occur.

12.1 Introduction to Jellyfish

Jellyfish are free-swimming zooplankton invertebrates that are made up of 97% water (Low et al. 2019). The word 'jellyfish' is universally used to portray cnidarians, ctenophores and salps (Haddock 2004). They are one of the oldest extant animal species, and their earliest existence was recorded in fossil snapshots approximately 500 million years ago (Cartwright et al. 2007). Hitherto, there are three major classes of jellyfish under the Phylum Cnidaria, such as Scyphozoa, Hydrozoa and

B. A. V. Maran (✉) · A. Jackson · T. Ananthan · M. Kumar
Borneo Marine Research Institute, Universiti Malaysia Sabah, Kota Kinabalu, Malaysia
e-mail: bavmaran@ums.edu.my

M. D. Shah et al. (eds.), *Marine Biotechnology: Applications in Food, Drugs and Energy*, https://doi.org/10.1007/978-981-99-0624-6_12

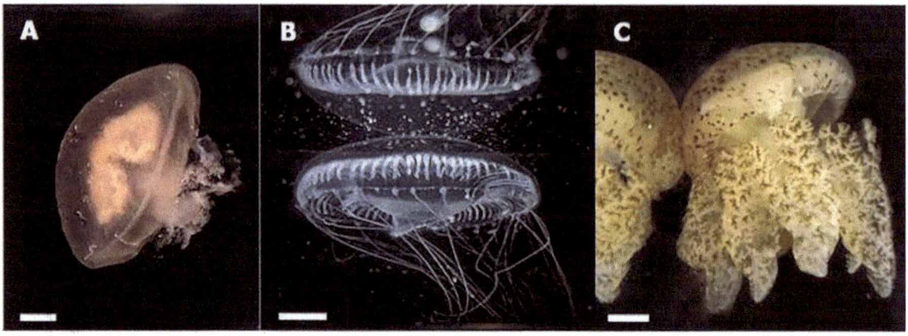

Fig. 12.1 (**a**) Rhizostome jellyfish (Scyphozoa); (**b**) *Aequorea* sp. (Hydrozoa); and (**c**) *Catostylus townsendii* (Scyphozoa) (Scale bar = 2 cm)

Cubozoa, also commonly known as true jellyfish, freshwater jellyfish and box jellyfish, respectively. They can range from as large as two meters in bell height and diameter or as small as 0.5 mm in bell height and diameter, and the representative species are *Cyanea capillata* (Linnaeus, 1758) and *Staurocladia* sp., from the classes Scyphozoa and Hydrozoa, respectively (Fig. 12.1).

Jellyfish can easily be recognized from its translucent, bell-shaped exumbrella, trailing tentacles and oral arms that are attached to the bottom of its bell (Russell 1958). Along the margin of the exumbrella are rounded lobes called lappets, which help to flex the bell. In some indentation of the lappets, there are sensory organs called rhopalia. These organs are able to perceive light and gravity and control the contraction pace of the swimming muscles of jellyfish (Meglitsch 1991). Scyphomedusae and Cubomedusae usually are marine and prefer estuaries, especially in oceans with consistent currents and low salinity (Arai 1997). Regardless, there are records of 'transportation' of invasive jellyfish into local waters from elsewhere, as the species are not known to occur in that area (Chuan et al. 2020, 2021). In some countries, jellyfish species such as *Mastigias papua* (Lesson, 1830) has been abundantly reported in lakes and called as jellyfish lakes (Dawson et al. 2001).

Jellyfish can reproduce both sexually and asexually. In the former, the male medusae releases sperm to be fertilized externally by the female medusae, which ingest the strings of sperm. As for the latter, once the ciliated larvae mature from the egg stage, they will settle on a suitable seabed or sea floor via a process called strobilation and develop into benthic polyps, before forming into ephyrae (Dawson 2003). The ephyrae then grow into a medusa, and the cycle goes on. When the temperature is warm and nutrients are present in abundance, jellyfish usually spawn in large quantities, commonly referred to as blooms (Lucas 2001; Dawson 2003). This can be supported by a study by Uye (2008) deduced that warmer sea waters often lead to higher polyps reproduction and increased jellyfish birth rates.

Because blooms are random and unpredictable, jellyfish are regarded by many as pests, and stakeholders are urged to use jellyfish in the medical, food and cosmetic industries. Since dense aggregations of Jellyfish blooms have also been known to cause imminent problems to the fishing industry, aquaculture, tourism, power plants and blocking or clogging waterways (Stoecker et al. 1987; Lucas 2001; Purcell et al. 2007). For instance, in November 2007, *Pelagia noctiluca* (Forsskål, 1775) swarmed aquaculture farms containing 100,000 salmon in Ireland, which resulted in damage worth 1 million euro (Smith 2007). Ten years after, history repeated itself as other fish farms in Ireland were invaded by blooms of the same species of jellyfish, killing off 80% of the farmed salmon stock (O'Sullivan 2017). In another case, fish stocks in the Adriatic Sea that bore the brunt of *Pelagia* sp. jellyfish blooms from 20 to 30 years ago have still not recovered from the damage (Boero 2013). Massive jellyfish blooms of *Aurelia aurita* (Linnaeus, 1758) have also clogged the power plant operations in China, causing a loss in electricity supply in one-third of the city (Dong et al. 2010). Some species of jellyfish are notoriously known to inflict symptoms like the Irukandji syndrome upon contact, which could lead to hypertension, and death in extreme cases. Aside from that, jellyfish are often a product of by-catch and are usually discarded as they can damage the nets that they are caught in accidentally.

Scyphozoan jellyfish are known to possess various types of compounds such as proteins, lipids and carbohydrates. Due to these compounds, various studies have cemented the significance of jellyfish in nutraceuticals (Hsieh et al. 2001), cosmeceuticals and biomedical fields (Leone et al. 2015, 2019; Kim et al. 2016). One of the most important developments in jellyfish applications is the discovery of the Green Fluorescent Protein (GFP) from the crystal jellyfish, *Aequorea* sp., as it was usable as a marker protein in organic structures for DNA studies (Shimomura et al. 1962). In some studies, crude venom in jellyfish was shown to contain antitumoral activities and is toxic against colon cancer and hepatoma cells in humans (Li et al. 2012; Lee et al. 2017), indicating their potential use in the medical field. Regardless, the utilization of jellyfish are fairly new in these industries issues such as small market size, high production costs and constraints regarding specifications of quality, safety assurance, extraction techniques, minimization of environmental impact and palatability issues (Gellenbeck 2012; Xiong et al. 2018; Camacho et al. 2019) should be taken into consideration when working with novel biomaterials so as to not pose a threat to human health and the ecosystem.

12.2 Collagen

The etymology of the word 'collagen' originates from Greek—'kola' is a term that means gum and 'gen' refers to the process of creating collagen. Collagen is one of the only proteins that can be found in high concentrations in animals, especially in mammals, as it makes up 25% to 30% of the body's total protein content (Müller 2003). Collagen is a major component of the endomysium, which is a network of connective tissues. They can also be present in a variety of tissues and organs, including the corneas, bones, blood vessels, cartilage, muscles and dentin of the teeth (Silvipriya et al. 2015). In fibrous tissues like the skin, tendons and ligaments, collagen appears as fibrils, and they are primarily produced by fibroblasts and a variety of different epithelial cells (Di Lullo et al. 2002; Kadler et al. 2007). In the case of invertebrates, collagen is available in the walls and cuticles of their bodies, while collagen in plants and unicellular organisms are close to non-existent, as polysaccharides and cellulose take over the functions within the organisms that are originally performed by collagen.

12.2.1 Collagen Structure

Collagen is a heteropolymer, and its core structural unit is comprised of three polypeptide chains arranged in a triple-helical configuration. Two of these chains are similar and the third has a chemical makeup that differs from the former two. Each chain is made up of 1050 amino acids that are twisted around each other in a 300 nm long, right-handed helical shape, has a molecular weight of 290,000 and a diameter of 1.5 nm. Collagen has a repeating sequence Glycine-X-Y, of which X and Y can be made up of any type of amino acids, but the most common types are proline and hydroxyproline, respectively. On the other hand, glycine is positioned at the third amino acid in the tropocollagen molecule to allow for a solid packing of the three chains (Fig. 12.2).

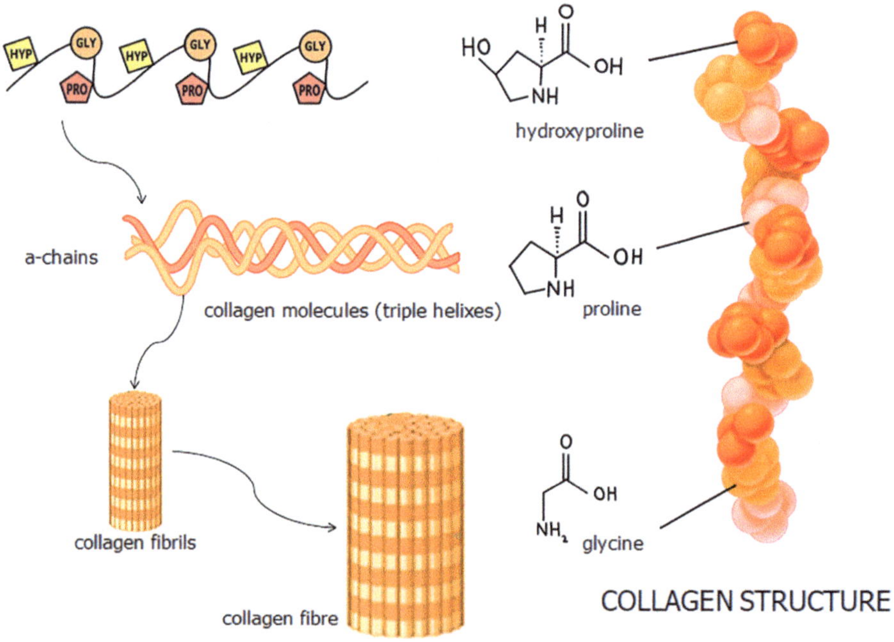

Fig. 12.2 Collagen fibrils are organized into fibre bundles. A zipper mechanism is used to weave individual alpha (α)-chains into triple helices. Fibrils are made up of bundles of triple helices that are aggregated into bigger fibres

12.2.2 Types of Collagen

Hitherto, there are 28 different types of collagen, and each one of them is made up of 46 separate polypeptide chains. While all collagen types have a triple helix in their structures, they all differ in their lengths, sizes and structures in each non-helical section (Miller et al. 1984). All collagen types are classified with numerical values, type I to type XXVIII, based on their chronological sequence of discovery. However, of all the different types of collagen, type I collagen is the most abundant and accounts for 90% of the collagen in the body due to its wide prevalence in almost all connective tissues (Table 12.1) (Cheah 1985).

12.2.3 Collagen in Marine Organisms

Due to its unique properties, collagen has been used in the growing industries of medicines, as well as nutritional supplements, cosmetics, food and beverages for many years (Huang 1988; Coppola et al. 2020; Barzideh et al. 2013). Collagen sources like porcine and bovine have long been used in industrial applications, especially in the medical field. However, some religious practitioners do not consume these products due to religious reasons. This, along with other environmental concerns, has prompted researchers to look for new collagen sources (Ogawa et al. 2004; Jongjareonrak et al. 2005; Song et al. 2006; Heu et al. 2010; Gómez-Guillén et al. 2011). Collagen from marine animals has the potential because it has a low risk of transmission of diseases, poses no religious obstacles to its consumption, provides an abundance of raw materials and yields a higher percentage of collagen compared to other sources (Senaratne et al. 2006).

Parts of sponges, jellyfish and fish, such as their extracellular matrix, connective tissues, bones, flesh, scales and fins all contain collagen (Ramshaw et al. 2009; Brunt and Burgess 2018; Trim et al. 2020). Numerous publications have documented the extraction of collagen from marine organisms, including black drums, cuttlefish, brownstripe red snappers, flatfish, skates, ocellate puffer fish, brown-banded bamboo sharks, Baltic cods and carps (Nagai et al. 2000; Sadowska et al. 2003; Ogawa et al. 2004; Jongjareonrak et al. 2005; Hwang et al. 2007; Duan et al. 2009; Heu et al.

Table 12.1 Types of collagen and their distribution within mammal tissues (adapted from Silvipriya et al. 2015; Coppola et al. 2020)

Types	Distribution of collagen
I	Skin, bone, teeth, tendon, ligament, vascular ligature organs (the main constituent of the organic part of bones)
II	Eyes and cartilage (the main constituent of cartilage)
III	Reticulate (the main constituent of reticular fibres), skin, muscle and blood vessels
IV	Forms the epithelium-secreted layer of the basement membrane and the basal lamina
V	Hair, cell surfaces and placenta
VI	Cornea, often associated with type I collagen

2010; Kittiphattanabawon et al. 2010). Cnidarian species like sea pen and sea anemone are also researched for their collagen, and studies have recorded alpha $\alpha 1$ and $\alpha 2$ chains, and $\alpha 3$ chains derived from their collagen molecules, respectively (Katzman and Kang 1972; Tillet-Barret et al. 1992) (Table 12.2).

12.3 Current Studies in Jellyfish Collagen

Jellyfish are composed of two layers of tissues, which are the epidermis and gastrodermis. The extracellular matrix in between the tissues is commonly described as the mesoglea and acts like a hydrostatic skeleton to the organism. Jellyfish mesoglea is a highly hydrated fibrous substance comprising mucopolysaccharides, collagen fibrils, protein-rich microfibrils to mammalian fibrillin, heparin sulphate proteoglycans and other structural proteins (Sarras et al. 1991; Gambini et al. 2012). One of the first studies to characterize the contents of mesoglea in jellyfish was the discovery of $\alpha 1\alpha 2\alpha 3$ protein molecules in its collagen (Miura and Kimura 1985). Subsequent studies documented the possibility of *Rhopilema asamushi* Uchida, 1927 to contain $\alpha 1\alpha 2\alpha 3\alpha 4$ heterotetramer molecules in their collagen, as well as a high yield of collagen from its exumbrella (Nagai et al. 1999). Other jellyfish were also found to possess collagen proteins, such as *Cyanea nozakii* Kishinouye, 1891, which was recorded to contain ($\alpha 1$) 3 molecules (Zhang et al. 2014). Jellyfish collagen has consistent batch-to-batch reproducibility. It is a promising alternative to mammalian collagen-based biomaterials for various medical applications involving bone and tissue regeneration. Nevertheless, the current state of knowledge in this area is still lacking, as little is known about the collagen content and properties in the mesoglea of jellyfish, especially its applications in the aforementioned industries.

Jellyfish collagen has been shown to differ from collagen derived from land animals and other marine collagens in terms of amino acid content, which affects other elements of collagen, such as thermal behaviour, isoelectric pH and solubility (Kimura et al. 1983; Miura and Kimura 1985; Nagai et al. 1999). Studies have found that the concentration of amino acids, which is generally low in marine sources, is even lower in jellyfish, resulting in decreased thermal stability of collagen (Kimura et al. 1983; Miura and Kimura 1985; Nagai et al. 1999). Despite the number of collagen studies in jellyfish (Table 12.2), most of the publications have only characterized collagen in Scyphomedusae, while there is little to no research on the collagen content in Hydromedusae and Cubomedusae.

12.4 Jellyfish Collagen Extraction Methods

Multiple extraction methods have been proposed and studied for different sources of jellyfish collagen (Table 12.3). These methods are derived from the basic method composed of three simple steps, which are preparation, extraction and recovery. The preparation step includes cleaning, separating animal parts and minimizing sample size by cutting or mincing the sample. Traditionally, when preparing jellyfish, the

Table 12.2 Applications of collagen derived from some of the marine organisms

Collagen source		Tissues	Applications	References
Common name	Scientific name (Order: Family)			
Chum Salmon	*Oncorhynchus keta* (Walbaum, 1792) (Salmoniformes: Salmonidae)	Skin	Promotes development of long bones in growing male rats.	Xu et al. (2010)
		Skin	Facilitates learning and memory in aged mice	Pei et al. (2010)
		Skin	Alleviates oxidative stress in aged skin, demonstrates protective effects on chronological skin ageing	Liang et al. (2010)
Hoki/ blue grenadier	*Macruronus novaezelandiae* (Hector, 1871) (Gadiformes: Macruronidae)	Skin	Citric-acid crosslinking in electrospun nanofibres	Cumming et al. (2018)
Belanger's croaker	*Johnius belangerii* (Cuvier, 1830) (Eupercaria/misc.: Sciaenidae)	Skin	Increases anti-oxidative enzyme levels in hepatoma cell	Mendis et al. (2005a)
Sea urchin	*Paracentrotus lividus* (Lamarck, 1816) (Camarodonta: Parechinidae)	Peristomial membrane	Characterization of collagen	Benedetto et al. (2014)
Shark	*Prionace glauca* (Linnaeus, 1758) (Carcharhiniformes: Carcharhinidae)	Skin	Production of scaffold for bone tissue engineering	Diogo et al. (2018)
Sponge	*Geodia cydonium* (Linnaeus, 1767) (Tetractinellida: Geodiidae)	–	Characterization of collagen	Diehl-Seifert et al. (1985)
	Chondrosia reniformis Nardo, 1847 (Chondrosiida: Chondrosiidae)	–	Characterization of collagen	Swatschek et al. (2002)
Cuttlefish	*Sepia lycidas* Gray, 1849 (Sepiida: Sepiidae)	Skin	Characterization of collagen	Nagai et al. (2001)
Octopus	*Callistoctopus ornatus (Gould, 1852)*	Arm	Characterization of collagen	Nagai et al. (2002a, b)

(continued)

Table 12.2 (continued)

Collagen source		Tissues	Applications	References
	[=*Callistoctopus arakawai* Taki, 1964] (Octopoda: Octopodidae)			
Squid	*Illex argentinus* (Castellanos, 1960) (Oegopsida: Ommastrephidae)	Skin	Characterization of collagen	Kolodziejska et al. (1999)
	Dosidicus gigas (d'Orbigny [in 1834–1847], 1835) (Oegopsida: Ommastrephinae)	Skin	Increases cell viability exposed to t-BHP	Mendis et al. (2005b)
			Reduces ferric	Alemán et al. (2011)
Diamondback squid	*Thysanoteuthis rhombus* Troschel, 1857 (Oegopsida: Thysanoteuthidae)	Skin	Characterization of collagen	Nagai (2004)
Nile tilapia	*Oreochromis niloticus* (Linnaeus, 1758) (Cichliformes: Cichlidae)	Skin	Wound healing	Hu et al. (2017)
			Protective effect against free radical-induced cellular and DNA damage in murine microglial cell	Vo et al. (2011)
Alaska Pollack	*Gadus chalcogrammus* Pallas, 1814 [=*Theragra chalcogramma* (Pallas, 1814)] (Gadiformes: Gadidae)	Skin	Increases cell viability exposed to t-BHP	Kim et al. (2001)
Pacific cod	*Gadus macrocephalus* Tilesius, 1810 (Gadiformes: Gadidae)	Skin	Protective effect against oxidation-induced DNA damage in mouse macrophages cell	Ngo et al. (2011)
			Protective effect against oxidation of membrane lipids, proteins and nuclear DNA in mouse monocyte cells	Himaya et al. (2012)
Seaweed pipefish	*Syngnathus schlegeli* Kaup,	–	Characterization of collagen	Khan et al. (2009)

(continued)

Table 12.2 (continued)

Collagen source		Tissues	Applications	References
	1856 (Syngnathiformes: Syngnathidae)			

Table 12.3 Three different jellyfish collagen extraction methods

Methods	References
Acid-solubilized collagen	Khong et al. (2016)
	Berillis (2015)
	Sionkowska et al. (2020)
	Kim et al. (2016)
	Jafari et al. (2020)
	Zhang et al. (2014)
	Addad et al. (2011)
Pepsin-solubilized collagen	Barzideh et al. (2013)
	Sionkowska et al. (2020)
	Mitura et al. (2020)
	Jafari et al. (2020)
	Calejo et al. (2009)
	Zhang et al. (2014)
	Addad et al. (2011)
Physical aided acid-extraction collagen	Khong et al. (2018)

oral arms are isolated from their umbrella, which is then divided into the mesoglea, the exumbrella and the sub-umbrella (Barzideh et al. 2013). These samples must be reduced in size to undergo chemical pre-treatments which remove non-collagenous proteins, colours or pigments, and lipids. The pre-treatment is done using sodium hydroxide (NaOH), which does not cause structural alteration to collagen chains, followed by alcohol, specifically butyl-alcohol or ethanol, and oxygen peroxide. Table salt (NaCl) can also be used as a substitute for NaOH for the removal of non-collagenous proteins. Throughout the extraction stage, warm water is used as collagen have poor solubility in cold water due to the presence of strong cross-links in its threefold helix structure.

Neutral salt-solubilized collagen, acid-solubilized collagen and pepsin-solubilized collagen are the three common methods that are used to extract collagen (Zhang et al. 2007). Acid-soluble and pepsin-soluble extraction methods are mostly used by researchers to extract collagen from jellyfish (Table 12.3), whereas neutral salt solutions are used to extract minimally cross-linked collagen molecules. Hitherto, a physical-aided acid-assisted extraction method has been developed, which is a new method of collagen extraction which results in a greater homogeneity and solubilization of dewatered jellyfish tissues (Khong et al. 2018).

The extraction processes were mostly derived and carried out in accordance with Miura and Kimura (1985) and Nagai et al. (2000) (Fig. 12.3). However, the chemical

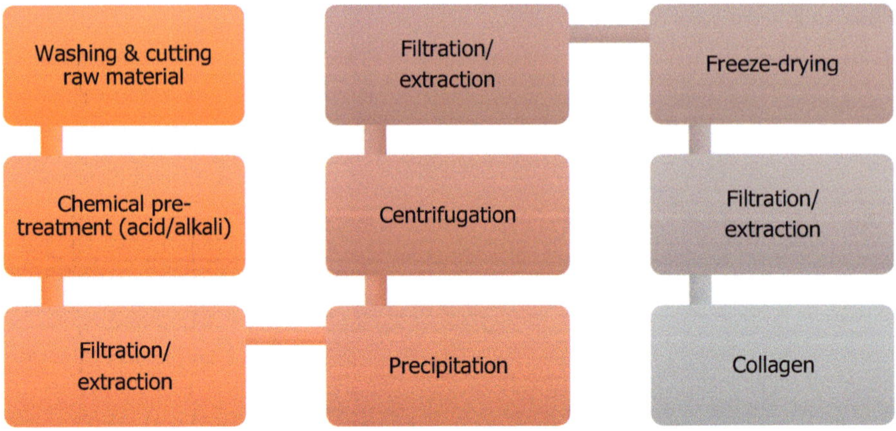

Fig. 12.3 Flow chart showing the process of jellyfish collagen extraction

composition, yield and properties of the extracted jellyfish collagen vary between these procedures. The entire extraction procedure takes place at 4 °C. The recovery product is known as acid-soluble collagen when extracted using only acid (ASC). When extracting collagen from marine animal tissues, the most commonly utilized diluted acid is acetic acid at a finishing concentration of 0.5 M, but lactic acid and citric acid are also used. Aside from that, citrate buffer (pH 2–3) and hydrochloric acid (HCl) (pH 2–3) are other examples of acidic solvents. Collagen is often extracted from jellyfish by solubilizing it in a 0.5 M acetic acid solution for 3 days, then salting it out with a disodium phosphate (Na_2HPO_4) solution. Centrifugation is used to separate the precipitated collagen, which is then solubilized in acetic acid and purified by re-precipitation with solid NaCl at a concentration of 0.9 M. As water makes up about 95% of marine animals like jellyfish, this impacts collagen solubility in acetic acid. To increase the collagen solubility in diluted acid and increase the extraction yield, jellyfish must be homogenized or freeze-dried. Atelocollagen can also be produced by digesting ASC with pepsin (Addad et al. 2011; Song et al. 2006).

12.4.1 Acid-Soluble Collagen (ASC)

Using the ASC method (Fig. 12.4), jellyfish tissues are extracted with 0.1 M NaOH after being rinsed with distilled water. The insoluble tissues are re-suspended in 0.5 M acetic acid (1:1000 w/v), and acid-soluble proteins are extracted for 3 days with 0.5 M acetic acid, which is repeated twice. The extracts are dialysed extensively with 0.02 M Na_2HPO_4 after filtering and compressing the insoluble mesoglea with cheesecloth. The collected precipitates are dissolved in 0.5 M acetic acid after centrifugation at 6000 g for 30 min at 4 °C. Solid NaCl needs to be added to the supernatant after centrifugation at 20,000 g for 1 h, which will measure to a

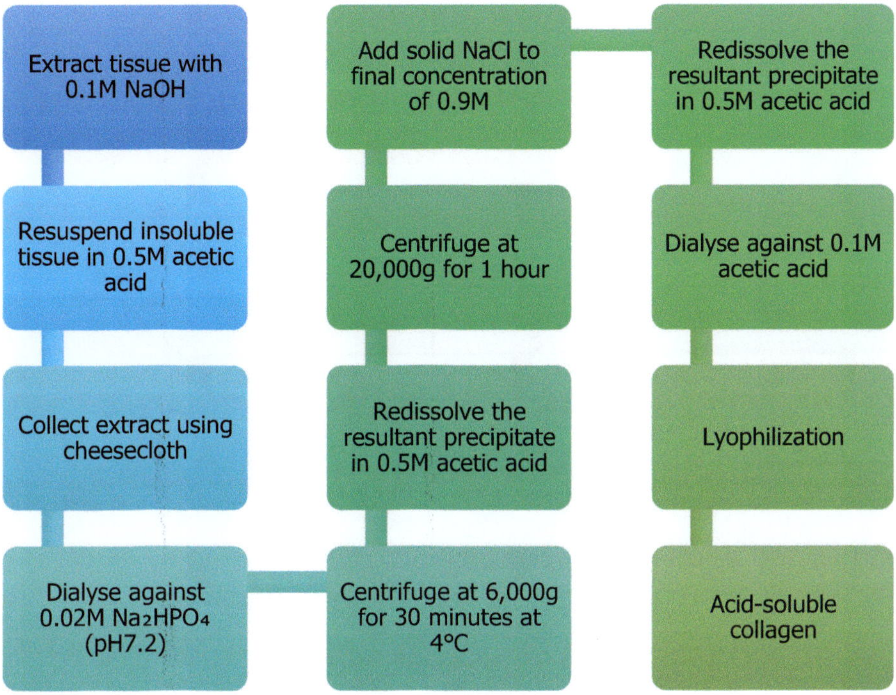

Fig. 12.4 Flow chart showing acid-soluble collagen extraction method

concentration of 0.9 M. The acid-soluble collagen fraction is dissolved in 0.5 M acetic acid, dialysed against 0.1 M acetic acid and lyophilized from the 0.9 M NaC1-precipitable fraction (Miura and Kimura 1985; Nagai et al. 2000).

12.4.2 Pepsin-Soluble Collagen (PSC)

When the enzyme pepsin is added during the extraction process, the isolated collagen is termed pepsin-soluble collagen (Fig. 12.5). Jellyfish tissues are washed in distilled water and then with 0.1 M NaOH. The insoluble particles are re-suspended in 0.5 M acetic acid (1:100 w/v) and digested with 10% (w/v) pepsin for 48 h at 4 °C. The pepsin-solubilized collagen is then centrifuged at 20,000 g for 1 h, then dialysed for 3 days against 0.02 M Na_2HPO_4 (pH 7.2). The resulting precipitate is dissolved in 0.5 M acetic acid and salted out with NaCl to a final concentration of 1.0 M after centrifugation at 20,000 g for 1 h. After that, the precipitate is dissolved in 0.5 M acetic acid, dialysed against 0.1 M acetic acid and then lyophilized, which will yield extremely high amounts of collagen based on its lyophilized dry weight (Miura and Kimura 1985; Nagai et al. 1999). With this method, the best collagen yield to date is achieved with the oral arms of *Rhizostoma pulmo* (Macri, 1778) (2–10 mg collagen/g of wet tissue) (Addad et al. 2011).

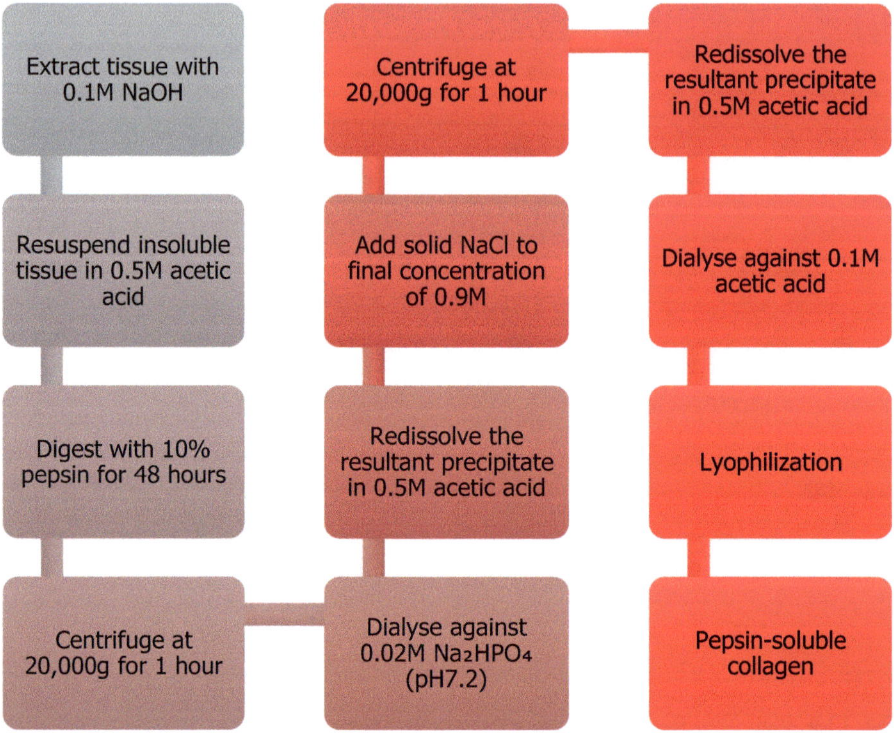

Fig. 12.5 Flow chart showing pepsin-soluble collagen extraction method

12.4.3 Other Known Methods

Yusoff et al. (2013) have presented a new approach for extracting collagen from aquatic animals that combine acidic treatments with a series of physical and mechanical treatments, including pH changes, homogenization, mixing and sonication (Fig. 12.6). The study found that the extraction yield increased dramatically when compared to traditional extraction methods after physical treatment. *Acromitus hardenbergi* Stiasny, 1934 was sampled to test this new collagen extraction procedure, which resulted in enhanced solubilization and homogeneity of dewatered jellyfish tissues (Khong et al. 2016). In brief, jellyfish tissues are rinsed using distilled water and extracted three times using 0.1 M of NaOH. The insoluble tissue is re-suspended and carefully mixed (1:1, w/v) in 0.5 M acetic acid. After that, the suspension is sonicated for 15 min, followed by 1 h of rigorous mixing at 4 °C. Soluble extracts are combined and dialysed extensively (1:10, v/v) with 0.02 M phosphate buffer at a pH value of 7.2. Collagen is salted off by resolving the precipitate in 0.5 M acetic acid and adding NaCl to a final concentration of 0.9 M. The precipitate is purified further by dissolving in 0.5 M acetic acid and dialysis against 0.1 M acetic acid, 0.05 M acetic acid and 0.025 M acetic acid, followed by

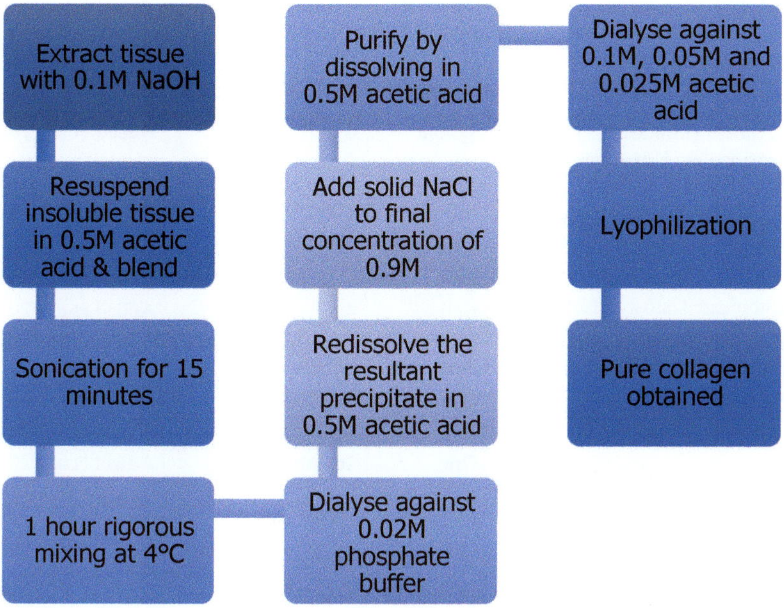

Fig. 12.6 Flow chart showing physical-induced acid-assisted collagen extraction method

0.1 M acetic acid, 0.05 M acetic acid and 0.025 M acetic acid. When the dialysed collagen slurry is lyophilized, pure collagen is recovered.

Enhanced acid-assisted collagen extraction from jellyfish *A. hardenbergi* using physical-assisted processes increased extraction efficiency (yield) by approximately seven folds (bell arms) and five folds (oral arms), and by approximately two folds from both bell and oral arms when compared to conventional PSC extraction. The enhanced process produced greater protein content, and hydroxyproline content results when compared to conventional methods, while also producing lower ash content. Previous studies have revealed that using this method, collagen extract is higher in purity than collagen extracted from other methods, aside from being devoid of heavy metal contamination (Khong et al. 2018).

According to several studies, jellyfish are made up of roughly 60% collagen, with species like *R. asamushi* and *S. meleagris*, which are scyphomedusae commonly consumed by Eastern populations, having the greatest collagen yield (Table 12.4). Collagen derived from *Catostylus tagi* (Haeckel, 1869) is already being investigated for commercial use, indicating that this species could be a good source of collagen. In vitro, jellyfish collagen is found to be toxic-free, heavy metal-free and potentially hypoallergenic (Yusoff et al. 2013). However, recovering collagen from jellyfish tissues is a difficult task due to the tissues' specialization in resisting solubilization, as a result of adaptability to the severe ocean environment. To date, the documented methods to extract jellyfish collagen have been arduous and time-consuming, requiring a lot of stages and time while obtaining poor yields. There are also studies reported to have found poor or fruitless collagen yields (Barzideh et al. 2014). With

Table 12.4 The collagen content on different species of jellyfish. Collagen content is reported as a percentage of total mass, either wet or dry ([a])

Species	Tissue	Collagen content (%)	Source
Aurelia aurita (Linnaeus, 1758) (Semaeostomeae: Ulmaridae)	Whole	0.01	Addad et al. (2011)
Chrysaora sp.	Bell	9–19	Barzideh et al. (2013)
Pelagia noctiluca (Forsskål, 1775) (Semaeostomeae: Pelagiidae)	Whole	0.07	Addad et al. (2011)
Catostylus tagi (Haeckel, 1869) (Rhizostomeae: Catostylidae)	Bell	2.7[a]	Calejo et al. (2009)
	Bell	4.5	Addad et al. (2011)
Cotylorhiza tuberculata (Macri, 1778) (Rhizostomeae: Cepheidae)	Oral arms	19.4	Addad et al. (2011)
	Bell	<10[a]	Addad et al. (2011)
	Bell	8.3–31.5	Addad et al. (2011)
Rhizostoma pulmo (Macri, 1778) (Rhizostomeae: Rhizostomatidae)	Oral arms	26–90	Addad et al. (2011)
	Bell	<10[a]	Addad et al. (2011)
Rhopilema asamushi Uchida, 1927 (Rhizostomeae: Rhizostomatidae)	–	35.2[a]	Nagai et al. (2000)
Rhopilema esculentum Kishinouye, 1891 (Rhizostomeae: Rhizostomatidae)	Mesoglea	0.28[a]	Cheng et al. (2017)
Stomolophus meleagris Agassiz, 1860 (Rhizostomeae: Stomolophidae)	Mesoglea	46.4[a]	Nagai et al. (1999)
Nemopilema nomurai Kishinouye, 1922 (Rhizostomeae: Rhizostomatidae)	Mesoglea	2.2[a]	Miura and Kimura (1985)

higher output and quality, jellyfish collagen could be a viable alternative to land-based collagen in the manufacturing of nutraceuticals, medicines, cosmeceuticals and biomaterials.

12.5 Toxicity of Jellyfish Collagen

Toxicity studies involve the evaluation of chemical safety, their toxicological properties and the determination of the highest exposure level, such as in doses or concentrations, and are fundamental for human health risk assessment. Measuring toxicity levels and defining the hazard can give a clear understanding of whether or not the changes caused by the chemical treatment could cause adverse or non-adverse effects on the test subject (Lewis et al. 2002). Studies on the toxicity of jellyfish collagen should be thoroughly conducted if it were to be used in health sciences, as it has been previously mentioned to have various effects, such as anti-oxidant (Yu et al. 2006; Ding et al. 2011), anti-hypertensive, anti-hyperlipidemic

(Liu et al. 2012), UV-protective (Zhuang et al. 2009a, b, c) and immunostimulant activities (Sugahara et al. 2006).

In vitro studies conducted on the properties of jellyfish, collagen has shown that it can be used as a carrier and scaffold for various cell types. The extracted collagen has also exhibited high levels of cell viability and is non-toxic (Song et al. 2006). Compared to other collagens, jellyfish collagen has better cell viability in multiple tissues, such as the gastrointestinal tract and bone. This was studied by observing the high porosity and interconnected pore structure of the freeze-dried collagen scaffolds made from the NHS and EDC cross-linked with a jellyfish during in vivo testing. The results of the study revealed that the animal's immune response was similar to that of a cattle's. The potential of this material was also apparent when compared to other bio-based scaffolds (Song et al. 2006; Addad et al. 2011).

Another study looked at the proliferation and expression of chondrogenic markers in human mesenchymal stem cells (hMSCs) using porous scaffolds made from jellyfish collagen. It was discovered that it has no cytotoxic impact and promotes hMSC proliferation in this investigation. On jellyfish collagen scaffolds, chondrogenic-induced hMSC stimulation exhibited viable cells and elevation of chondrogenic markers at the mRNA level from day 1 to 21 (Hoyer et al. 2014). Khong et al. (2018) stated that jellyfish collagen can be safely used even at a concentration of around 1000 g/mL for 3 days. The non-toxicity of the collagens was observed in the various tests performed on it. However, the results of the LDH and MTT assays could not estimate the LC50 values for all the collagens.

The results of the study were consistent and showed that cells treated with jellyfish collagen exhibited better cell viability than those that were treated with other natural materials such as gelatin and bovine collagen. The researchers also noted that the vitality of the cells treated with the collagen did not decrease after 10 days (Song et al. 2006). In vitro, the results of the study showed that the use of jellyfish collagen did not cause adverse effects. However, the cells treated with collagen exhibited varying levels of vitality. The elevated levels of LDH, which are associated with a compromised plasma membrane, were also observed in the cells that were treated with collagen (Khong et al. 2018).

Human studies have shown that the release of LDH from collagen can trigger allergic reactions and inflammation. According to Khong et al. (2018), IASC method significantly decreased the release of LDH from collagen. This method is considered to be non-toxic and hypoallergenic and can be used for the treatment of various conditions.

12.6 Applications of Jellyfish Collagen

12.6.1 Nutraceuticals

Due to their popularity as a food source in Eastern countries, several studies were conducted on the traditional claims that consuming jellyfish can cure arthritis, gout and ulcers, decrease hypertension, treat bronchitis, alleviate back pain and stimulate blood flow, aside from improving digestion, remedying fatigue, aiding weight loss, reducing skin swelling and treating cancer (Rudloe 1992; Hsieh and Rudloe 1994; Jones and Rudloe 1995; Hsieh et al. 2001; You et al. 2016).

In Asia, semi-dried jellyfish are a multi-billion-dollar seafood industry (Hsieh et al. 2001). Traditionally processing techniques use a multi-phase processing approach by combining salt (NaCl) and alum ($AIKSO_4H.12H_20$), which is also known as aluminium sulphate with many uses, such as industrial flocculation. This combination is used to reduce water content, lower pH and firm the texture. Jellyfish are usually freshly harvested and processed lightly to improve durability and simplify subsequent processing, including cleaning and cutting (Abdullah et al. 2015).

Subsequent studies have also found that collagen from jellyfish has been reported to have a lower content of harmful elements, such as arsenic, lead and mercury, compared to the national standard for food industries (Zhang et al. 2014). Furthermore, the collagen content and nutritional composition of jellyfish are described to have low calorific values, low-fat content, immunostimulant and anti-oxidant properties, and high concentrations of proteins and minerals, making jellyfish a potential source of food, feed and functional food (Hsieh et al. 2001; Sugahara et al. 2006; Leone et al. 2015; Khong et al. 2016). For instance, isolated collagen from *Chrysaora* sp. showed considerable anti-oxidant and angiotensin-converting enzyme (ACE) inhibitory activities (Barzideh et al. 2014), which prevents the body from producing angiotensin II. The substance narrows blood vessels in humans, which may lead to hypertension, heart failure and other diseases (Peng et al. 2005). Aside from that, jellyfish are also a good source of protein, low in calories, carbs and crude fat, and are good to consume as a supplementary diet (Hsieh et al. 2001; Purcell et al. 2007; Khong et al. 2016).

Current new techniques have been developed for the stabilization and processing of jellyfish into semi-finished food items without the use of alum (Bleve et al. 2021). This is due to human health concerns, which include digestion problems, irritation to respiratory and gastrointestinal tracts, and eyes when consuming alum (Ahirrao 2008). To monitor the suggested methods and characterize the jellyfish-derived products, a set of safety and quality metrics, as well as technical and nutritional features, should be employed to further utilize this marine source in the nutraceutical industries.

12.6.2 Cosmeceuticals

Collagen is the principal component that contributes to the production of skin, muscle tissues, cartilage and bone (Hsieh and Rudloe 1994). It is widely marketed in the cosmeceutical industry as it largely contributes to the skin's elasticity and strength, the growth of tissues and organs, and the protection of the skin by preventing toxins and pathogens from being absorbed through the skin's pores (Fratzl 2008; Silvipriya et al. 2015). Because of its moisturizing, renewing and film-forming qualities, collagen is one of the most commonly used ingredients in cosmetic compositions, as it addresses dermatological problems such as degraded skin quality and density (Pachence 1987). It is a natural ingredient with humectant and skin moisturizer properties that can bind water, while also preventing trans-epidermal water loss (TEWL).

Skin ageing occurs as a result of collagen fibres becoming degraded over time, resulting in a reduction in the thickness and strength of the tissues. Hence, many people use collagen supplements as part of their anti-ageing skincare routine to give off a lush, young appearance (Rodríguez et al. 2017). One of the treatments of anti-ageing is wrinkle fillers. These are collagen-based fillers that are used to minimize the appearance of wrinkles. They are completely safe, natural, biocompatible, allergy-free, easy to remove and biodegradable.

In the cosmetics industry, collagen is normally available in the form of creams and nutritional supplements (Silvipriya et al. 2015). In the past, only bovine and swine skins were employed as the primary sources of collagen for industrial applications. However, the surge in diseases stemming from using these products, such as bovine spongiform encephalopathy (BSE), transmissible spongiform encephalopathy (TSE) and foot and mouth disease (FMD), has called for a re-evaluation of the use of bovine and swine as a source of collagen (Ogawa et al. 2004; Song et al. 2006). Due to these issues, there has been a recent surge in the use of marine-derived chemicals as cosmetic ingredients due to their chemical and biological diversity, and also their adaptations for surviving in harsh marine conditions. However, studies on jellyfish collagen in the cosmetic industry are still lacking. There are not enough studies to describe the collagen content, and the novelty of the product will also cause people to hesitate in using it. Jellyfish collagen contains a high number of anti-oxidants, which can protect the human skin from getting exposed to ultraviolet (UV) radiation from pro-oxidative damage due to UV-induced reactive oxygen species (ROS), such as hydroxyl radicals, superoxide radicals and hydrogen peroxide (H_2O_2) (Fan et al. 2013; Guillerme et al. 2017). In a cytotoxicity study by Aziz et al. (2021), collagen extract from the jellyfish *R. hispidum* seems to have no cytotoxic effects on the mouse embryonic 3 T3 fibroblast cell line and may have the potential to be developed as an active drug in cosmeceutical applications.

Polypeptide isolates from *R. pulmo* collagen have been reported to reduce cell damage, which is suitable for reducing damage to the skin. Furthermore, in *R. esculentum*, anti-fatigue collagen peptides and melanogenesis-inhibitory functions have also been revealed (Ding et al. 2011; Zhuang et al. 2009a, b, c).

Mucin, a glycoprotein molecule consisting of protein and saccharide, produced by jellyfish have a tonne of potential in the cosmetic industry as it can maintain the condition of the skin, aside from containing numerous hydrating benefits (Joseph 2017). A study found the moisturizing effects of jellyfish collagen reported a natural moisturizing protein called as filaggrin, that can able to increase the water-holding capacity of the skin. This protein amounted to 211.7% in *Nemopilema nomurai* Kishinouye, 1922 (Kim et al. 2016). Other proteins reported in the same study were hyaluronan synthase-3 (HAS-3) (139.9%) and aquaporin-3 (AQP-3) (212.5%), which is involved in the synthesis of hyaluronic acid, and transportation of water and glycerol through the membrane, respectively (Kim et al. 2016). Another study focused on the whitening effects of the same species of jellyfish, where the extract of *N. nomurai* is hydrolysed through commercial proteolytic enzymes. It was found that neutrase-treated collagen extract inhibited melanin synthesis by 89.9%, while at the same time decreasing the expression of tyrosinase, an oxidase responsible for controlling melanin production (Zhuang et al. 2009a, b, c; Lee et al. 2014). Additionally, in cosmetics, collagen can also be used as a thickening agent in hair, skin and oral care (Sionkowska et al. 2020).

One of the commercial jellyfish products contains 1000 ppm of jellyfish extract, along with other moisturizing ingredients such as hyaluronic acid, xanthan gum, glycerine, butylene glycol and water. Its key features include replenishing the skin's vitality and making it firmer due to high levels of collagen, proteins and glycoproteins; softening and smoothing out lines and wrinkles; and providing long-lasting hydration (Michalun and DiNardo 2014). Another product on the biodegradable cellulose sheet mask includes jellyfish extract, glycerine, butylene glycol, lecithin, etc. The product claims to be loaded with *R. esculentum* extract to strengthen the skin's barrier, boost skin's moisture, improve skin suppleness and promote healthier and younger-looking skin (personal observation of all authors).

On another note, after the discovery of the immortal jellyfish, *Turritopsis dohrnii* (Weismann, 1883), the cosmetic industry was inspired by its ability to reverse its life cycle from its medusa form to its polyp form and proceeded to mimic its properties and implement the idea of anti-ageing into a skin care product. Hence, an anti-ageing peptide was developed based on the idea and the peptide has been shown to be effective in cell rejuvenation and DNA damage reduction, where it reverses DNA damage, and aids with cell repair and longevity. A recent study looked at the genomics behind the species' rejuvenation abilities, and it was found that *T. dohrnii* can able to express genes that are involved in DNA replication, DNA repair, and apoptosis, and exhibit telomere maintenance activities, which are vital for the survival of cells and in cell proliferation and rejuvenation. Other than that, the study also proposed that the species contain pluripotency inducers, which is involved in the species' capability in differentiating into many cell types, and this is one of the factors of why *T. dohrnii* can able to skip death in its life cycle (Pascual-Torner et al. 2022). In a nutshell, collagen is one of the most sought-after products in the cosmetic industry, and studies have shown that jellyfish contain a considerable amount of collagen, which has moisturizing and whitening properties especially from species such as *Rhopilema asamushi, R. esculentum, R. hispidum, Rhizostoma pulmo, Stomolophus meleagris, Catostylus tagi, Nemopilema nomurai, Cotylorhiza tuberculata* and *Chrysaora sp.* (Nagai et al. 1999, 2000; Calejo et al. 2009; Addad et al. 2011; Lee et al. 2014; Zhang et al. 2014; Kim et al. 2016).

12.6.3 Pharmaceuticals

One of the preliminary studies on jellyfish collagen in the biotechnology sector started with a study on gelatin. A study on the physicochemical characteristics of jellyfish collagen revealed that the gelatin of *Rhopilema hispidum* consists of α1-chain, α2-chain, β-chain, γ-chain and glycine, proline, hydroxyproline and alanine amino acids (Cho et al. 2014). The same study also showed that jellyfish collagen also varies by species in terms of solubility. For instance, *R. hispidum* collagen has a high solubility at acidic pH but a low solubility at neutral pH, but the collagen of *A. aurita* can be retrieved at neutral pH (Miki et al. 2015), allowing its properties to be studied for future applications in tissue engineering.

A study to better understand the properties of jellyfish collagen and its potential as a substitute for collagen derived from mammalian sources. The study's findings revealed that jellyfish collagen may offer a reliable in vitro surrounding for the ovarian cancer cells OvCa-3 and SKOV-3, promoting their proliferation and migration in particular. Cancer cell attachment, shape and indicators of the epithelial to mesenchymal transition were all maintained and supported by moulded *R. pulmo* sponges. As a result, it has been demonstrated to be acceptable for advancing growing techniques and is an excellent substitute for mammalian protein supplies for human cell culture (Paradiso et al. 2019). The jellyfish *Rhizostoma luteum* (Quoy and Gaimard 1827) seems to be an outstanding candidate for future use in marine-derived drugs, as it has the most significant anti-oxidant activity, as well as a high content of polyunsaturated fatty acids (PUFAs), essential fatty acid linoleic and amino acid (Paradiso et al. 2019), which can improve metabolic patterns and reduce systemic inflammation (Cicero et al. 2012).

Another study has shown the haemostatic qualities of collagen derived from the mesoglea of the jellyfish *R. esculentum*, which was found to have haemostatic capabilities for wound healing (Cheng et al. 2017; Felician et al. 2019). Aside from that, jellyfish collagen has been claimed to have anti-fatigue, anti-cancer and anti-oxidant properties (Ding et al. 2011; Lee et al. 2014). The application of jellyfish collagen in industries such as medicinal, nutraceutical and biotechnological, however, requires sufficient funding from the government to mass process the technology required for industrial utilization (Kim et al. 2014).

According to Mearns-Spragg et al. (2020), jellyfish collagen can be utilized as a biomaterial, namely a cell matrix for culturing a type of glial cell located throughout parts of the brain and spinal cord, iPSC-derived Microglia (iMGL). Thus, iPSC is derived from skin or blood cells that have been reprogrammed back into an embryonic-like state that enables the development of human cells for therapeutic purposes, whereas microglia is a cell located in the brain and spinal cord that acts as the main form of active immune defence in the central nervous system. Further, they deduced that jellyfish collagen displays a superior option to rat tail mammalian type 1 collagen (RTC) when tested on the biological impact of human cells, as it did not cause an increase in cell clumping and cell deaths (Mearns-Spragg et al. 2020).

12.7 Future Perspectives

One of the most sought-after products in the pharmaceutical, nutraceutical and especially cosmetic industries is collagen, due to its anti-ageing properties. Jellyfish contain a considerable amount of collagen that has moisturizing properties, along with anti-oxidants and immunostimulants, and has potential in cell culture and regenerative medicine. With jellyfish blooms frequenting the waters, especially in areas where fisheries are important, scientists are looking for ways to utilize them beneficially. Due to the growing demand for skin care, food security and advancement of technology in the medicinal field, the usage of jellyfish collagen will not be limited to just functional food and sheet masks in the future. Hair care, bone tissue regeneration, wound dressing and mini pellets are industries that should be looking into harvesting and utilizing jellyfish collagen as a sustainable source. However, precautions should be taken so that jellyfish exploitation will not occur.

Acknowledgement We acknowledge the financial support received from the International Grant: LPA2007 from MABIK, Korea.

References

Abdullah A, Nurjanah N, Hidayat T, Aji DU et al (2015) Fatty acid profile of jellyfish (Aurelia aurita) as a source raw material of aquatic result rich benefit. Int J Chem Biomol Sci 1(1):12–16

Addad S, Exposito JY, Faye C, Ricard-Blum S, Lethias C et al (2011) Isolation, characterization and biological evaluation of jellyfish collagen for use in biomedical applications. Mar Drugs 9: 967–983

Ahirrao SD (2008) Effect of a common water purifier potash alum on drinking water quality and human health. Nat Environ Pollut Technol 7(1):79–81

Alemán A, Giménez B, Pérez-Santín E, Gómez-Guillén MC, Montero P et al (2011) Contribution of Leu and Hyp residues to antioxidant and ACE-inhibitory activities of peptides sequences isolated from squid gelatine hydrolysate. Food Chem 125:334–341

Arai MN (1997) A functional biology of Scyphozoa. Chapman & Hall, Netherlands

Aziz NA, Salim N, Zarei M, Saari N, Yusoff FM et al (2021) Extraction, anti-tyrosinase, and antioxidant activities of the collagen hydrolysate derived from Rhopilema hispidum. Prep Biochem Biotechnol 51(1):44–53

Barzideh Z, Latiff AA, Gan C-Y, Benjakul S, Karim AA et al (2013) Isolation and characterisation of collagen from the ribbon jellyfish (Chrysaora sp.). Int J Food Sci Technol 49(6):1490–1499

Barzideh Z, Latiff AA, Gan CY, Abedin M, Alias AK et al (2014) ACE inhibitory and antioxidant activities of collagen hydrolysates from the ribbon jellyfish (Chrysaora sp.). Food Technol Biotechnol 52(4):495–504

Benedetto CD, Barbaglio A, Martinello T, Alongi V, Fassini D, Cullorà E, Patruno M, Bonasoro F, Barbosa MA, Carnevali MDC, Sugni M et al (2014) Production, characterization and biocompatibility of marine collagen matrices from an alternative and sustainable source: the sea urchin Paracentrotus lividus. Mar Drugs 12(9):4912–4933

Berillis P (2015) Marine collagen: extraction and applications. Res Trends Biochem Mol Biol Microbiol 1–13

Bleve G, Ramires FA, De Domenico S, Leone A et al (2021) An alum-free jellyfish treatment for food applications. Front Nutrition 559

Boero F (2013) Review of jellyfish blooms in the Mediterranean and Black Sea. FAO, Rome, pp 1–35

Brunt EG, Burgess JG (2018) The promise of marine molecules as cosmetic active ingredients. Int J Cosmet Sci 40(1):1–15

Calejo MT, Morais ZB, Fernandes AI et al (2009) Isolation and biochemical characterisation of a novel collagen from Catostylus tagi. J Biomater Sci 20(14):2073–2087

Camacho F, Macedo A, Malcata F et al (2019) Potential industrial applications and commercialization of microalgae in the functional food and feed industries: a short review. Mar Drugs 17(312)

Cartwright P, Halgedahl SL, Hendricks JR, Jarrard RD, Marques AC, Collins AG, Lieberman BS et al (2007) Exceptionally preserved jellyfishes from the Middle Cambrian. PLoS One 2(10): e1121

Cheah KS (1985) Collagen genes and inherited connective tissue disease. Biochem J 229(2):287

Cheng X, Shao Z, Li C, Yu L, Raja MA, Liu C et al (2017) Isolation, characterization and evaluation of collagen from jellyfish Rhopilema esculentum Kishinouye for use in hemostatic aplications. PLoS One 12(1):e0169731

Cho S, Ahn JR, Koo JS, Kim SB et al (2014) Physicochemical properties of gelatin from jellyfish Rhopilema hispidum. Fish Aquatic Sci 17(3):299–304

Chuan CH, Venmathi Maran BA, Yap TK, Cheong KC, Hussein MAS, Saleh E, Tan SH et al (2020) First record of jellyfish Anomalorhiza shawi Light, 1921 (Cnidaria: Scyphozoa) and its associated organisms in Sabah, Malaysia. Reg Stud Mar Sci 35:1–7

Chuan CH, Venmathi Maran BA, Yap TK, Cheong KC, Hussein MAS, Saleh E et al (2021) New records of cubozoan and scyphozoan jellyfish from Sabah waters. Malaysia Diversity 13:420

Cicero AFG, Reggi A, Parini A, Borghi C et al (2012) Application of polyunsaturated fatty acids in internal medicine: beyond the established cardiovascular effects. Arch Med Sci 8(5):784–793

Coppola D, Oliviero M, Vitale GA, Lauritano C, D'Ambra I, Iannace S, de Pascale D et al (2020) Marine collagen from alternative and sustainable sources: extraction, processing and applications. Mar Drugs 18(4):214

Cumming MH, Leonard AR, LeCorre-Bordes DS, Hofman K et al (2018) Intra-fibrillar citric acid crosslinking of marine collagen electrospun nanofibres. Int J Biol Macromol 114:874–881

Dawson MN (2003) Macro-morphological variation among cryptic species of the moon jellyfish, Aurelia (Cnidaria: Scyphozoa). Mar Biol 143(2):369–379

Dawson MN, Martin LE, Lolita K, Penland K et al (2001) Jellyfish swarms, tourists, and the Christ-child. Hydrobiologia 451:131–144

Di Lullo GA, Sweeney SM, Korkko J, Ala-Kokko L, San Antonio JD et al (2002) Mapping the ligand-binding sites and disease-associated mutations on the most abundant protein in the human, type I collagen. J Biol Chem 277(6):4223–4231

Diehl-Seifert B, Kurelec B, Zhan RK, Dorn A, Jericevic B et al (1985) Attachment of sponge cells to collagen substrata: effects of a collagen assembly factor. J Cell Sci 79:271–285

Ding JF, Li YY, Xu JJ, Su XR, Gao X, Yue FP et al (2011) Study on effect of jellyfish collagen hydrolysate on anti-fatigue and anti-oxidation. Food Hydrocoll 25:1350–1353

Diogo GS, López-Senra EL, Pirraco RP, Canadas RF, Fernandes EM, Serra J, Pérez-Martín RI, Sotelo CG, Marques AP, González P, Moreira-Silva J, Silva TH, Reis RL et al (2018) Marine collagen/apatite composite scaffolds envisaging hard tissue applications. Mar Drugs 16(8):269

Dong Z, Liu D, Keesing JK (2010) Jellyfish blooms in China: dominant species, causes and consequences. Mar Pollut Bull 60(7):954–963

Duan R, Zhang J, Du X, Yao X, Konno K et al (2009) Properties of collagen from skin, scale and bone of carp (Cyprinus carpio). Food Chem 112(3):702–706

Fan J, Zhuang Y, Li B et al (2013) Effects of collagen and collagen hydrolysate from jellyfish umbrella on histological and immunity changes of mice photoaging. Nutrients 5(1):223–233

Felician FF, Yu RH, Li MZ, Li CJ, Chen HQ, Jiang Y, Xu HM et al (2019) The wound healing potential of collagen peptides derived from the jellyfish Rhopilema esculentum. Chin J Traumatol 22(1):12–20

Fratzl P (2008) Collagen: structure and mechanics. Springer, New York, pp 1–496

Gambini C, Abou B, Ponton A, Cornelissen AJ et al (2012) Micro and macrorheology of jellyfish extracellular matrix. Biophys J 102(1):1–9

Gellenbeck KW (2012) Utilization of algal materials for nutraceutical and cosmeceutical applications-what do manufacturers need to know? J Appl Phycol 24:309–313
Gómez-Guillén MC, Giménez B, López-Caballero MA, Montero MP et al (2011) Functional and bioactive properties of collagen and gelatin from alternative sources: a review. Food Hydrocoll 25(8):1813–1827
Guillerme JB, Couteau C, Coiffard L et al (2017) Applications for marine resources in cosmetics. Cosmetics 4(3):35
Haddock SHD (2004) A golden age of gelata: past and future research on planktonic ctenophores and cnidarians. Hydrobiologia 530-531:549–556
Heu MS, Lee JH, Kim HJ, Jee SJ, Lee JS, Jeon YJ, Kim JS et al (2010) Characterization of acid-and pepsin-soluble collagens from flatfish skin. Food Sci Biotechnol 19(1):27–33
Himaya SWA, Ngo DH, Ryu B, Kim SK et al (2012) An active peptide purified from gastrointestinal enzyme hydrolysate of Pacific cod skin gelatine attenuates angiotensin-1 converting enzyme (ACE) activity and cellular oxidative stress. Food Chem 132(4):1872–1882
Hoyer B, Bernhardt A, Lode A, Heinemann S, Sewing J, Klinger M, Notbohm H, Gelinsky M (2014) Jellyfish collagen scaffolds for cartilage tissue engineering. Acta Biomater 10(2):883–892. https://doi.org/10.1016/j.actbio.2013.10.022
Hsieh Y-HP, Rudloe J (1994) Potential of utilizing jellyfish as food in Western countries. Trends Food Sci Technol 5:225–229
Hsieh Y-HP, Leong FM, Rudloe J et al (2001) Jellyfish as food. Hydrobiologia 45(1):11–17
Hu Z, Yang P, Zhou C, Li S, Hong P et al (2017) Marine collagen peptides from the skin of Nile tilapia (Oreochromis niloticus): characterization and would healing evaluation. Mar Drugs 15: 102
Huang YAW (1988) Cannonball jellyfish (Stomolophus meleagris) as a food resource. J Food Sci 53(2):341–343
Hwang JH, Mizuta S, Yokoyama Y, Yoshinaka R et al (2007) Purification and characterization of molecular species of collagen in the skin of skate (Raja kenojei). Food Chem 100(3):921–925
Jafari H, Lista A, Siekapen MM, Ghaffari-Bohlouli P, Nie L, Alimoradi H, Shavandi A (2020) Fish collagen: extraction, characterization, and applications for biomaterials engineering. Polymers (Basel) 12(10):2230
Jones RP, Rudloe J (1995) Harvesting & processing of Florida cannonball jellyfish. Report to the Florida international affairs commission, Florida Department of Commerce. Southeastern Fisheries Association, Tallahassee, p 28
Jongjareonrak A, Benjakul S, Visessanguan W, Nagai T, Tanaka M et al (2005) Isolation and characterisation of acid and pepsin-solubilised collagens from the skin of Brownstripe red snapper (Lutjanus vitta). Food Chem 93(3):475–484
Joseph A (2017) Abode of nutraceuticals, pharmaceuticals and biotoxins. In: Joseph A (ed) Investigating seafloors and oceans. Elsevier, Oxford, pp 493–554
Kadler KE, Baldock C, Bella J, Boot-Handford RP et al (2007) Collagens at a glance. J Cell Sci 120(12):1955–1958
Katzman RL, Kang AH (1972) The presence of fucose, mannose, and glucosamine-containing heteropolysaccharide in collagen from the sea anemone Metridium dianthus. J Biol Chem 247(17):5486–5489
Khan SB, Qian Z-J, Ryu B, Kim S-K et al (2009) Isolation and biochemical characterizatio of collagens from seaweed pipefish, Syngnathus schlegeli. Biotechnol Bioprocess Eng 14:436–442
Khong NM, Yusoff FM, Jamilah B, Basri M, Maznah I, Chan KW, Nishikawa J et al (2016) Nutritional composition and total collagen content of three commercially important edible jellyfish. Food Chem 196:953–960
Khong NM, Yusoff FM, Jamilah B, Basri M, Maznah I, Chan KW, Nishikawa J et al (2018) Improved collagen extraction from jellyfish (Acromitus hardenbergi) with increased physical-induced solubilization processes. Food Chem 251:41–50

Kim SK, Kim YT, Byun HG, Nam KS, Joo DS, Shahidi F et al (2001) Isolation and characterization of antioxidative peptides from gelatine hydrolysate of Alaksa. J Agric Food Chem 49:1984–1989

Kim DY, Lee J, Kim D et al (2014) A study on direction of industrial utilization for jellyfish in Korea. Fish Ocean Educ Res 26(3):587–596

Kim DW, Park TS, Kim JY, Choi SG, Lee DW et al (2016) Moisturising effect of jellyfish collagen extract. J Soc Cosmet Sci Korea 42(2):153–162

Kimura S, Miura S, Park YH et al (1983) Collagen as the major edible component of jellyfish (Stomolophus nomural). J Food Sci 48(6):1758–1760

Kittiphattanabawon P, Benjakul S, Visessanguan W, Kishimura H, Shahidi F et al (2010) Isolation and characterisation of collagen from the skin of brownbanded bamboo shark (Chiloscyllium punctatum). Food Chem 119(4):1519–1526

Kolodziejska I, Sikorski ZE, Niecikowska C et al (1999) Parameters affecting the isolation of collagen from squid (Illex argentinus) skins. Food Chem 66:153–157

Lee HY, Kim YH, Um SY, Jung US, Chang MS, Lee NH et al (2014) Melanogenesis inhibition effects of Nemopilema nomurai hydrolysed extracts. J Soc Cosmet Sci Korea 40(4):341–347

Lee H, Bae SK, Kim M, Pyo MJ, Kim M, Yang S, Won C-K, Yoon WD, Han CH, Kang C et al (2017) Anticancer effect of Nemopilema nomurai jellyfish venom on hepg2 cells and a tumor xenograft animal model. Evidence-Based Complementary and Alternative Medicine:1–12

Leone A, Lecci RM, Durante M, Meli F, Piraino S et al (2015) The bright side of gelatinous blooms: nutraceutical value and antioxidant properties of three Mediterranean jellyfish (Scyphozoa). Mar Drugs 13(8):4654–4681

Leone A, Lecci RM, Milisenda G, Piraino S et al (2019) Mediterranean jellyfish as novel food: effects of thermal processing on antioxidant, phenolic, and protein contents. Eur Food Res Technol 245(8):1611–1627

Lewis WR, Billington R, Debryune E, Gamer A, Lang B, Carpanini F et al (2002) Recognition of adverse and nonadverse effects in toxicity studies. Toxicol Pathol 30(1):66–74

Li C, Li P, Feng J, Li I, Yu H et al (2012) Cytotoxicity of the venom from the nematocysts of jellyfish Cyanea nozakii Kishinouye. Toxicol Ind Health 28:186–192

Liang J, Pei X, Zhang Z, Wang N, Wang J, Li Y et al (2010) The protective effects of long-term oral administration of marine collagen hydrolysate from chum salmon on collagen matrix homeostasis in the chronological aged skin of Sprague-Dawley male rats. J Food Sci 75(8):H230–H238

Liu X, Zhang M, Zhang C, Liu C et al (2012) Angiotensin converting enzyme (ACE) inhibitory, antihypertensive and antihyperlipidaemic activities of protein hydrolysates from Rhopilema esculentum. Food Chem 134:2134–2140

Low LB, Syazwan WM, Rizman-Idid M et al (2019) A morphological evaluation of Chrysaora chinensis of peninsular Malaysia and distinguishing its populations using geometric morphometrics. Acta Oceanol Sin 38(10):67–74

Lucas CH (2001) Reproduction and life history strategies of the common jellyfish, Aurelia aurita, in relation to its ambient environment. Hydrobiologia 451:229–246

Mearns-Spragg A, Tilman J, Tams D, Barnes A et al (2020) The biological evaluation of jellyfish collagen as a new research tool for the growth and culture of iPSC derived microglia. Front Mar Sci 7:689

Meglitsch PA (1991) Invertebrate zoology. Oxford University Press, p 86

Mendis E, Rajapakse N, Kim SK (2005a) Antioxidant properties of a radical-scavenging peptide purified from enzymatically prepared fish skin gelatine hydrolysate. J Agric Food Chem 53: 581–587

Mendis E, Rajapakse N, Byun HG, Kim SK et al (2005b) Investigation of jumbo squid (Dosidicus gigas) skin gelatin peptides for their in vitro antioxidant effects. Life Sci 77:2166–2178

Michalun MV, DiNardo JC (2014) Skin care and cosmetic ingredients dictionary. Milady, New York, pp 1–339

Miki A, Inaba-Inoue S, Baba T, Kihira K, Fukada H, Oda M et al (2015) Structural and physical properties of collagen extracted from moon jellyfish under neutral pH conditions. Biosci Biotech Bioch 79(10):1–5

Miller EJ, Piez K, Reddi A et al (1984) Biomedical and industrial application of collagen. Elsevier, New York, pp 41–81

Mitura S, Sionkowska A, Jaiswal A et al (2020) Biopolymers for hydrogels in cosmetics. J Mater Sci Mater Med 31(6):1–14

Miura S, Kimura S (1985) Jellyfish mesogloea collagen – characterization of molecules as α1α2α3 heterotrimers. J Biol Chem 260:15352–15356

Müller WE (2003) The origin of metazoan complexity: Porifera as integrated animals. Integr Comp Biol 43(1):3–10

Nagai T (2004) Collagen from diamondback squid (Thysanoteuthis rhombus) outer skins. Z Naturforsch C 59:271–275

Nagai T, Ogawa T, Nakamura T, Nakamura H et al (1999) Collagen of edible jellyfish exumbrella. J Sci Food Agr 79:855–858

Nagai T, Worawattanamateekul W, Suzuki N, Nakamura T, Ito T, Fujiki K, Nakao M, Yano T et al (2000) Isolation and characterization of collagen from rhizostomous jellyfish (Rhopilema asamushi). Food Chem 70:205–208

Nagai T, Yamashita E, Taniguchi K, Kanamori N, Suzuki N et al (2001) Isolation and characterization of collagen from the outer skin waste material of cuttlefish (Sepia lycidas). Food Chem 72:425–429

Nagai T, Araki Y, Suzuki N et al (2002a) Collagen of the skin of ocellate puffer fish (Takifugu rubripes). Food Chem 78:173–177

Nagai T, Nagamori K, Yamashita E, Suzuki N et al (2002b) Collagen of octopus Callistoctopus arakawai arm. Int J Food Sci Technol 37:285–289

Ngo DH, Ryu B, Vo TS, Himaya SWA, Wijesekara I, Kim SK et al (2011) Free radical scavenging and angiotensin-I converting enzyme inhibitory peptides from Pacific cod (Gadus macrocephalus) skin gelatine. Int J Biol Macromol 49(5):1110–1116

O'Sullivan K (2017) Stinger jellyfish swarms wipe out farmed salmon in west of Ireland. The Irish Times: https://www.irishtimes.com/news/environment/stinger-jellyfish-swarms-wipe-out-farmed-salmon-in-west-of-ireland-1.3247245. Accessed 5 Dec 2020

Ogawa M, Portier RJ, Moody MW, Bell J, Schexnayder MA, Losso JN et al (2004) Biochemical properties of bone and scale collagens isolated from the subtropical fish black drum (Pogonia cromis) and sheepshead seabream (Archosargus probatocephalus). Food Chem 88(4):495–501

Pachence JM (1987) Collagen: its place in the medical device industry. Med Device Diagnostic Ind 9:49–55

Paradiso F, Fitzgerald J, Yao S, Barry F, Taraballi F, Gonzalez D, Francis L et al (2019) Marine collagen substrates for 2D and 3D ovarian cancer cell systems. Front Bioeng Biotechnol 7:343

Pascual-Torner M, Carrero D, Perez-Silva JG, Alvarez-Puente D, Roiz-Valle D, Bretones G, Rodrigues D, Maeso D, Mateo-Gonzalez E, Espanol Y, Marino G, Acuna JL, Quesada V, Lopez-Otin C et al (2022) Comparative genomics of mortal and immortal cnidarians unveils novel keys behind rejuvenation. Dev Biol 119(36):e2118763119

Pei X, Yang R, Zhang Z, Gao L, Wang J, Xu Y, Zhao M, Han X, Liu Z, Li Y et al (2010) Marine collagen peptide isolated from chum salmon (*Oncorhynchus keta*) skin facilitates learning and memory in aged C57BL/6J mice. Food Chem 118(2):333–340

Peng H, Carretero OA, Vuljaj N, Liao T-D, Motivala A, Peterson EL, Rhaleb N-E et al (2005) Angiotensin-converting enzyme inhibitors: a new mechanism of action. Circulation 112:2436–2445

Purcell JE, Uye S, Lo W et al (2007) Anthropogenic causes of jellyfish blooms and their direct consequences for humans: a review. Mar Ecol Prog Ser 350:153–174

Ramshaw JA, Peng YY, Glattauer V, Werkmeister JA et al (2009) Collagens as biomaterials. J Mater Sci Mater Med 20(1):3–8

Rodríguez MIA, Barroso LGR, Sánchez ML et al (2017) Collagen: a review on its sources and potential cosmetic applications. J Cosmet Dermatol 17(1):20

Rudloe J (1992) Jellyfish: a new fishery for the Florida panhandle. Report prepared for the Apalachee regional planning council, Blounstown. U.S. Department of Commerce, Economic Development Administration, EDA, Florida, pp 35

Russell FS (1958) The Medusae of the British Isles Vol. I: Anthomdedusae, Leptomedusae, Limnomedusae, Trachymedusae, and Narcomeusae. Iver: Cambridge University Press

Sadowska M, Kołodziejska I, Niecikowska C et al (2003) Isolation of collagen from the skins of Baltic cod (*Gadus morhua*). Food Chem 81(2):257–262

Sarras MP, Madden ME, Zhang X, Gunwar S, Huff JK, Hudson BG et al (1991) Extracellular matrix (mesoglea) of *Hydra vulgaris*. Dev Biol 148(2):481–494

Senaratne LS, Park PJ, Kim SK et al (2006) Isolation and characterization of collagen from brown backed toadfish (*Lagocephalus gloveri*) skin. Bioresour Technol 97(2):191–197

Shimomura O, Johnson FH, Saiga Y et al (1962) Extraction, purification and properties of aequorin, a bioluminescent protein from the luminous hydromedusan, *Aequorea*. J Cell Comp Physiol 59(3):223–239

Silvipriya KS, Kumar KK, Bhat AR, Kumar BD, John A, Iakshmanan P et al (2015) Collagen: animal sources and biomedical applications. J Appl Pharmaceut Sci 5(3):123–127

Sionkowska A, Adamiak K, Musial K, Gadomska M et al (2020) Collagen based materials in cosmetic applications: a review. Materials 13:4217

Smith K (2007) Jellyfish attack wipes out N. Ireland salmon farm. https://www.reuters.com/article/us-irish-jellyfish-idUSL2241858320071122. Accessed 5 Dec 2020

Song E, Kim SY, Chun T, Byun HJ, Lee YM et al (2006) Collagen scaffolds derived from a marine source and their biocompatibility. Biomaterials 27(15):2951–2961

Stoecker DK, Michaels AE, Davis LH et al (1987) Grazing by the jellyfish, *Aurelia aurita*, on microzooplankton. J Plankton Res 9(5):901–915

Sugahara T, Ueno M, Goto Y, Shiraishi R, Doi M, Akiyama K, Yamauchi S et al (2006) Immunostimulation effect of jellyfish collagen. Biosci Biotechnol Biochem 70(9):2131–2137

Swatschek D, Schatton W, Muller WEG, Kreuter J et al (2002) Microparticles derived from marine sponge collagen (SCMPs): preparation, characterization and suitability for dermal delivery of all-trans retinol. Eur J Pharm Biopharm 54:123–133

Tillet-Barret E, Franc J-M, Franc S, Garrone R et al (1992) Characterization of heterotrimeric collagen molecules in a sea-pen (Cnidaria, Octocorallia). Eur J Biochem 203(1–2):179–184

Trim SA, Wandrey F, Trim CM et al (2020) Beauty from the deep: cnidarians in cosmetics. Nova Science Publishers

Uye S (2008) Blooms of the giant jellyfish *Nemopilema nomurai*: a threat to the fisheries sustainability of the East Asian marginal seas. Plankton Benthos Res 3:125–131

Vo TS, Ngo DH, Kim JA, Ryu B, Kim SK et al (2011) An antihypertensive peptide from tilapia gelatin diminishes free radical formation in murine microglial cells. J Agric Food Chem 59(22): 12193–12197

Xiong JQ, Kurade MB, Jeon BH et al (2018) Can microalgae remove pharmaceutical contaminants from water? Trends Biotechnol 36:30–44

Xu Y, Han X, Li Y (2010) Effect of marine collagen peptides on long bone development in growing rats. J Sci Food Agric 90(9):1485–1491

You K, Bian Y, Ma C, Chi X, Liu Z, Zhang Y et al (2016) Study on the carry capacity of edible jellyfish fishery in Liaodong Bay. J Ocean Univ China 15(3):471–479

Yu HH, Liu XG, Xing RE, Liu S, Guo ZY, Wang PB, Li CP, Li PC et al (2006) In vitro determination of antioxidant activity of proteins from jellyfish *Rhopilema esculentum*. Food Chem 95:123–130

Yusoff FM, Bakar J, Basri M, Ismail M, Khong NMH et al (2013) A method for extracting collagen from aquatic animals, collagen and products containing it. Patent pending

Zhang Y, Liu W, Li G, Shi B, Miao Y, Wu X et al (2007) Isolation and partial characterization of pepsin-soluble collagen from the skin of grass carp (*Ctenopharyngodon idella*). Food Chem 103(3):906–912

Zhang J, Duan R, Huang L, Song Y, Regenstein JM et al (2014) Characterisation of acid-soluble and pepsin-solubilised collagen from jellyfish (*Cyanea nozakii* Kishinouye). Food Chem 150: 22–26

Zhuang Y, Zhao X, Li B et al (2009a) Optimization of antioxidant activity by response surface methodology in hydrolysates of jellyfish (*Rhopilema esculentum*) umbrella collagen. J Zhejiang Univ Sci B 10(8):572–579

Zhuang Y, Sun L, Zhao X, Wang J, Hou H, Li B (2009b) Antioxidant and melanogenesis-inhibitory activities of collagen peptide from jellyfish (*Rhopilema esculentum*). J Sci Food Agric 89(10): 1722–1727

Zhuang Y, Hou H, Zhao X, Zhang Z, Li B et al (2009c) Effects of collagen and collagen hydrolysate from jellyfish (*Rhopilema esculentum*) on mice skin photoaging induced by UV irradiation. J Food Sci 74(6):183–188

Application of Biotechnology in White Syndrome Coral Disease Identification

13

Khodzori Fikri Akmal, Saad Shahbudin, Zepri Abdul Muhaimin, Muhammad Dawood Shah, and Wei Sheng Chong

Abstract

Coral disease is one of the major threats affecting Caribbean reefs over the past four decades. In the Indo-Pacific, an increasing number of disease signs and syndromes have been documented at various reef localities, with white syndrome (WS) being the most common. Hence, most efforts are directed towards understanding the causal microbial pathogens using biotechnological applications. Based on preliminary coral surveys at selected reef sites around Tioman Island, Malaysia, six coral diseases and eight signs of compromised health were identified with the WS commonly found afflicting the *Acropora* and *Montipora* corals. In light of this, further examination was done to identify the potential of microbial pathogens from the apparently recorded WS coral disease using biochemical, molecular and histological methods. The data presented constitute baseline information on the status of coral disease that could be used by the relevant authorities to improve management and regulatory approaches in this marine-protected area.

K. F. Akmal (✉) · Z. Abdul Muhaimin · M. D. Shah · W. S. Chong
Borneo Marine Research Institute, Universiti Malaysia Sabah, Jalan UMS, Kota Kinabalu, Sabah, Malaysia
e-mail: mfakmal@ums.edu.my

S. Shahbudin
Department of Marine Science, Kulliyyah of Science, International Islamic University Malaysia, Jalan Sultan Ahmad Shah, Kuantan, Pahang, Malaysia

M. D. Shah et al. (eds.), *Marine Biotechnology: Applications in Food, Drugs and Energy*, https://doi.org/10.1007/978-981-99-0624-6_13

13.1 Introduction

Malaysia's coral reefs cover nearly 3600 km^2 (Wilkinson 2008) and contain more than 500 hermatypic coral species (Waheed 2016). However, they are increasingly exposed to multiple threats. On the east coast of Peninsular Malaysia, extensive coastal development is the major threat to coral reefs (Akmal et al. 2019; Akmal and Shahbudin 2021), while sedimentation has contributed to poor reef development on the west coast of the peninsular (Safuan et al. 2018; Akmal et al. 2018). This is exacerbated by bleaching events in 1998 and 2010 which resulted in widespread coral bleaching around the islands of Peninsular Malaysia (Tun et al. 2010; Tan and Heron 2011). The bleaching phenomenon puts the persistence of coral reefs under pressure and may lead to the spread of coral disease (Manzello 2015; Walton et al. 2018).

Coral disease is defined as any impairment that affects coral health, resulting in physiological and morphological dysfunctions (Woodley et al. 2016). It can also be described as any abnormal sign, jeopardizing coral health (Beeden et al. 2008; Raymundo et al. 2008). Disease emergence is caused by the effect of ocean warming and the associated anthropogenic stressors (Mydlarz et al. 2009; Woodley et al. 2016). According to the US National Oceanic and Atmospheric Administration (NOAA), the average global ocean temperature is expected to rise by approximately 1 to 4 °C in 2100 (Collins et al. 2013). This contributes to the bleaching events and coral disease outbreaks either independently or synergistically with other environmental drivers such as sedimentation and land-based pollution (Pollock et al. 2014; Maynard et al. 2015).

The coral disease is widely distributed, affecting reefs throughout the Indo-Pacific and the Caribbean (Weil et al. 2012; Woodley et al. 2016). About 40 coral diseases have been identified globally (Bruckner 2015), with the Caribbean accounting for more than 35 diseases (Weil 2004; Work and Aeby 2006). Meanwhile, coral diseases varied across the Indo-Pacific. Seven diseases, for example, have been identified in the Great Barrier Reefs (GBR) (Willis et al. 2004) and ten identified along the reefs in Japan (Wada et al. 2018), with the white syndrome (WS) being common in both reef regions. WS is characterized as any distinctive patches or bands of the exposed white coral skeleton (Work and Aeby 2011), afflicting the most common *Acropora*, *Montipora* and *Pachyseris* corals (Sussman et al. 2008). The epizootic is notably known for the mass hard coral mortality (Willis et al. 2004). Environmental stressors such as rising seawater temperatures and nutrient pollution have been linked to WS outbreaks in the GBR (Bruno et al. 2007; Pollock et al. 2014) and other Indo-Pacific coral assemblages (Ruiz-Moreno et al. 2012; Couch et al. 2014; Aeby et al. 2016; Weil et al. 2019).

Despite having widespread distribution and causing regional reef decline, the aetiology of WS is yet to be comprehensively determined (e.g. Sussman et al. 2008; Luna et al. 2010; Zhenyu et al. 2013; Ushijima et al. 2014; Nugraha et al. 2019). Hence, most efforts are directed towards understanding the causative microbial pathogen using several disease diagnostic tools, including culturing biochemical techniques (Ushijima et al. 2012; Zhenyu et al. 2013; Pollock et al. 2014; Nugraha et al. 2019) and emerging molecular tools such as fluorescent in situ hybridization (Pollock et al. 2014), denaturing gradient gel electrophoresis (Bourne 2005), microarray (Edge et al. 2005) and metagenomic (Okamura et al. 2010) as well as a variety of histological techniques (Pollock et al. 2011; Ushijima et al. 2012; Work and Meteyer 2014; Wada et al. 2018).

In Malaysia, the study of coral disease is still in its infancy stage. The reef survey is primarily focusing on the status of coral conditions (e.g. Toda et al. 2007; Shahbudin et al. 2017; Safuan et al. 2018; Hanapiah et al. 2019; Akmal et al. 2018, 2019; Akmal and Shahbudin 2021) rather than to prioritize on coral disease prevalence (e.g. Miller et al. 2015; Akmal and Shahbudin 2020) due to the lack of a coral disease taxonomist. Therefore, this study aims to identify the coral disease and signs of compromised coral health at selected reef sites around Tioman Island, Malaysia. The present study also attempts to examine the microbial pathogens of the common Indo-Pacific WS using biochemical, molecular and histological methods. The findings of this study contribute to the baseline information on the coral disease, including the possible causal agents associated with WS for future conservation plans in the Marine Park of Tioman Island.

13.2 Tioman Island Marine Park

Tioman Island is the largest island, situated on the east coast of Peninsular Malaysia (02°48′52.1″N and 104°10′29.3″E) off the South China Sea. It covers approximately 21 km in length and 12 km in width. The island's topography is covered by nearly 58% of rocky headlands and cliffs, while the remaining 42% is covered by sandy beach coastal areas (Department of Marine Park Malaysia 2011). It is estimated that approximately 3200 people live in nine villages including three (Tekek, Air Batang and Salang villages) on the northwest coast, four (Lanting, Genting, Paya and Nipah villages) on the southeast coast, one (Mukut) on the south coast and one (Juara) on the east coast of Tioman Island (Omar et al. 2015). Tekek serves as the capital village, providing many facilities, including a health clinic, bank, jetty complex, marina, several retail stores and an airport.

Tioman Island has rapidly evolved from a remote fishing village to a thriving tourist destination (Omar et al. 2015). From 2000 to 2017, this Marine Park attracted more than 200 thousand local and foreign tourists per year (Department of Marine Park Malaysia 2017). Besides, approximately 72 resorts and 34 dive centres have been developed and the majority are concentrated along the west coast site of Tioman Island (Reef Check Malaysia 2018). However, increased coastal development and tourism in adjacent coastal areas have potentially contributed to negative impacts which lead to low coverage of live corals at many of the reef sites reported on the west coast site of Tioman (Toda et al. 2007; Shahbudin et al. 2017; Akmal et al. 2019). Furthermore, some of the reef sites were also reported to have experienced a bleaching event in 2010, which resulted in the partial mortality of approximately 50% of the live corals (Tan and Heron 2011). Such natural and human-induced disturbances might contribute to the spread of coral disease, which could also be linked to a loss of tourism revenue for local communities.

13.3 Coral Disease Survey and Sample Collection

In this study, coral disease surveys were done four times in 2 years (July 2018, October 2018, March 2019 and June 2019) at three selected reef sites namely Sanggit, Salang and Bakau Bays around Tioman Island (Fig. 13.1). Diseased and compromised coral colonies observed in the study area were photographed using an Olympus TG-4 underwater camera. For sample collection, six coral fragments (1–2 cm in diameter) displaying the WS sign were collected from a depth between 3 and 15 m at each reef site. The targeted corals for WS were tabular species of *Acropora* (*A. cytherea* and *A. hyacinthus*) and foliose species of *Montipora aequituberculata*. Other samples, including seawater, sediment and algae found in adjacent infected coral colonies were also collected in the study area. In total, 27 samples were collected ($n = 9$ per site) for each sampling month. Throughout the sampling period, a total of 108 samples were collected including 72 infected corals, 12 water, 12 sediment and 12 algae. All samples were collected under permit from the section of Marine Park Malaysia (permit number: JTLM 630–7 Jld. 8–49).

13.4 Coral Disease and Signs of Compromised Health Recorded in Tioman Island

Based on the lesion characteristics including type, pattern, colour and/or progression rate (Beeden et al. 2008; Raymundo et al. 2008), field observations identified six coral diseases in Tioman Island, namely white syndrome (Fig. 13.2), skeletal eroding band (Fig. 13.3), yellow band disease (Fig. 13.4), unexplained growth anomalies (Fig. 13.5), ulcerative white spots (Fig. 13.6) and atramentous necrosis (Fig. 13.7). Furthermore, eight signs of coral compromised health were identified in the study area, including focal bleaching (Fig. 13.8), pigmentation response (Fig. 13.9), algal and sponge overgrowth (Fig. 13.10), predation scars (Fig. 13.11), sediment necrosis (Fig. 13.12), physical damage (Fig. 13.13), explained growth anomalies (Fig. 13.14) and trematodiasis (Fig. 13.15). Coral surveys also indicated that white syndrome (WS) is the most common disease and is susceptible to *Acropora* (*A. hyacinthus* and *A. cytherea*) and *Montipora* (*M. aequituberculata* and *M. foliosa*) corals. Hence, further examination was done to determine the causal microbial agents on the occurrence of WS recorded in Tioman Island Marine Park.

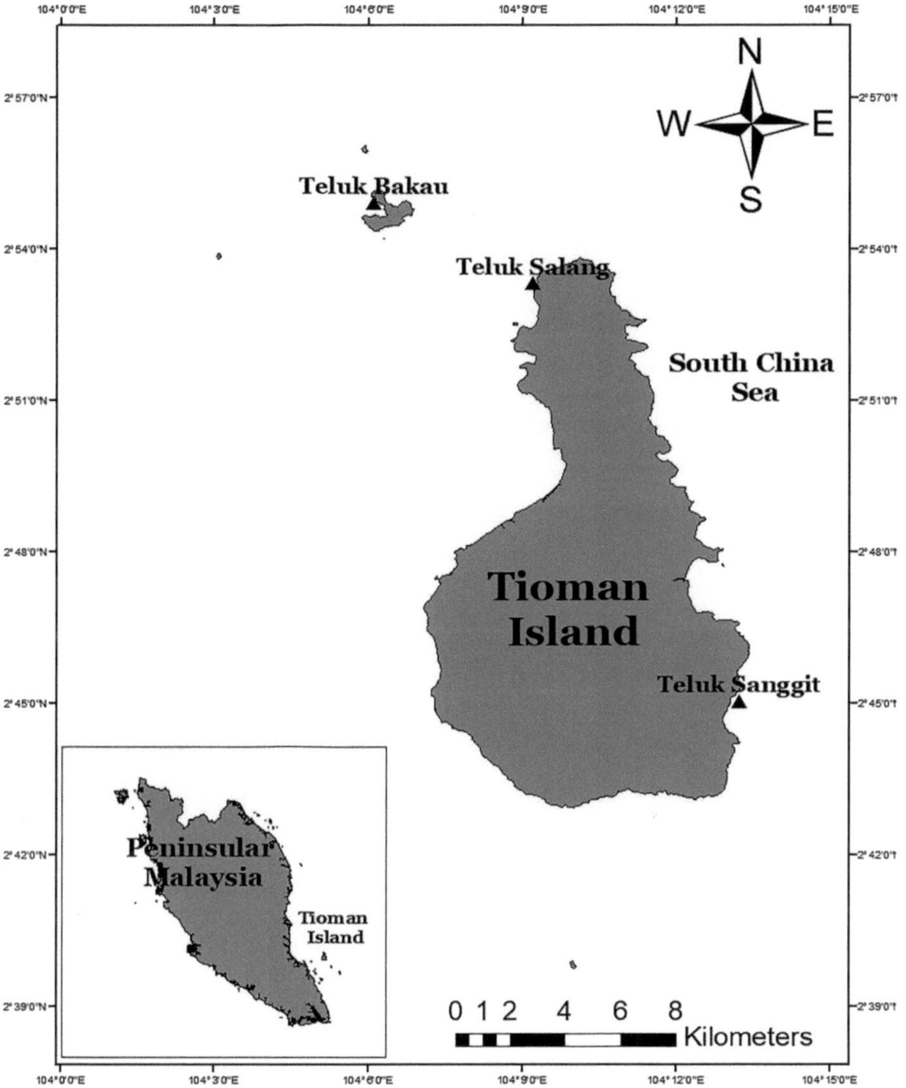

Fig. 13.1 Location of the 3 reef sites (Sanggit Bay: 02° 45′ 08.6″ N, 104° 13′ 13.8″ E, Salang Bay: 02° 53′ 34.4″ N, 104° 09′ 24.0″ E and Bakau Bay: 02° 54′ 61.4″ N, 104° 06′ 74.6″ E) for coral disease survey and sample collection around Tioman Island

Fig. 13.2 Infected corals by white syndrome exhibit diffused patterns of tissue loss, leaving them exposed with a bare and intact white skeleton. Tissue loss may progressively moderate to rapid, leaving white colour in active lesion fronts and yellow-brown as fouling develops over the exposed skeleton. It commonly affects the tabular *Acropora* (**a**) and foliose *Montipora* (**b**) corals

Fig. 13.3 Infected corals by skeletal eroding band exhibit diffused patterns of tissue loss, exposed to an eroded skeleton and covered by ciliates known as *Halofolliculina corallasia*. Relatively slow progression rate of tissue loss, leaving speckled black or dark green colours of ciliates at low or high coverage densities. It commonly affects the massive *Dipsastraea* (**a**) and *Goniastrea* (**b**) corals

13.5 Biochemical Technique

13.5.1 Bacterial Isolation and Biochemical Identification

The biochemical analysis was done through Gram staining, oxidase test and catalase test, followed by identification using the API 20NE Kit. Beforehand, the bacteria isolation was done on both Tryptic Soy Agar (TSA) and Thiosulfate Citrate Bile Salt-sucrose (TCBS) media using the streaking technique. Uniform yellow and green colonies (2–4 mm in diameter) were selected as the *Vibrio* isolates based on Mustapha et al. (2013). The pure colonies of *Vibrio* isolates were then retrieved through further sub-culture on TSA media supplemented with 10% NaCl. Following the Gram staining technique, these colonies were then morphologically observed

Fig. 13.4 Infected corals by yellow band disease exhibit diffused and multifocal patterns of tissue loss, exposed to a bare skeleton. The slow rate of tissue loss results in the pale yellow-white colour of blotches or patches of circular rings. It commonly affects the free-living fungid corals such as *Fungia* (**a**), *Lithophyllon* (**b**) and *Ctenactis* (**c**, **d**) corals

Fig. 13.5 Unexplained growth anomalies are known as tumour-like diseases. Infected corals have focal or multifocal patterns of abnormal whitish skeletal elements such as ridges and valleys. The skeletal elements appeared to be smaller or larger than healthy coral corallites and protruded on the surface of the colonies. It commonly affects the tabular (**a**) and branching (**b**) *Acropora* corals

under the light microscope to classify them as either Gram-positive or Gram-negative bacteria (reference). Additionally, the positive result reliability was also indicated by purple colour formation, while no colour changes indicated a negative result following the oxidase test. Further catalase test was done using 3% hydrogen

Fig. 13.6 Infected corals by ulcerative white spots exhibit a multifocal pattern of tissue loss, exposed with a bare and intact white skeleton. Lesions are typically small, less than 1 cm in diameter and form a regular ovoid shape. It commonly affects the foliose *Montipora* (**a**) and massive *Porites* (**b**) corals

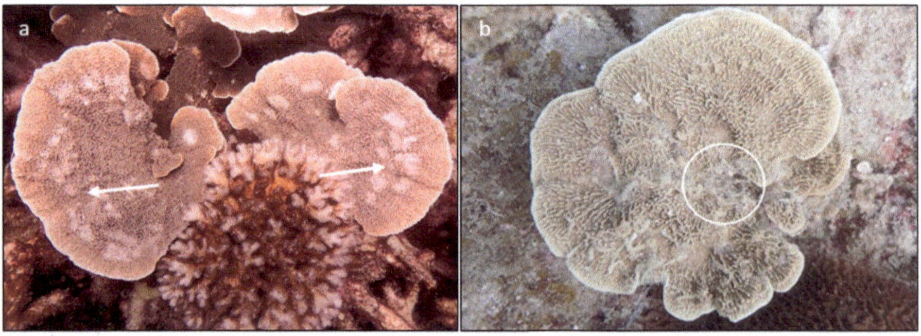

Fig. 13.7 Infected corals by atramentous necrosis exhibit multifocal and irregular patterns of tissue loss, exposing bare white skeletons and subsequently colonized by a fouling community over the exposed skeleton. . Slow the rate of tissue loss, resulting in a grey or black colour as fouling develops. It commonly affects the foliose Montipora corals (**a**, **b**)

peroxide (H_2O_2) which had been dropped over the colony attached to the microscope slide, where the presence (indicates a positive result) or absence (indicates a negative result) of bubbles or foams were observed. With the obtained morphological information of the bacteria colonies, the biochemical identification using the API 20NE Kit was carried out, involving several assimilation tests and numerical profiling using the APIWEB™ system.

13.5.2 Bacterial Identification Based on API20NE Kit

From a total of 100 Gram-negative *Vibrio* isolates, 50 identical pure colonies were chosen. All pure colonies were found to be oxidase-positive (as evidenced by the development of purple colour in the oxidase strip) and catalase-negative (indicated by gas production in 3% H_2O_2). Following their phenotypic features, two *Vibrio* spp.

Fig. 13.8 Focal bleaching can be characterized as focal or non-focal bleaching. Infected corals have multifocal (focal) or diffuse (non-focal) patterns of white bleached tissue. There is no eroded skeleton or colonized algae since coral polyps are still alive and visible. It commonly affects the massive *Porites* corals (**a**, **b**)

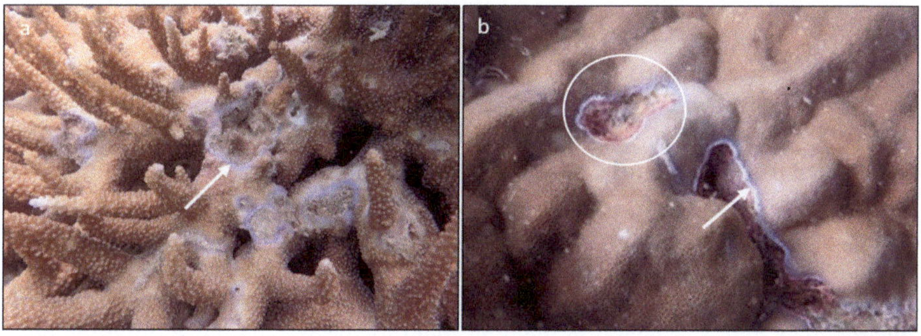

Fig. 13.9 The sign pigmentation response is a healing response of a coral host to multiple stressors such as algal abrasion, boring animals and fish bites. Infected corals have multifocal and diffuse patterns of tissue loss displayed with brightly pink, purple or blue tissue discolouration. It commonly affects the branching *Acropora* (**a**) and massive *Porites* (**b**) corals

(*V. vulnificus* and *V. alginolyticus*) and one *Photobacterium damselae* were identified using the API 20NE identification system (Table 13.1). The results of biochemical reactions revealed that *Vibrio* spp. reduced nitrate to nitrite and produce indole. They were also indicated to be glucose fermentation-positive and arginine dihydrolase-negative. However, *V. vulnificus* (GEL +/ PNPG + /GLU –) and *V. alginolyticus* (GEL –/ PNPG - /GLU +) showed differences in the processes of gelatin hydrolysis (GEL), β-galactosidase activity (PNPG) and assimilation of glucose reaction (GLU). Like *Vibrio* spp., *P. damselae* also demonstrated a positive reaction to nitrate reduction and glucose fermentation. Even so, it was different from *Vibrio* spp. by having indole production-negative and arginine dihydrolase-positive.

Fig. 13.10 Infected corals by algal and sponge overgrowth have been colonized and overgrown by turf or fine algae (**a**) and encrusting sponges, mainly from cyanobacteriosponge (*Terpios hoshinota*) (**b**). They appeared to be smothered over the surface of the coral colony, resulting in coral tissue necrosis. Algae commonly overshadow the tabular *Acropora*, while both algae and sponges commonly overgrown branching *Acropora*, *Isopora*, *Porites* and *Pocillopora* corals

13.5.3 Biochemical Identification of *Vibrio* spp.

About 100 isolates of *Vibrio* belonging to the family Vibrionaceae were successfully cultured from WS diseased corals and their adjacent seawater, sediment and algae. The presence of Gram-negative bacilli with curve and rod forms in *Vibrio* isolates (Fig. 13.16), as well as the ability to form green or yellow colonies on TCBS medium, verified their morphological traits. The ability to grow on TSA medium with 10% NaCl revealed a pervasive halophilic trait of *Vibrio* spp. that contributes towards their high-salt adaption, helping them to survive the harsh environmental conditions (Jayasinghe et al. 2010; Gomathi et al. 2013; Kalburge et al. 2014).

Based on the biochemical reactions, two *Vibrio* spp. (*V. vulnificus* and *V. alginolyticus*) and one *Photobacterium damselae* (previously known as *V. damsela*) were discovered to be 99.9% identical. Coastal locations are densely inhabited by these bacterial species. Over 80 *Vibrio* species have been found globally, with *V. cholera*, *V. vulnificus*, *V. alginolyticus* and *V. parahaemolyticus* being the most clinically implicated in human illnesses, including severe diarrheal sickness (vibriosis) connected to polluted water and raw seafood consumption. Furthermore, the pathogenicity of *V. alginolyticus* and *P. damselae* in the aquaculture business might result in the enormous death of produced fish, oysters and other bivalves. Since numerous *Vibrio* spp. are known to cause significant mortality in aquatic species (e.g. Naka et al. 2011; Hashem and El-Barbary 2013; Mirbakhsh et al. 2014), their presence in Tioman Island might have aided the onset of WS coral disease.

Despite that, the species identified using the biochemical test did not confirm the sequence findings from the molecular 16S rRNA gene samples. It is because the quick detection of clinical and environmental bacterial species from the miniaturized test of API 20NE kit is ineffectual in detecting specific Vibrionaceae family (Colodner et al. 2004; Martinez-Urtaza et al. 2006). To allow the accurate detection

Fig. 13.11 Infected corals by predation scars have distinct feeding or grazing scars, exposed with a bare white skeleton. Feeding scars may have various diffuse, focal or multifocal patterns of tissue loss depending on coral predators. The *Acanthaster* starfish feeding scars may leave a diffuse pattern of rapid tissue loss and naturally occur from the colony base for branching *Acropora* (**a**) or colony edge for tabular *Acropora* (**b**) and massive *Goniastrea* (**c**) corals. Meanwhile, fish predators may leave various regular and irregular patterns of feeding scars depending on fish bites such as scars left by parrotfish (**d**), trigger or pufferfish (**e**) and damselfish (**f**)

of the bacteria family, the temperature, incubation period and media composition need to be optimized (Ottaviani et al. 2003). In addition, the APIWEBTM system database contains just five *Vibrio* spp. (*V. vulnificus*, *V. alginolyticus*, *V. cholera*, *V. metschnikovii* and *V. parahaemolyticus*) and a single species of *P. damselae*. As a result, bacterial species found using a biochemical test may be misidentified. Hence, the 16S rRNA primer was further utilized to identify *Vibrio* spp. through molecular

Fig. 13.12 Infected corals by sediment necrosis exhibit a diffuse pattern of tissue loss accumulated with fine sediment, filamentous algae, mucus secretion and pigmentation response. It more commonly affects the massive compared to branching corals such as *Dipsastraea* (**a**) and *Porites* (**b**) corals

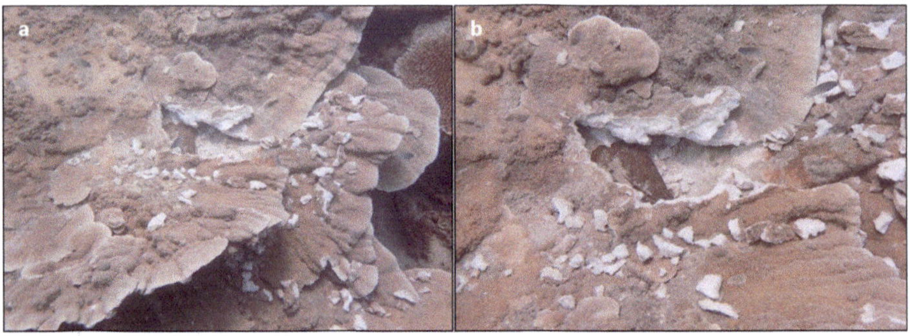

Fig. 13.13 Infected corals by physical damage exhibit fragments that have been damaged due to physical impacts caused by direct contact with fins, anchors or boat propellers. It commonly affects the fragile branching *Acropora* and foliose *Montipora* (**a**, **b**) corals

Fig. 13.14 Infected corals by explaining growth anomalies exhibit focal to multifocal skeletal deformations that appear to be associated with the presence of other invertebrates such as crabs and barnacles within or above the colony surface. It commonly affects the massive *Porites* (**a**) and *Platygyra* (**b**) corals

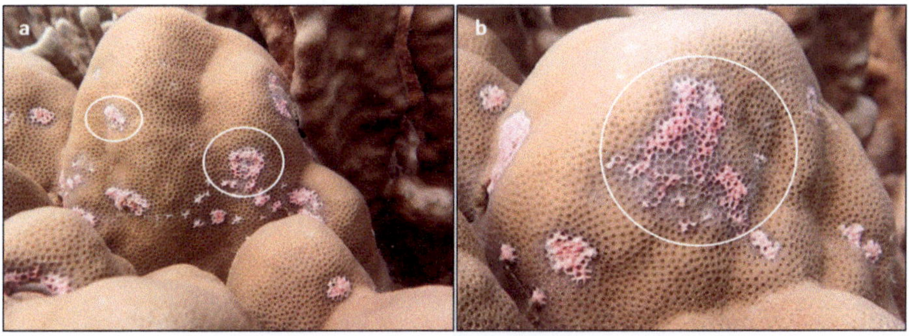

Fig. 13.15 Infected corals by trematodiasis exhibit distinct and multifocal patterns of swelling tissue over the colony surface. Swollen nodules develop a distinct white to pink colour in response to encysted parasitic trematodes. Trematode cysts are typically small (1–2 mm) and protrude in an irregular cluster shape. It commonly affects the massive *Porites* coral (**a**, **b**)

Table 13.1 Phenotypic characterization of bacterial isolates using API 20NE Kit

	Vibrio isolates		
Phenotypic characteristics	*Vibrio vulnificus*	*Vibrio alginolyticus*	*Photobacterium damselae*
Samples ($n = 50$)	$n = 15$	$n = 15$	$n = 20$
Gram's stain	–	–	–
Cell morphology	Cocci	Cocci	Cocci
Oxidase	+	+	+
Catalase	–	–	–
Nitrate reduction (NO_3)	+	+	+
Indole production (TRP)	+	+	–
Glucose fermentation (GLF)	+	+	+
Arginine dihydrolase (ADH)	–	–	+
Hydrolysis of:			
Aesculin (ESC)	+	+	–
Gelatin (GEL)	+	–	–
Enzyme activity of β-Galactosidase (PNPG)	+	–	+
Assimilation of:			
Glucose (GLU)	–	+	+
Mannose (MNE)	–	–	–
Malate (MLT)	+	+	+

Note. +: indicates a positive result, –: indicates a negative result

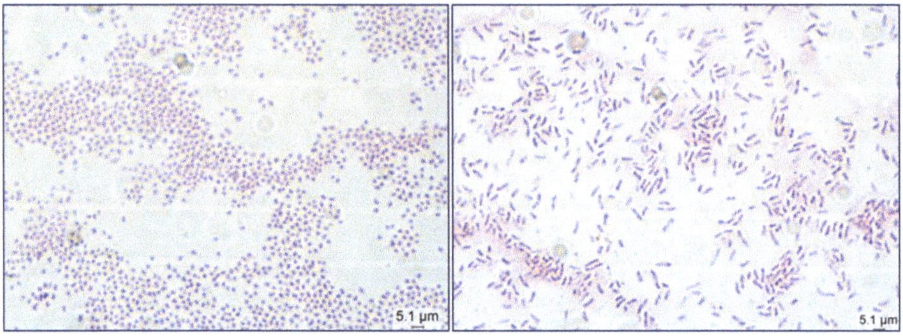

Fig. 13.16 Morphology of Gram-negative bacilli with curve and rod shapes for *Vibrio* isolates after gram staining under a light microscope with 40x objective

approach in the study area. Supportively, the 16S rRNA gene has been shown to offer the most accurate phylogenetic categorization of microbes (Kang et al. 2003; Kim and Bang 2008) and is widely used to identify new microbial diseases (Clarridge 2004).

13.6 Molecular Technique

13.6.1 Bacterial Identification Based on 16S rRNA Gene Sequencing

The molecular analysis was done through several procedures of DNA extraction, polymerase chain reaction (PCR) amplification, cyclic conditions, gel electrophoresis and phylogenetic analysis. Genomic DNA was extracted from pure cultured *Vibrio* isolates using the Wizard Genomic DNA kit (Promega, USA), following the manufacturer's instructions. The extracted DNA template was quantified using the Nanodrop quantification method (Desjardins and Conklin 2011) and stored at −20 °C for longer storage (Ahmed et al. 2014). PCR amplification of the 16S rRNA gene was performed using universal primers 8F and 1492R; 5-′-AGAGTTTGATCCTGGCTCAG-3′ and 5′-GGTTACCTTGTTACGACTT-3′ (Lane 1991). Cycling conditions consisted of 3 min for initial denaturation at 94 °C followed by 30 cycles of denaturation at 94 °C for 1 min, annealing at 55 °C for 1 min, extension at 72 °C for 1 min and concluded by 10 min extension step at 72 °C. After that, the PCR products were visualized through gel electrophoresis using a 1% agarose gel under the UV Transilluminator (Alphaimager™ 2200, Germany). All successful single-band PCR products were then sent to the First Base, Malaysia for the DNA sequencing process. Followed by that, the phylogenetic analysis was performed for all sample sequences using Sequence Scanner vers. 1.0 (Gene Codes Corporation), BioEdit ver. 7 (Hall et al. 2011) and ClustalX ver. 2.1 (Larkin et al. 2007). Next, sequence data were aligned with the most similar sequences retrieved from the Genbank database using the NCBI BLAST program

(Altschul et al. 1997) and further analysed using the EzTaxon-e database (Kim et al. 2012) (Table 13.2). A total of 20 sample sequences from seven bacterial strains were identified and further submitted to the Genbank database (Table 13.3). Besides that, 12 reference sequences of 16S rRNA gene retrieved from the Genbank database were also used to construct the phylogenetic tree under the following accession numbers: NR117887 (*Vibrio brasiliensis*), NR025491 (*V. hepatarius*), NR118093 (*V. tubiashii*), NR113784 (*V. harveyi*), NR121709 (*V. alginolyticus*), NR114632 (*V. parahaemolyticus*), NR113782 (*V. campbelli*), NR036888 (*V. vulnificus*), NR025476 (*V. neptunius*), NR117892 (*V. coralliilyticus*), NR156028 (*V. ishigakensis*) and NR042343 (*Photobacterium rosenbergii*). *Agrobacterium tumefaciens* (NR041396) was used as an outgroup sequence. In total, 33 nucleotide sequences (samples, reference and outgroup sequences) were aligned using Clustal W incorporated in Molecular Evolutionary Genetics Analysis (MEGA) software ver. 6 using default settings (Tamura et al. 2013). MEGA ver. 6 software was also used to construct the phylogenetic tree based on neighbour-joining (NJ) and maximum likelihood (ML) evolutionary distance analyses. NJ was computed using Tamura-3 parameter (Tamura 1992). Meanwhile, ML was computed using General Time Reversible + Gamma + Proportion Invariant (GTR + G + I) (Nei and Kumar 2000) following the best-fit model estimated using MrModeltest ver. 2.3 program (Nylander 2004). Both evolutionary distance analyses were performed with a 1000 bootstrap (Tamura 1992).

13.6.2 Molecular Phylogenetic Tree of Bacterial Species

Based on the nearly complete 16S rRNA gene (1500 bp), the molecular phylogenetic tree is made of 33 nucleotide sequences that comprise sample, reference and outgroup (Fig. 13.17). *A. tumefaciens* sequence was utilized as an outgroup for the tree topology. All sample sequences were clustered with seven coral pathogen strains, which are *V. brasiliensis* (NR117887), *V. hepatarius* (NR025491), *V. tubiashii* (NR025491), *V. campbellii* (NR113782), *V. coralliilyticus* (NR117892), *V. ishigakensis* (NR156028) and *P. rosenbergii* (NR042343). Six of the 20 sample sequences exhibited 99% sequence similarities with *V. brasiliensis*, which has been referenced to the marine sponge *Scleritoderma cyanea* (Hoffmann et al. 2012). Meanwhile, four sample sequences from WS-infected corals were 99% identical to a previously identified coral pathogen *V. corallilyticus*, which causes bleaching in *Pocillopora damicornis* and WS disease in *Acropora*, *Montipora* and *Pachyseris* corals (Ben-Haim et al. 2003). (Sussman et al. 2008). Furthermore, three sediment samples were found to be 100% identical to *P. rosenbergii*, which is related to bleaching corals (Thompson et al. 2005). Other *Vibrio* spp. found in this study (*V. hepatarius*, *V. tubiashi*, *V. campbellii* and *V. ishigakensis*) were also engaged in disease aetiology of aquatic species such as corals, fish and bivalves (Thompson et al. 2003a, 2003b).

Table 13.2 Bacterial species identified following Genbank and Eztaxon databases

No.	Sample ID	Strains	Genbank	BLAST similarity (%)	EzTaxon	Similarity (%)
1	TSGP1	ATCC BAA-450	*V. Coralliilyticus*	100%	*V. Coralliilyticus*	99.4%
2	TSGP2	ATCC BAA-450	*V. Coralliilyticus*	100%	*V. Coralliilyticus*	99.4%
3	TSGP3	JCM 19231	*V. Ishigakensis*	99.3%	*V. Ishigakensis*	99.4%
4	TSGP4	LMG 20546	*V. brasiliensis*	99.9%	*V. brasiliensis*	99.7%
5	TSGP5	LMG 20546	*V. brasiliensis*	99.9%	*V. brasiliensis*	99.7%
6	TSGP6	CAIM 519	*V. campbellii*	99.7%	*V. campbellii*	100%
7	TBKP7	CAIM 519	*V. campbellii*	99.7%	*V. campbellii*	100%
8	TBKP8	LMG 22223	*P. rosenbergii*	99.6%	*P. rosenbergii*	100%
9	TBKP9	LMG 20546	*V. brasiliensis*	99.9%	*V. brasiliensis*	99.7%
10	TBKP10	LMG 20546	*V. brasiliensis*	99.9%	*V. brasiliensis*	99.7%
11	TBKP11	ATCC BAA-450	*V. Coralliilyticus*	100%	*V. Coralliilyticus*	99.4%
12	TBKP12	JCM 19231	*V. Ishigakensis*	99.3%	*V. Ishigakensis*	99.4%
13	TBKP13	LMG 20546	*V. brasiliensis*	99.3%	*V. brasiliensis*	99.7%
14	TSKP14	CAIM 519	*V. campbellii*	99.7%	*V. campbellii*	100%
15	TSLP15	LMG 22223	*P. rosenbergii*	99.6%	*P. rosenbergii*	100%
16	TSLP16	LMG 20546	*V. brasiliensis*	99.9%	*V. brasiliensis*	99.7%
17	TSLP17	LMG 22223	*P. rosenbergii*	99.5%	*P. rosenbergii*	99.9%
18	TSLP18	LMG 20362	*V. Hepatarius*	99.8%	*V. Hepatarius*	100%
19	TSLP19	ATCC BAA-450	*V. Coralliilyticus*	100%	*V. Coralliilyticus*	99.4%
20	TSLP20	ATCC 19109	*V. Tubiashii*	99.7%	*V. Tubiashii*	99.7%

Table 13.3 Sample ID, location, isolation source and accession no. of sample sequences

No.	Sample ID	Location	Isolation source	Accession No.
1	TSGP1	Sanggit Bay	WS *A. cytherea*	MN339960
2	TSGP2	Sanggit Bay	WS *A. cytherea*	MN339962
3	TSGP3	Sanggit Bay	WS *M. aequituberculata*	MN339964
4	TSGP4	Sanggit Bay	Seawater	MN339954
5	TSGP5	Sanggit Bay	Sediment	MN339953
6	TSGP6	Sanggit Bay	Algae	MN339957
7	TBKP7	Bakau Bay	Seawater	MN339958
8	TBKP8	Bakau Bay	Sediment	MN339950
9	TBKP9	Bakau Bay	Algae	MN339951
10	TBKP10	Bakau Bay	WS *M. aequituberculata*	MN339956
11	TBKP11	Bakau Bay	WS *A. hyacinthus*	MN339961
12	TBKP12	Bakau Bay	WS *A. hyacinthus*	MN339965
13	TBKP13	Bakau Bay	WS *M. aequituberculata*	MN339955
14	TSKP14	Salang Bay	Seawater	MN339959
15	TSLP15	Salang Bay	Sediment	MN339949
16	TSLP16	Salang Bay	Algae	MN339952
17	TSLP17	Salang Bay	Sediment	MN339948
18	TSLP18	Salang Bay	WS *M. aequituberculata*	MN339967
19	TSLP19	Salang Bay	WS *M. aequituberculata*	MN339963
20	TSLP20	Salang Bay	WS *A. hyacinthus*	MN339966

13.6.3 Molecular Identification and Prevalence of Vibrio Bacterial Pathogens

Six *Vibrio* spp. (*V. coralliilyticus*, *V. hepatarius*, *V. brasiliensis*, *V. tubiashi*, *V. campbellii* and *V. ishigakensis*) and one *Photobacterium rosenbergii* were identified using the nearly complete 16S rRNA gene sequences (1500 bp). These bacterial species were 99.3% to 100% linked to bacterial pathogens associated with severe diseases in coral and other aquatic species, according to the Genbank and Eztaxon. At 96% bootstrap value, a taxonomic cluster revealed that all sequences recovered from WS-infected corals were phylogenetically related to *V. coralliilyticus*. This *Vibrio* sp. is a widespread coral pathogen found on Indo-Pacific reefs (Ben-Haim and Rosenberg 2002; Ben-Haim et al. 2003; Sussman et al. 2008).

V. coralliilyticus has previously been identified as a specific temperature-dependent pathogen given its rapid destruction of coral tissue just within 2 weeks when water temperatures increase beyond 25 °C (Ben-Haim and Rosenberg 2002). It has also been identified as an etiological agent of for *Pocillopora* coral bleaching (e.g. Ben-Haim et al. 2003) and possibly for WS signs in *Acropora*, *Pachyseris* and *Montipora* corals (e.g. Sussman et al. 2008). Furthermore, the high prevalence of *V. coralliilyticus* has been linked to gorgonian mortality in the Mediterranean Sea (Bally and Garrabou 2007), diseased oyster larvae (*Crassostrea gigas*) in the United

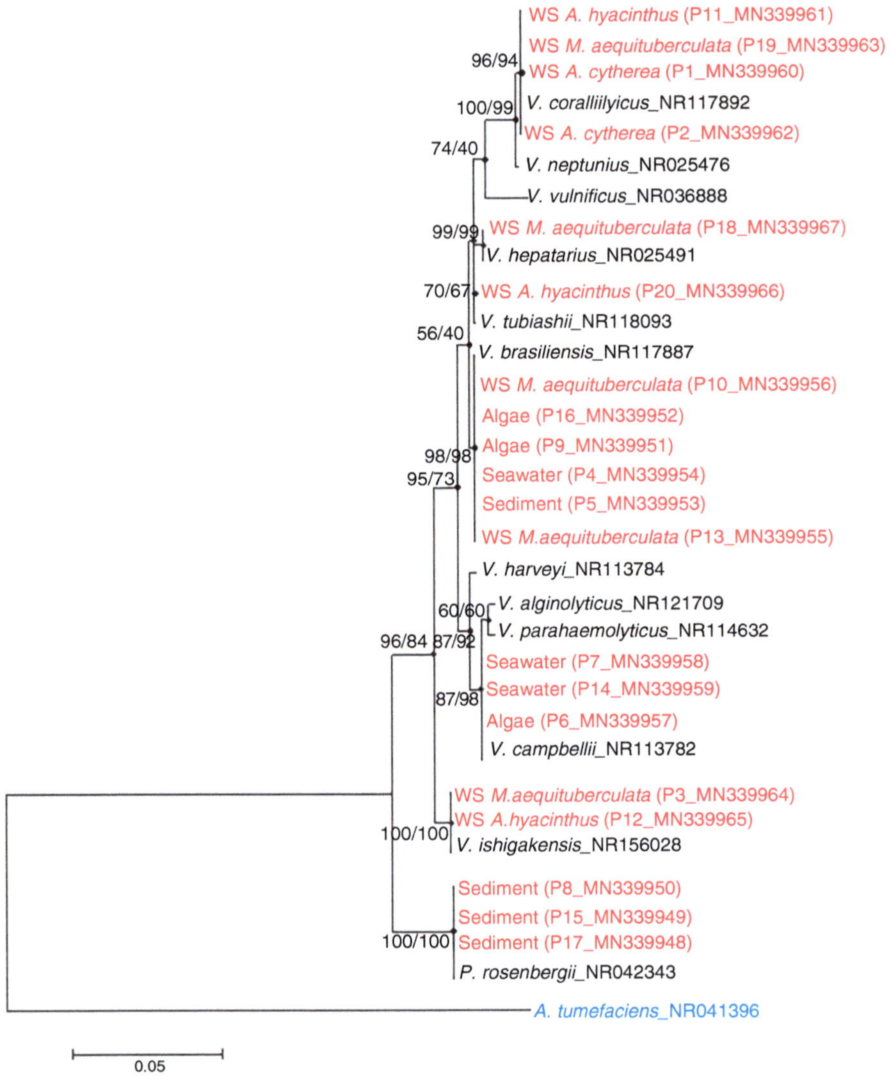

Fig. 13.17 The evolutionary distance ML phylogenetic tree is based on 16S rRNA gene sequences of isolates obtained in this study. Coral pathogens are marked in red and labelled from P1 until P20. Reference strains are marked in black and outgroup, marked in blue. Nodes represent bootstrap values ≥50% based on 1000 replicates generated from ml and neighbour-joining (NJ) analyses. Scale bar corresponds to a 5% estimated sequence divergence

Kingdom and scallop (*Nodipecten nodosus*) in Brazil (Ben-Haim et al. 2003). In this study, *V. coralliilyticus* also had the highest mean prevalence (45.4% ± 6.7) and can be found at all selected reef sites in TIMP.

The molecular phylogenetic tree also indicated that certain WS-infected coral sequences were classified as *V. hepatarius* (99% bootstrap value), *V. brasiliensis*

(98%), *V. campbellii* (87%) and *V. ishigakensis* (100%). These *vibrio* spp. were also found at all reef locations, covering more than 5% of the mean prevalence. *V. hepatarius* and *V. brasiliensis* have previously been identified as the causative microbial agents for diseased aquatic organisms such as bivalves, fish, rotifers and shrimps (Thompson et al. 2003a; Thompson et al. 2003b), whereas *V. campbellii* and *V. ishigakensis* have been found to predominate in the reef sediments and seawater (Thompson et al. 2004; Gao et al. 2016). Additionally, *V. campbellii* is also an infectious pathogen for corals, prawns and oysters (Lin et al. 2010). On the other hand, the molecular phylogenetic tree obtained from the 16S rRNA gene has grouped all sediment sequences with the previously identified coral pathogen *P. rosenbergii* with 100% bootstrap support (Thompson et al. 2005). This *Photobacterium* sp. was isolated from both diseased and healthy corals on Magnetic Island in Australia (Thompson et al. 2005).

13.7 Histological Method

13.7.1 WS Coral Tissue Analysis

The histological analysis was done on both healthy and diseased (displaying sign of WS disease) coral tissues, which involved tissue collection, fixation, decalcification, processing, embedding, sectioning, slide staining and image analysis. Samples from both healthy and diseased coral fragments (1–2 cm in diameter) of *A. cytherea* were collected in triplicates at every reef site by using a hammer and chisel. All coral fragments were immediately fixed in 10% seawater formalin solution and stored at room temperature for 24 hours. Then, coral fragments were transferred to a decalcifying solution in 10% buffered hydrochloric acid (HCl) to remove their skeleton. The decalcification process takes about 4 to 24 h depending on the dissolution of the calcium carbonate structure (Maboloc et al. 2016). After that, all decalcified coral fragments were cut vertically and placed in standard processing cassettes as replicates. These samples underwent tissue processing using a Thermo Scientific automated tissue processor (Excelsior™, AS) for 19 h, followed by several hours of dehydration and infiltration processes. Following that, the tissue embedding was done by filtering coral tissues beforehand into the paraffin dispenser at 58–60 °C while waiting for the cassette to be filled with wax. The tissues were then embedded onto the cassette with heated forceps before being filled with molten paraffin at 65 °C. The paraffin blocks were allowed to freeze on the cold plate (Leica, USA). The blocks were kept refrigerated at 4 °C overnight after the paraffin was firm. Prior to the tissue sectioning, the blocks were kept cool and trimmed using the block trimmer (Thermo Scientific, Malaysia). Later, 5 μm thick sections were cut using a Thermo Scientific microtome machine and placed onto the microscopic slide. Later, the slides were dried on the hot plate (Marshall Scientific, USA) to ensure the tissues were completely affixed to the slides. The slide sections were then submitted for a staining procedure by the staining protocol described by Culling et al. (2014).

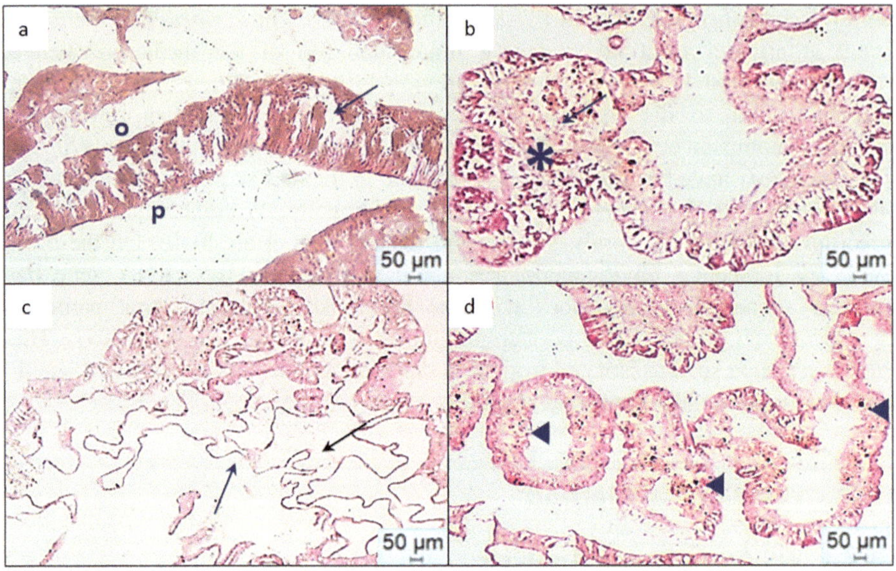

Fig. 13.18 Histological structures of WS coral tissue at a magnification of 40x objective. (**a**) reduction of zooxanthellae algae (arrow) within necrotic tissues of the epidermis (o) and gastrodermis (p), (**b**) necrosis and dissociation (arrow) of mesenterial filaments (asterisk), (**c**) hyperplasia of the body wall (arrows) and (**d**) the presence of bacterial colonies (arrowheads) within the basal body wall

Finally, the slides were examined under a NIKON digital microscope with NIS Element ver. 3 imaging software at a magnification of 10x and 40x objectives.

13.7.2 Histological Comparisons of WS Diseased and Healthy Coral

Distinctive histological structures of WS-infected (Fig. 13.18) and healthy (Fig. 13.19) *A. cytherea* coral tissues were observed as they were stained with hematoxylin and eosin. The upper body wall of WS tissue showed separation of the epidermis, gastrodermis and thin layer connective tissue of mesoglea, as well as a loss of coral symbiotic zooxanthellae algae inside the epidermis and gastrodermis layers. Additionally, within the basal body wall of WS tissue, necrotic mesenterial filaments, hyperplasia development and the presence of bacterial colonies were all detected. Healthy coral tissue, on the other hand, has intact epidermis, gastrodermis, mesoglea and mesenterial filaments within its upper and basal body walls. Furthermore, zooxanthellae algae and nematocyst battery containing nematocysts were also intact within the basal body wall of healthy tissue.

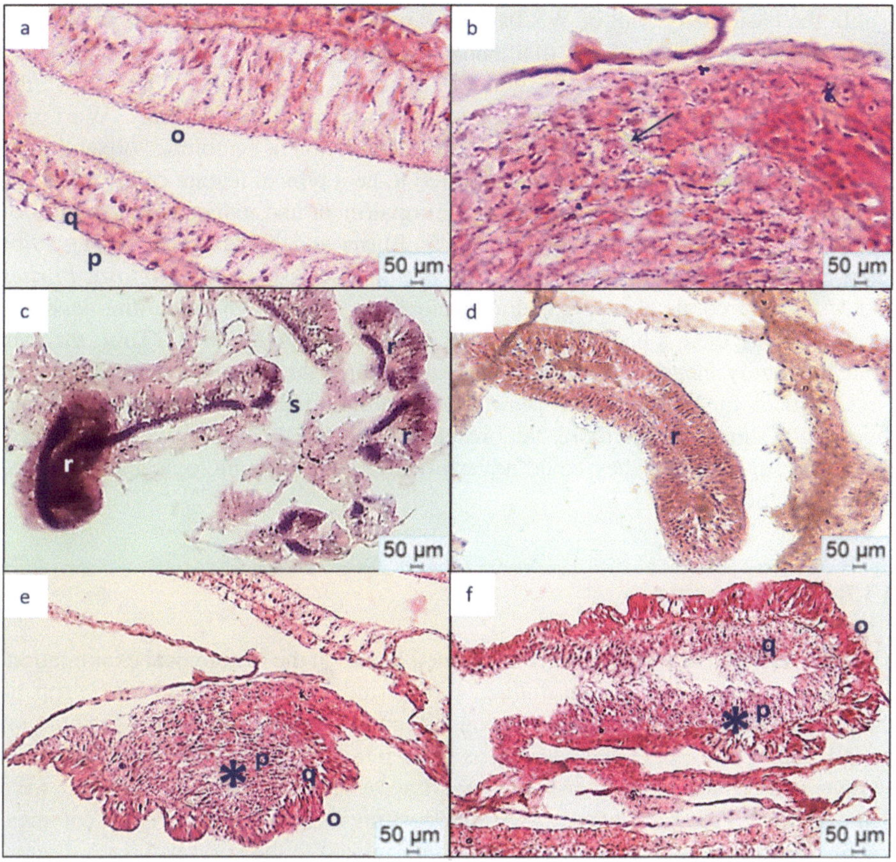

Fig. 13.19 Histological structures of healthy coral tissue at a magnification of 40x objective. (**a**) upper body wall consists of the epidermis (o) and gastrodermis (p), separated by a thin connective tissue of mesoglea (q), (**b**) basal body wall consists of nematocyst within coiled sporocysts (arrow), (**c, d**) mesenterial filaments (r) with cnidoglandular cap projecting into the gastrovascular canal (s), (**e, f**) gastrodermis (p) replete with red-staining zooxanthellae algae (asterisks) separated from the epidermis (o) by the mesoglea (q)

13.7.3 Histological Differences Between WS Diseased and Healthy Coral Tissue

Histological studies are critical for detecting abnormalities in coral physiological systems (Weil et al. 2006; Williams et al. 2011). The abnormalities may include skeletal density loss and coral symbiotic disruption, as well as hyperplasia development inside the basal body wall (Woodley et al. 2008; Bourne et al. 2009). This study investigated the histological section of WS tissue loss in *Acropora* coral acquired from TIMP reef sites. In the upper body wall, there was a loss of symbiotic zooxanthellae algae within the epidermis and gastrodermis thin layers. Meanwhile, within the basal body wall of WS tissue, nematocyst depletion, as well as necrosis and dissociation of mesenterial filaments, was detected.

Histological findings are consistent with previous research which has shown the altered feeding, digesting and defence mechanisms of diseased corals (Work and Rameyer 2005; Burns and Takabayashi 2011). Depletion of symbiotic zooxanthellae inside the upper body wall has been reported to be a typical feature of WS diseased coral tissue, producing metabolic capacity impairment and reduced growth rates of the coral hosts (Work and Rameyer 2005; Burns and Takabayashi 2011). Prior research of histological characteristics on the disease lesions of *Acropora*, *Porites* and *Montipora* corals also revealed a considerable drop in symbiotic dinoflagellate density, giving them a bleached appearance (Work and Aeby 2011). Thus, the data from this study clearly show that WS lesions alter the cellular composition and physiological functions of *Acropora* coral, including their defence, feeding and digesting systems. Furthermore, the presence of bacterial colonies within the basal body wall suggests that they could adversely affect the coral biological and physiological functions.

13.8 Conclusion

The infected WS coral tissues were confirmed based on the histological examination. The biochemical and molecular findings of this study also highlighted *Vibrio* spp. as the primary microbial pathogens causing WS coral disease. Throughout the cultural period, six *Vibrio* spp. (*V. coralliilyticus, V. hepatarius, V. brasiliensis, V. tubiashi, V. campbellii* and *V. ishigakensis*) and one *Photobacterium rosenbergii* were detected, with *V. coralliilyticus* being the most prevalent in infected coral colonies. *Vibrio coralliilyticus* was shown to be closely related to the coral pathogen discovered in the Indo-Pacific reefs. Accordingly, additional bacterial species discovered on Tioman reefs were also found to be closely related to lethal bacterial diseases in aquatic creatures at various reef localities.

Overall, this study provides useful data for identifying healthy and diseased coral colonies, thereby providing a good metric for coral health assessment. The identification result also provides a method for distinguishing between human impacts (e.g. sedimentation and nutrient pollution) and other biological stressors (e.g. predation by *Acanthaster* starfish and corallivorous fishes). Moreover, the incorporation of biotechnological applications provides more comprehensive data and information on potential bacterial agents associated with the occurrence of coral disease in the study area.

In addition, this study clearly illustrated the negative impact of increased tourism and coastal development on the occurrence of coral disease around Tioman Island. Therefore, it highlights the importance of effective regulatory and management approaches to reduce the stressors associated with the rapidly growing tourism industry. Several significant steps must be taken to minimize the rapid pace of coastal development and the increasing number of tourists visiting Marine Park. For example, through implementation and rigorously enforcing the existing guidelines and policies on ecotourism, notably in marine reserves, by coordinating and controlling the carrying capacity of reef recreational activities such as diving, snorkelling and boating. Spatial containment of these tourism activities can be applied by dispersing snorkelers and divers over extensive reef sites to avoid the intensifying levels of use at a specific site that may cause physical damage and injury to corals. As a consequence, the occurrence of diseased and compromised corals may be extended, and some coral species may become extinct if these activities are not meticulously managed.

Funding This research work was funded by the Skim Penyelidikan Lantikan Baharu (SLB2214) from the Universiti Malaysia Sabah and partially supported by the Fundamental Research Grant Scheme (FRGS/1/2022/WAB05/UMS/02/4; FRG0574–1/2022) under the Ministry of Higher Education Malaysia.

References

Aeby GS, Callahan S, Cox EF, Runyon C, Smith A, Stanton FG, Ushijima B, Work TM (2016) Emerging coral diseases in Kāne 'ohe bay, O 'ahu, Hawai 'i (USA): two major disease outbreaks of acute *Montipora* white syndrome. Dis Aquat Org 119(3):189–198

Ahmed OB, Asghar AH, Elhassan MM (2014) Comparison of three DNA extraction methods for polymerase chain reaction (PCR) analysis of bacterial genomic DNA. Afr J Microbiol Res 8(6): 598–602

Akmal KF, Saad S, Mohammad-Noor N (2018) Coral community structure in payar Island marine park, Malaysia. J Sustain Sci Manag 14:29–39

Akmal KF, Shahbudin S, Faiz MHM, Hamizan YM (2019) Diversity and abundance of scleractinian corals in the east coast of peninsular Malaysia: a case study of Redang and Tioman Islands. Ocean Science Journal 54(3):435–456

Akmal KF, Shahbudin S (2020) Baseline assessment of coral health and disease in Tioman Island Marine Park, Malaysia. *Commun Ecol* 21(3):285–301

Akmal KF, Shahbudin S (2021) Spatial distribution of coral communities on fringing reefs at Tioman Island Marine Park, Malaysia. Zoology Ecol 31(2):158–168

Altschul SF, Madden TL, Schäffer AA, Zhang J, Zhang Z, Miller W, Lipman DJ (1997) Gapped BLAST and PSI-BLAST: a new generation of protein database search programs. Nucleic Acids Res 25(17):3389–3402

Bally M, Garrabou J (2007) Thermodependent bacterial pathogens and mass mortalities in temperate benthic communities: a new case of emerging disease linked to climate change. Glob Chang Biol 13(10):2078–2088

Beeden R, Willis B, Raymundo L, Page C, Weil E (2008) Underwater cards for assessing coral health on indo-pacific reefs. In: Coral Reef Targeted Research and Capacity Building for Management Program. Currie Communications, Melbourne

Ben-Haim Y, Rosenberg E (2002) A novel vibrio sp. pathogen of the coral *Pocillopora damicornis*. Mar Biol 141(1):47–55

Ben-Haim Y, Zicherman-Keren M, Rosenberg E (2003) Temperature-regulated bleaching and lysis of the coral *Pocillopora damicornis* by the novel pathogen vibrio coralliilyticus. Appl Environ Microbiol 69(7):4236–4242

Bourne DG (2005) Microbiological assessment of a disease outbreak on corals from Magnetic Island (great barrier reef, Australia). Coral Reefs 24(2):304–312

Bourne DG, Garren M, Work TM, Rosenberg E, Smith GW, Harvell CD (2009) Microbial disease and the coral holobiont. Trends Microbiol 17(12):554–562

Bruckner AW (2015) History of coral disease research. In: Woodley CM, Downs CA, Bruckner AW, Porter JW, Galloway SB (eds) Diseases of coral. Wiley, Hoboken, pp 52–84

Bruno JF, Selig ER, Casey KS, Page CA, Willis BL, Harvell CD, Sweatman H, Melendy AM (2007) Thermal stress and coral cover as drivers of coral disease outbreaks. PLoS Biol 5(6):e124

Burns JH, Takabayashi M (2011) Histopathology of growth anomaly affecting the coral, *Montipora capitata*: implications on biological functions and population viability. PLoS One 6(12):e28854

Clarridge JE (2004) Impact of 16S rRNA gene sequence analysis for identification of bacteria on clinical microbiology and infectious diseases. Clin Microbiol Rev 17(4):840–862

Colodner R, Raz R, Meir I, Lazarovich T, Lerner L, Kopelowitz J, Keness Y, Sakran W, Ken-Dror S, Bisharat N (2004) Identification of the emerging pathogen *Vibrio vulnificus* biotype 3 by commercially available phenotypic methods. J Clin Microbiol 42(9):4137–4140

Collins M, Knutti R, Arblaster J, Dufresne JL, Fichefet T, Friedlingstein P, Gao X, Gutowski WJ, Johns T, Krinner G, Shongwe M (2013) Long-term climate change: projections, commitments and irreversibility. In: Stocker TF, Qin D, Plattner GK, Tignor M, Allen SK, Boschung AJ, Nauels A, Xi Y, Bex V, Midgley PM (eds) Climate change 2013: the physical science basis: contribution of working group I to the fifth assessment report of the intergovernmental panel on climate change. University Press, Cambridge, pp 1029–1136

Culling CFA, Allison RT, Barr WT (2014) Cellular pathology technique. Elsevier

Couch CS, Garriques JD, Barnett C, Preskitt L, Cotton S, Giddens J, Walsh W (2014) Spatial and temporal patterns of coral health and disease along leeward Hawai'i Island. Coral Reefs 33(3): 693–704

Department of Marine Park Malaysia (2011) Handbook of the living marine resources of Malaysia Marine Park. Ministry of Natural Resources and Environment, Putrajaya, p 23

Department of Marine Park Malaysia 2017 Total number of visitors in marine parks from year 2000 to year 2017. Web service available online at: http://marinepark.dof.gov. Last Accessed January 2018

Desjardins PR, Conklin DS (2011) Microvolume quantitation of nucleic acids. Curr Protoc Mol Biol 93(1):A-3J

Edge SE, Morgan MB, Gleason DF, Snell TW (2005) Development of a coral cDNA array to examine gene expression profiles in Montastraea faveolata exposed to environmental stress. Mar Pollut Bull 51(5–7):507–523

Gao F, Al-saari N, Amin AR, Sato K, Mino S, Suda W, Oshima K, Hattori M, Ohkuma M, Hargreaves PI, Meirelles PM (2016) *Vibrio ishigakensis* sp. nov., in *Halioticoli* clade isolated from seawater in Okinawa coral reef area, Japan. Syst Appl Microbiol 39(5):330–335

Gomathi RS, Vinothkumar R, Arunagiri K (2013) Isolation and identification vibrios from marine seafood samples. Int J Curr Microbiol App Sci 2(2):36–43

Hall T, Biosciences I, Carlsbad C (2011) BioEdit: an important software for molecular biology. GERF Bull Biosci 2(1):60–61

Hanapiah MFM, Saad S, Ahmad Z, Yusof MH, Khodzori MFA (2019) Assessment of benthic and coral community structure in an inshore reef in Balok, Pahang, Malaysia. J Biol Div 20(3): 872–877

Hashem M, El-Barbary M (2013) *Vibrio harveyi* infection in Arabian surgeon fish (*Acanthurus sohal*) of Red Sea at Hurghada, Egypt. Egyptian J Aquatic Res 39(3):199–203

Hoffmann M, Monday SR, Allard MW, Strain EA, Whittaker P, Naum M, McCarthy PJ, Lopez JV, Fischer M, Brown EW (2012) Vibrio caribbeanicus sp. nov., isolated from the marine sponge Scleritoderma cyanea. Int J Syst Evol Microbiol 62(8):1736–1743

Jayasinghe CVL, Ahmed SBN, Kariyawasam MGIU (2010) The isolation and identification of *vibrio* species in marine shrimps of Sri Lanka. J Food Agric 1(1):36–44

Kalburge SS, Whitaker WB, Boyd EF (2014) High-salt preadaptation of *Vibrio parahaemolyticus* enhances survival in response to lethal environmental stresses. J Food Prot 77(2):246–253

Kang HS, Heo MS, Lee JH (2003) Use of 16S-23S rRNA intergenic spacer region for rapid detection of Vibrio fluvialis. Korean J Environ Biol 21:77–85

Kim MN, Bang HJ (2008) Detection of marine pathogenic bacterial *vibrio* species by multiplex polymerase chain reaction (PCR). J Environ Biol 29(4):543–546

Kim OS, Cho YJ, Lee K, Yoon SH, Kim M, Na H, Park SC, Jeon YS, Lee JH, Yi H, Won S (2012) Introducing EzTaxon-e: a prokaryotic 16S rRNA gene sequence database with phylotypes that represent uncultured species. Int J Syst Evol Microbiol 62(3):716–721

Lane DJ (1991) 16S/23S rRNA sequencing. Nucl Acid Tech Bacterial Systematics:115–175

Larkin MA, Blackshields G, Brown NP, Chenna R, McGettigan PA, McWilliam H, Valentin F, Wallace IM, Wilm A, Lopez R, Thompson JD (2007) Clustal W and Clustal X version 2.0. Bioinformatics 23(21):2947–2948

Lin B, Wang Z, Malanoski AP, O'Grady EA, Wimpee CF, Vuddhakul V, Alves N Jr, Thompson FL, Gomez-Gil B, Vora GJ (2010) Comparative genomic analyses identify the *Vibrio harveyi* genome sequenced strains BAA-1116 and HY01 as Vibrio campbellii. Environ Microbiol Rep 2(1):81–89

Luna GM, Bongiorni L, Gili C, Biavasco F, Danovaro R (2010) *Vibrio harveyi* as a causative agent of the white syndrome in tropical stony corals. Environ Microbiol Rep 2(1):120–127

Maboloc EA, Jamodiong EA, Villanueva RD (2016) Reproductive biology and larval development of the scleractinian corals *Favites colemani* and *F. abdita* (Faviidae) in northwestern Philippines. Invertebr Reprod Develop 60(1):1–11

Manzello DP (2015) Rapid recent warming of coral reefs in the Florida keys. Sci Rep 5:16762

Martinez-Urtaza J, Lozano-Leon A, Viña-Feas A, de Novoa J, Garcia-Martin O (2006) Differences in the API 20E biochemical patterns of clinical and environmental *Vibrio parahaemolyticus* isolates. FEMS Microbiol Lett 255(1):75–81

Maynard J, Van Hooidonk R, Eakin CM, Puotinen M, Garren M, Williams G, Heron SF, Lamb J, Weil E, Willis B, Harvell CD (2015) Projections of climate conditions that increase coral disease susceptibility and pathogen abundance and virulence. Nat Clim Chang 5(7):688–694

Miller J, Sweet MJ, Wood E, Bythell J (2015) Baseline coral disease surveys within three marine parks in Sabah, Borneo. PeerJ 3:e1391

Mirbakhsh M, Afsharnasab M, Khanafari A, Razavi MR (2014) Molecular identification of *Vibrio harveyi* from larval stage of Pacific white shrimp (*Litopenaeus vannamei*) Boone (Crustacea: Decapoda) by polymerase chain reaction and 16S rDNA sequencing. Iran J Fish Sci 13(2): 384–393

Mustapha S, Mustapha EM, Nozha C (2013) *Vibrio alginolyticus*: an emerging pathogen of food borne diseases. Int J Sci Technol Res 2:302–309

Mydlarz LD, Couch CS, Weil E, Smith G, Harvell CD (2009) Immune defenses of healthy, bleached and diseased *Montastraea faveolata* during a natural bleaching event. Dis Aquat Org 87(1–2):67–78

Naka H, Dias GM, Thompson CC, Dubay C, Thompson FL, Crosa JH (2011) Complete genome sequence of the marine fish pathogen *vibrio anguillarum* harboring the pJM1 virulence plasmid and genomic comparison with other virulent strains of *V. anguillarum* and *V. ordalii*. Infect Immun 79(7):2889–2900

Nei M, Kumar S (2000) Molecular evolution and phylogenetics. Oxford University Press

Nugraha WA, Handoko A, Effendy M (2019) Prevalence and bacteria associated with white band disease on *Acropora* sp. from Gili Labak Island Sumenep District Indonesia. In: IOP conference series: earth and environmental science, vol 276. IOP Publishing, p 012063

Nylander JAA (2004) MrModeltest v2. Program distributed by the author. Evolutionary Biology Centre, Uppsala University, p 2

Okamura Y, Kimura T, Yokouchi H, Meneses-Osorio M, Katoh M, Matsunaga T, Takeyama H (2010) Isolation and characterization of a GDSL esterase from the metagenome of a marine sponge-associated bacteria. Mar Biotechnol 12(4):395–402

Omar SI, Othman AG, Mohamed B, Bahauddin A (2015) Coastal resort life cycle: an overview of Tioman Island, Malaysia. Tourism Planning Develop 12(3):266–280

Ottaviani D, Masini L, Bacchiocchi S (2003) A biochemical protocol for the isolation and identification of current species of *vibrio* in seafood. J Appl Microbiol 95(6):1277–1284

Pollock FJ, Morris PJ, Willis BL, Bourne DG (2011) The urgent need for robust coral disease diagnostics. PLoS Pathog 7(10):e1002183

Pollock FJ, Lamb JB, Field SN, Heron SF, Schaffelke B, Shedrawi G, Bourne DG, Willis BL (2014) Sediment and turbidity associated with offshore dredging increase coral disease prevalence on nearby reefs. PLoS One 1(11):e0165541

Raymundo LJ, Couch CS, Harvell CD, Raymundo J, Bruckner AW, Work TM, Weil E, Woodley CM, Jordan-dahlgren E, Willis BL, Sato Y (2008) Coral disease handbook guidelines for assessment, monitoring & management. In: Coral Reef Targeted Research & Capacity Building for Management Program. Currie Communications, Melbourne

Reef Check Malaysia (2018) Status of coral reefs in Malaysia 2017. In: Reef Check Malaysia Survey Report. Reef Check Malaysia

Ruiz-Moreno D, Willis BL, Page AC, Weil E, Cróquer A, Vargas-Angel B, Jordan-Garza AG, Jordán-Dahlgren E, Raymundo L, Harvell CD (2012) Global coral disease prevalence associated with sea temperature anomalies and local factors. Dis Aquat Org 100(3):249–261

Safuan CDM, Ali A, Zainol Z, Ali A, Akhir MFM, Muslim AM, Bachok Z (2018) A baseline assessment of coral reef in Malacca Straits, Malaysia. Ocean Sci J 53(2):275–283

Shahbudin S, Akmal KF, Faris S, Normawaty MN, Mukai Y (2017) Current status of coral reefs in Tioman Island, Peninsular Malaysia. Turkish J Zoology 41(2):294–305

Sussman M, Willis BL, Victor S, Bourne DG (2008) Coral pathogens identified for white syndrome (WS) epizootics in the indo-Pacific. PLoS One 3(6):e2393

Tamura K (1992) Estimation of the number of nucleotide substitutions when there are strong transition-transversion and G+ C-content biases. Mol Biol Evol 9(4):678–687

Tamura K, Stecher G, Peterson D, Filipski A, Kumar S (2013) MEGA6: molecular evolutionary genetics analysis version 6.0. Mol Biol Evol 30(12):2725–2729

Tan CH, Heron SF (2011) First observed severe mass bleaching in Malaysia, greater coral triangle, Galaxea. J Coral Reef Stud 13(1):27–28

Thompson FL, Thompson CC, Hoste B, Vandemeulebroecke K, Gullian M, Swings J (2003a) *Vibrio fortis* sp. nov. and *vibrio hepatarius* sp. nov., isolated from aquatic animals and the marine environment. Int J Syst Evol Microbiol 53(5):1495–1501

Thompson FL, Li Y, Gomez-Gil B, Thompson CC, Hoste B, Vandemeulebroecke K, Rupp GS, Pereira A, De Bem MM, Sorgeloos P, Swings J (2003b) *Vibrio neptunius* sp. nov., *Vibrio brasiliensis* sp. nov. and *vibrio xuii* sp. nov., isolated from the marine aquaculture environment (bivalves, fish, rotifers and shrimps). Int J Syst Evol Microbiol 53(1):245–252

Thompson JR, Randa MA, Marcelino LA, Tomita-Mitchell A, Lim E, Polz MF (2004) Diversity and dynamics of a North Atlantic coastal *vibrio* community. Appl Environ Microbiol 70(7): 4103–4110

Thompson FL, Thompson CC, Naser S, Hoste B, Vandemeulebroecke K, Munn C, Bourne D, Swings J (2005) *Photobacterium rosenbergii* sp. nov. and *Enterovibrio coralii* sp. nov., vibrios associated with coral bleaching. Int J Syst Evol Microbiol 55(2):913–917

Toda T, Okashita T, Maekawa T, Alfian BAAK, Rajuddin MKM, Nakajima R, Chen W, Takahashi KT, Othman BHR, Terazaki M (2007) Community structures of coral reefs around peninsular Malaysia. J Oceanogr 63(1):113–123

Tun K, Chou LM, Low J, Yeemin T, Phongsuwan N, Setiasih N, Wilson J, Amri AY, Adzis KEA, Lane D (2010) A regional overview on the 2010 coral bleaching event in Southeast Asia. In: Kimura T, Tun KC (eds) Status *of coral reefs in east Asian seas region: 2010*. Ministry of the Environment, Tokyo, pp 16–27

Ushijima B, Smith A, Aeby GS, Callahan SM (2012) *Vibrio owensii* induces the tissue loss disease *Montipora* white syndrome in the Hawaiian reef coral *Montipora capitata*. PLoS One 7(10): e46717

Ushijima B, Videau P, Burger AH, Shore-Maggio A, Runyon CM, Sudek M, Aeby GS, Callahan SM (2014) *Vibrio coralliilyticus* strain OCN008 is an etiological agent of acute *Montipora* white syndrome. Appl Environ Microbiol 80(7):2102–2109

Wada N, Ohdera A, Mano N (2018) Coral disease in Japan. In: Coral reef studies of Japan. Springer, Singapore, pp 41–62

Waheed Z (2016) Patterns of coral species richness and reef connectivity in Malaysia. PhD Dissertation Leiden University, Nederland

Walton CJ, Hayes NK, Gilliam DS (2018) Impacts of a regional, multi-year, multi-species coral disease outbreak in Southeast Florida. Front Mar Sci 5:323

Weil E (2004) Coral reef diseases in the wider Caribbean. In: Rosenberg E, Loya Y (eds) Coral health and disease. SpringerVerlag, Berlin, pp 35–68

Weil E, Smith G, Gil-Agudelo DL (2006) Status and progress in coral reef disease research. Dis Aquat Org 69(1):1–7

Weil E, Irikawa A, Casareto B, Suzuki Y (2012) Extended geographic distribution of several indo-Pacific coral reef diseases. Dis Aquat Org 98(2):163–170

Weil E, Hernández-Delgado EA, Gonzalez M, Williams S, Suleimán-Ramos S, Figuerola M, Metz-Estrella T (2019) Spread of the new coral disease "SCTLD" into the Caribbean: implications for Puerto Rico. Reef Encounter 34(1):38–43

Wilkinson C (2008) Status of coral reefs of the world: 2008. Global Coral Reef Monitoring Network and Reef and Rainforest Research Centre, Townsville

Williams GJ, Work TM, Aeby GS, Knapp IS, Davy SK (2011) Gross and microscopic morphology of lesions in Cnidaria from Palmyra atoll, Central Pacific. J Invertebr Pathol 106(2):165–173

Willis BL, Page CA, Dinsdale EA (2004) Coral disease on the great barrier reef. In: Rosenberg E, Loya Y (eds) Coral health and disease. Springer, Berlin, Heidelberg, pp 69–104

Woodley CM, Bruckner AW, McLenon AL, Higgins JL, Galloway SB, Nicholson JH (2008) Field manual for investigating coral disease outbreaks. In: NOAA Technical Memorandum NOS NCCOS 80 and CRCP 6. National Oceanic and Atmospheric Administration, Silver Spring, MD, p 85

Woodley CM, Downs C, Bruckner AW, Porter JW, Galloway SB (eds) (2016) Diseases of coral. Wiley Blackwell

Work TM, Rameyer RA (2005) Characterizing lesions in corals from American Samoa. Coral Reefs 24(3):384–390

Work TM, Aeby GS (2006) Systematically describing gross lesions in corals. Dis Aquat Org 70(1–2):155–160

Work TM, Aeby GS (2011) Pathology of tissue loss (white syndrome) in *Acropora* sp. corals from the Central Pacific. J Invertebr Pathol 107(2):127–131

Work T, Meteyer C (2014) To understand coral disease, look at coral cells. EcoHealth 11(4): 610–618

Zhenyu X, Shaowen K, Chaoqun H, Zhixiong Z, Shifeng W, Yongcan Z (2013) First characterization of bacterial pathogen, *vibrio alginolyticus*, for Porites andrewsi white syndrome in the South China Sea. PLoS One 8(9):e75425

The Synergy of Remote Sensing in Marine Invasion Science 14

Wei Sheng Chong, Khodzori Fikri Akmal, and Muhammad Dawood Shah

Abstract

Changes in invasion understanding have been advocated for by researchers to better forecast invasions and lessen the consequences of invasive species on the environment and socioeconomics. This chapter aims to generate new ideas and promote research on remote sensing applications to advance robust invasion science and management. Remote sensing techniques can be used to examine and identify invasive species by combining a synergistic understanding of biological invasions associated with the aquaculture and shipping industries, invasive species detection aspects, limitations of marine remote sensing for invasion science, specific invasion metrics and change detection. Using these synergies is crucial for developing long-term management strategies based on interdisciplinary collaboration among academics, policymakers and communities. By monitoring and mapping the existence and distribution of marine invasive species, remote sensing can aid in ecosystem-based management of damaged coastal zones.

14.1 Introduction

Invasive species, defined as organisms introduced outside of their normal geographical area and beyond their natural dispersion capacity, are a crucial factor in defining ecosystem function and modifying local livelihoods and well-being globally (Bellard et al. 2016; Gallardo et al. 2016; Katsanevakis et al. 2014a, b; Vaz et al.

W. S. Chong (✉) · K. F. Akmal · M. D. Shah
Borneo Marine Research Institute, Universiti Malaysia Sabah, Jalan UMS, Kota Kinabalu, Sabah, Malaysia
e-mail: chong@ums.edu.my

M. D. Shah et al. (eds.), *Marine Biotechnology: Applications in Food, Drugs and Energy*, https://doi.org/10.1007/978-981-99-0624-6_14

Fig. 14.1 (**a**) *Pterois volitans* – lionfish (**b**) *Acanthaster plancii* – Crown-of-Thorns starfish, recorded in the shallow water

2018; Vilà and Hulme 2017). Invading alien species in marine ecosystems may have significant detrimental impacts on native biodiversity (e.g. native species being replaced and biodiversity degraded owing to environmental alteration, changes in community structure and adjustments in ecosystem services) (Katsanevakis et al. 2014b; Russell and Blackburn 2017; Vergés et al. 2016). Invasive species might rival indigenous species (Fig. 14.1) and change local benthic ecosystems, possibly causing subtidal community depletion (Klein and Verlaque 2008; Matijević et al. 2013). Also, invasive species' presence may cause seagrass meadows' deterioration, which has a devastating effect on coastal protection (Katsanevakis et al. 2014b).

Researchers have advocated for changes in invasion understanding to predict invasions better and mitigate the effects of invading alien species on the environment, socioeconomics and human health (Essl et al. 2017; Ricciardi et al. 2017; Vaz et al. 2017). Evolving biotechnological, environmental and ideological concerns and possibilities can transform invasion science in the future (Dehnen-Schmutz et al. 2018; Ricciardi et al. 2017). For example, the Sustainable Development Goals (SDG) 14 and the Intergovernmental Science-Policy Platform on Biodiversity and Ecosystem Services (IPBES) now acknowledge the technical development and advantages of thematic assessments on invasive alien species.

Considering the thematic assessment for invasion science, the authors want to inspire ideas and stimulate advanced research on remote sensing and Geographic Information System (GIS) applications in developing robust invasion science and management. Remote sensing is gathering data regarding an object without physical contact with it. Remote sensing has been especially beneficial for identifying and mapping flora and fauna invaders and predicting their present and prospective dispersals and effects (Hellmann et al. 2017; Müllerová et al. 2017; Rocchini et al. 2015; Safonova et al. 2019). Also, it has grown more vital for marine biological monitoring as data and GIS tools have become more available and ubiquitous, resulting in a marine remote sensing revolution (Vaz et al. 2019). Remote sensing is often used to obtain data on marine environments using remote sensors (installed on board satellites, aircraft or drones) and additional *in-situ* survey methods.

Why choose remote sensing? Remote sensing methods employing optical sensors (visible (VIS) wavelengths to shortwave infrared (SWIR) wavelengths) enable researchers to record the scale of benthic or intertidal environments and detect significant marine organisms (Gorelick et al. 2017). Satellite sensors with a medium spatial resolution (30–1000 m) (e.g. MODIS, Sentinel-3) give ocean colour data that may be correlated to factors such as chlorophyll or pigment content. They can aid in the identification and monitoring of marine invasive species in the riparian zones (e.g. riparian, emergent), at the water surface (e.g. phytoplankton, floating macrophytes) or in the water column (e.g. submerged macrophytes) (Gorelick et al. 2017).

Mapping invasive species by remote sensing in coastal and shallow waters may need platforms with better spatial resolution and appropriate sensors (e.g. Landsat) or extremely high-resolution platforms with optical sensors (e.g. Quickbird, World-View and unmanned aerial vehicles (UAVs)). The high-resolution satellites can enable repeated monitoring of emerging phenomena with a ground resolution of fewer than 30 metres (e.g. Landsat-8, Sentinel-2). The ability to collect ad-hoc data using low-cost equipment and software opens up new opportunities to improve our understanding of the marine ecosystem (Casella et al. 2017; Hamylton 2017). UAVs, for instance, may map fine-scale species heterogeneity (i.e. 1 m scale) without the orbital time, weather and cloud coverage restrictions of conventional remote sensing platforms (Chong et al. 2021; Hedley et al. 2016; Zaki et al. 2022). UAVs attempt to bridge the gap between conventional remote sensing platforms and marine habitat field monitoring methodologies. UAVs can collect data at a lower height, providing a higher spatial resolution and lowering expenses (Feng et al. 2015). The benefits of UAVs are the quick gathering of very high-resolution data at precisely set time intervals, making them beneficial for conservation and invasion checking (Jung et al. 2017; Kislik et al. 2018; Wang et al. 2012; Xu et al. 2018).

Remote sensing imaging and its derivatives provide a potential means of capturing bio-optical data and supporting image classification and change detection (Yao et al. 2019). These two methodologies are considered useful for identifying species richness and distribution in shallow waters and monitoring their spatial and temporal changes (Chong et al. 2021; Xiang et al. 2019). However, it is advised that remote sensing methods be paired with field measurements of the target species' biophysical characteristics to facilitate the correct data pre-processing (Hedley et al. 2016). The data may be utilized for the multi-scale or multi-temporal methodical inspection and mapping of the growing prevalence of invasive species in marine environments that are impacted by the increasing effects of shipping, eutrophication, pollution and climate change (Mannino et al. 2021).

With the development of the remote sensing field and the current trajectory of invasion research (Dehnen-Schmutz et al. 2018; Ricciardi et al. 2017), this chapter aims to investigate the remote sensing significance in addressing some of the important difficulties in advance of invasion research in the following sections. The authors' narrative aims to endorse the various functions of the field in invasion science by highlighting the specific matters for which remote sensing has turned out to be an advantage, namely in the following domains: (a) biological invasions connected with aquaculture and shipping industries; (b) invasive species detection aspects; (c) limitations of marine remote sensing for invasion science; (d) specific metrics of invasions and (e) change detection.

14.1.1 Biological Invasions Associated with Aquaculture and Shipping Industries

The shipping sector and modern aquaculture are well-known for introducing alien species across the globe, resulting in increased propagule pressure and invasion concerns (Bolch et al. 2020; Vaz et al. 2019). In addition to these activities, there is an increase in the transmission of algae, diseases, viruses and weeds. Changes in species composition and harmful algal blooms in coastal waters may have cascading impacts on community assembly for marine birds, corals, marine mammals, fish, shellfish and benthic ecosystems and remain a continual source of worry for environmentalists and conservationists (Anderson et al. 2002).

Invasive phytoplankton have been proven to propagate through ballast water (Carlton 1994; Olenin et al. 2000), and the species abundance observed in ballast water is increasing (Olenina et al. 2010). For millennia, shipping has served as an invasion channel, making it impossible to assess the biogeographical level of species transported to other maritime locations many years ago, several of which may have progressively become global (Olenina et al. 2010). Assessing a species' native or alien status is vital for ecological research and management; hence, target species detection elements, recognizing the limits of new technologies, and effective information exchange are urgently required.

In modern aquaculture, remote sensing has been employed for many applications. These applications span from mapping marine alien macroalgae to detecting and monitoring invading riparian species (Andrew and Ustin 2008; Hamada et al. 2007; Laba et al. 2008; West et al. 2017), as well as assessing harmful algal blooms (HAB) in marine environments (Hestir et al. 2012; Kudela et al. 2015). The employment of LiDAR (e.g. Riegl laser scanner) and also hyperspectral sensors, whether on an airborne platform (e.g. CASI, HyMap), a satellite (e.g. Hyperion, AVIRIS) or an aerial platform (e.g. TetraCAM), has proved very beneficial. Another example is the employment of a complicated system that incorporates an autonomous surface vehicle and a UAV to remove algal blooms and reduce their negative environmental consequences. Using electrocoagulation and flotation, Jung et al. (2017)) achieved a near-perfect eradication of cyanobacteria (99.53%).

Remote sensing may detect foreign algae, weeds (Mannino et al. 2021; Petrocelli et al. 2015), pests and diseases (Lin et al. 2014; Techy et al. 2010) in aquaculture locations. Remote sensing may be used to discover species that spread from aquaculture sites and anticipate their potential invasion regions by detecting algal blooms and using them in statistical modelling methodologies (Kudela et al. 2015). New research from the Yellow Sea indicates the possibility of utilizing UAVs to detect and even compute the biomass of green algae clinging to rafts with a red algae *Porphyra yezoensis* and adversely impacting *P. yezoensis* aquaculture. (Xu et al.

2018) demonstrated that the optimum algorithm for detecting green algae in aquaculture is the normalized green–red difference index. The authors presented a technique for detecting green algae and estimating biomass based on UAV-derived imagery, field surveys and Sentinel-2 satellite data.

14.1.2 Aspects of Invasive Species Detection

Improving *in-situ* survey techniques and field protocols is critical for reducing identification uncertainty and increasing target species detection. *In-situ* surveys, such as SCUBA diving and snorkelling for visual inspection (Safuan et al. 2018) or target species capture using fishing equipment, are often used to monitor marine invasive species. Because not all target species can be visually identified by observers or captured with fishing equipment, these approaches have a substantial disadvantage that might create data bias or impair data detectability (Katsanevakis et al. 2012). In biological community research, poor data detectability may lead to an underestimating of the population status variable or, worse, the invasive species richness. Numerous basic procedures in invasion science research, like mark-recapture, distance sampling and repeated presence–absence surveys for occupancy estimates, have been adjusted to circumvent the target species detection feature (Issaris et al. 2012; Katsanevakis et al. 2012). As a result, remote sensing and GIS technologies may be combined with *in-situ* surveys to increase target species identification.

A low-altitude airborne remote sensing platform, such as a drone, may easily acquire exact information on invasive species, allowing for accurate textual and geometric analysis across a far broader range than a single place examined by an *in-situ* survey (Colomina and Molina 2014; Toro and Tsourdos 2018). However, an increase in data resolution does not imply an increase in data interpretation since invading species features may increase the complexity of within-class texture, which commonly leads to misinterpretation. Furthermore, when the sizes of invasive species vary widely in a scene, multi-scale solutions are advised to avoid superfluous processing, especially for high-resolution imagery, and finding an appropriate scale range may be particularly challenging.

What can be done by the synergistic interplay of remote sensing and *in-situ* marine invasive species surveys is limited. The variable composition of sub-metre scale marine invasive species and their spectral and structural diversity introduce critical uncertainties into the relationship between species cover and above-water reflectance (Hedley et al. 2012). When mapping a toxic algal bloom on a sediment substrate as contrasted to a reef with a more compositionally organized structure, the species composition may limit feasible mapping accuracy (Lim et al. 2009; Mumby et al. 2004). The context of the marine environment also influences classification accuracy; for example, deep muddy settings may enhance the ambiguity of benthic mapping (Hedley et al. 2012; Mumby et al. 1997). These conditions must be considered when using thematic maps to control marine invasive species by giving precise information on the quality of features during the planning phase.

14.1.3 Marine Remote Sensing Limitations in Invasion Science

Air-water interactions such as atmospheric condition, water quality (depth and clarity) and sun glint limit the data quality for remote sensing in marine invasion research. If remotely sensed data are pre-processed, some of these limiting limitations may be more suitable to detect biophysical properties. Pre-processing approaches include radiometric, atmospheric and geometric adjustments, correction for sun glint, use of depth-invariant bands, and correction for the air–water interface (Hedley et al. 2005a, b; Kay et al. 2009; Mumby et al. 1998; Muslim et al. 2019; Pierrot Deseilligny and Clery 2012; Purkis and Pasterkamp 2004; Purkis 2005). The primary goal of remote sensing is to map regionally diverse or complex phenomena under changing environmental circumstances. Consequently, obtaining exact spectrum imaging findings on invasive species unaffected by sensor characteristics or meteorological or topographic conditions is critical for the results' dependability and repeatability.

The water column effects cause signal attenuation, with fewer than 10% of the submerged marine population signal detected at the top of the atmosphere. Consequently, sensor configuration, atmospheric, geometric and radiometric qualities are particularly crucial for the water column and submerged macrophytes (Muller-Karger et al. 2018). Satellite sensors intended to fit such conditions remain focused on waters, with a resolution of 250 to 1000 m, well beyond the spatial resolution required for mapping macrophytes. SPOT 7, and Landsat-8, feature greater signal-to-noise ratios and enhanced calibration procedures (Hossain et al. 2015). Submerged macrophytes mapping may thus be possible, but specific species mapping is expected to remain difficult without high spectral resolution data.

Sun glint is often inevitable in remote sensing imagery. The specular signal triggers excessive illumination in imagery, decreases the signal-to-noise ratios and dramatically reduces meaningful observations (Hedley et al. 2005a, b; Kutser et al. 2009). Kay et al. (2009) provide a thorough examination of the theoretical aspects of sun glint and its correction. Airborne and satellite imagery cannot be utilized to establish the exact amount and distribution of invasive species in regions impacted by sun glint due to inaccurate observations at the research site. As a result, the quality of map outputs may deteriorate, and radiometric analysis may result in erroneous classification approaches.

Remote sensing data collection is subject to atmospheric and cloud conditions, which weaken electromagnetic waves and produce data degradation and loss. Aerosol sensors or a physical-based multi-layer scattering model might be used when the number of particles or haze is high (Huang et al. 2016; Zarco-Tejada et al. 2013). Aerial platforms, such as UAVs, may significantly reduce their cloud and atmospheric effects by flying closer to their goals. UAV remote sensing can acquire high-quality data with various information that improve image interpretation for studying marine species' invasions. As a result, atmospheric adjustments are no longer mandatory for UAV image processing (Adão et al. 2017; Minařík et al. 2019).

In multi-spectral and hyperspectral satellite sensors, radiometric corrections are well developed, and atmospheric correction techniques, including empirical and radiative transfer modelling (RTM) algorithms, are well-defined (Aspinall et al. 2002; Dinguirard and Slater 1999; Smith and Milton 1999). Multi-spectral UAV sensors (e.g. TetraCAM, DJI multi-spectral) based on converted cameras are installed with little or no calibration, and their radiometric attributes are unknown (Aasen et al. 2015; Crusiol et al. 2017; Minařík et al. 2019). The next generation of multi-spectral, multi-array sensors (e.g. Parrot Sequoia, MicaSense RedEdge) allows for automatic correction measures based on capturing a single calibration target before and after the flight. However, user-designed correction algorithms are necessary for successful data gathering, processing and comparison utilizing many sensors (Assmann et al. 2019). Multi-spectral satellite and aerial sensors with well-defined sensor feature and data pre-processing techniques (such as Landsat and Sentinel) can be employed in invasion research. On the other hand, UAV multi-spectral sensors are manufactured in a broad range of series with minor parameter differences across versions. As a result, any UAV spectroscopy inquiry for invasion science must include at least basic radiometric and atmospheric data validation to ensure the imaging results' reliability and accuracy.

To summarize, remote sensing of marine invasive species needs high spatial resolution, spectral resolution and radiometric resolution for submerged marine invasive species. By allowing time-based techniques, advancements in comprehensive mapping with climate-related missions might enhance invasive marine mapping. The fast-growing scene of UAV platforms and sensors further boosts the possibility of mapping invasive species utilizing textural differences or object-based mapping employing segmentation methods, particularly when the region being surveyed is limited.

14.1.4 Specific Metrics of Invasion Science

Remote sensing and GIS technologies may be utilized to gather various useful metrics for the research of marine invasions, depending on the location or abundance of an invasive species (Wilson et al. 2014). It is critical to assess an invasive species' relative abundance, geographic distribution and degree of aggregation when identifying dominating invasive species, sensitive ecosystems and determining the trajectory of an invasion. If recruitment occurs beyond the parent population, the invasion has advanced to the point where it is driven by dispersal stress from invading metapopulations and is much more difficult to regulate (Higgins et al. 2001; Langdon et al. 2010; Rouget and Richardson 2003). This may be accomplished by calculating the extent to which an invasive species has spread from its originating inhabitants (Catford et al. 2012). These metrics may be produced in terms of approach by exporting vector files representing found invasive species to programmes such as R and Python. Visual interpretation, vector file delimitation and GIS measuring techniques may be utilized to determine the relative separation between the parent population and those that have propagated to surrounding native species. However, georeferencing may be required since these measures rely on the positional precision of remote sensing imagery.

14.1.5 Change Detection in Invasion Science

In-situ surveys should surely be conducted in addition to remote sensing observations, particularly when fresh invasions are recorded. Rather than expecting a foreign species to turn out to be too widespread, early identification of newly emerging invasions and prediction models (Le Louarn et al. 2017) are frequently more cost-effective, adding to vulnerability evaluations and increasing the likelihood of eradicating an invader (Vicente et al. 2013, 2016). However, the usefulness of remotely sensing may be reduced if a species is secretive, tiny, dispersed or extremely migratory. It is possible to have false positives, particularly within locations where invasive and indigenous species have comparable visual traits.

Researchers may employ remote sensing and GIS to discover a range of prospective locations for management prioritization, such as selecting the most or least invaded areas for urgent clearance operations. Furthermore, by submitting the study location's coordinates to geodatabases, other researchers can access them. Instances include the utilization of "black marble" night-time remote sensing for discovering human demographic trends that impact species behaviours (Mazor et al. 2013) or the maritime traffic detection that may lead to species relocation throughout oceans (Marino et al. 2015).

Time-series data may be used to track the spread of species through time and location. Spread distances measured on many occasions may be used to compute spread rates, which can then be coupled with other invasive species features (e.g. morphological quality, residential duration) to build prediction models (Vaz et al. 2019). It may also be feasible to determine combinations of environmental elements that contribute to invasion occurrences by watching when an invasive species first expand away from its native population. Many observations help identify the key invasion causes for a certain species. (Kudela et al. 2015) Using time-series analysis of field hyperspectral data demonstrated that cyanobacteria might replace phytoplankton blooms in a short period. Similar fast changes in cyanobacteria were observed by (Hestir et al. 2012) using hyperspectral data. Phytoplankton blooms in Tampa Bay can be developed over 2 to 3 days (Chen et al. 2010). Even so, the 13 years of Long Island Sound monitoring revealed that monthly data is inadequate to characterize episodic plankton blooms (Dierssen et al. 2015).

14.2 Invasion Science Information Sharing

The social realm nowadays becomes more solution-oriented as technology advances, and remote sensing at the intersection of human geography and ecology becomes appealing for invasion science (Dehnen-Schmutz et al. 2018). With increased off-shelves UAVs and the big data, coupled with readily accessible remotely sensed data, precise visions about invasions may be provided at exceptional geographical and temporal resolutions. Developing citizen science surveillance programmes and repositories based on remote sensing and GIS (e.g. educational websites and mobile applications) could be a promising approach in the future of invasion science and management. Indeed, using citizen science for participatory invasion reporting might significantly advance the invasive species' early detection, monitoring and control (Roy et al. 2018).

GIS has been extensively employed by biologists, surveyors, social scientists and palaeontologists for various missions (e.g. mapping marine foreign algae, identifying invasive goldenrod (Ishii and Washitani 2013), detecting changes in leafy spurge blooming (Mitchell and Glenn 2009) and disaster response and management (http://giscorps.org). However, it can be a toolkit for investigating the whole spectrum of invasions.

Four potential applications of remote sensing and GIS technology in marine invasion science are as follows: (a) to emphasize information or awareness to focus on the vivid alterations caused by species incursions; (b) to track the spread of invasive species; (c) to predict the spread of invasive species and (d) to create a platform for data sharing and networking among invasion scientists. GIS has been extensively utilized to present attractive, informative and useful information. GIS software or an online GIS platform such as WebGIS or Google Earth Engine (GEE) may export data in the raster KML format to show spatial data layers (Visser et al. 2014). Cloud-based GIS (e.g. GeoServer, ArcGIS server), which combine the GEE features with advanced GIS tools and are all accessible online, may be used to monitor and communicate data on species invasions. Although it demands a high-level specialized competence and is outside the scope of this chapter, this GIS approach has enormous potential to obtain and distribute data on invasive species.

14.3 Conclusions

The preceding non-exhaustive list demonstrates that marine biological invasions are a continuing phenomenon, with numerous invasive species extending their spread along the coast, with significant consequences for the ecosystem, human well-being and the financial system. As a result, monitoring the spread of marine invasive species is critical to protecting and managing marine ecosystems and executing relevant regulations (Katsanevakis et al. 2015).

Rapid and reliable technologies for monitoring and mapping the quantity and distribution of marine invasive species are necessary to assist in ecosystem-based sustainable management of impacted coastal regions. Other cost-efficient techniques for examining and identifying invasive species, particularly in marine protected areas (MPAs), are currently available, including remote sensing airborne or satellite platforms and *in-situ* measurements.

To advance remote sensing in marine invasion science, researchers must (a) understand the aspects of invasive species detection, (b) identify the limiting factors of marine remote sensing and (c) devote themselves to improving or developing methods, metrics and instruments for invasion science and (d) report findings together with its uncertainties. Using these synergies is becoming more important to provide successful and long-term management strategies based on interdisciplinary cooperation between academics, policymakers and communities.

Funding Information This study was funded by the Malaysian Ministry of Higher Education's Fundamental Research Grant Scheme (FRGS/1/2022/WAB05/UMS/02/1; FRG0568–1/2022) and the Universiti Malaysia Sabah's Skim Penyelidikan Lantikan Baharu (SLB2229) to Wei Sheng Chong.

References

Aasen H, Burkart A, Bolten A, Bareth G (2015) Generating 3D hyperspectral information with lightweight UAV snapshot cameras for vegetation monitoring: from camera calibration to quality assurance. ISPRS J Photogramm Remote Sens 108:245–259. https://doi.org/10.1016/j.isprsjprs.2015.08.002

Adão T, Hruška J, Pádua L, Bessa J, Peres E, Morais R, Sousa JJ (2017) Hyperspectral imaging: A review on UAV-based sensors, data processing and applications for agriculture and forestry. Remote Sens 9(11). https://doi.org/10.3390/rs9111110

Anderson DM, Glibert PM, Burkholder JM (2002) Harmful algal blooms and eutrophication: nutrient sources, composition, and consequences. Estuaries 25(4):704–726. https://doi.org/10.1007/BF02804901

Andrew ME, Ustin SL (2008) The role of environmental context in mapping invasive plants with hyperspectral image data. Remote Sens Environ 112(12):4301–4317

Aspinall RJ, Marcus WA, Boardman JW (2002) Considerations in collecting, processing, and analysing high spatial resolution hyperspectral data for environmental investigations. J Geograph Syst 4:15. https://doi.org/10.1007/s101090100071

Assmann JJ, Kerby JT, Cunliffe AM, Myers-Smith IH (2019) Vegetation monitoring using multispectral sensors — best practices and lessons learned from high latitudes. J Unmanned Vehicle Syst 7:54. https://doi.org/10.1139/juvs-2018-0018

Bellard C, Cassey P, Blackburn TM (2016) Alien species as a driver of recent extinctions. Biol Lett 12(2):20150623. https://doi.org/10.1098/rsbl.2015.0623

Bolch EA, Santos MJ, Ade C, Khanna S, Basinger NT, Reader MO, Hestir EL (2020) Remote detection of invasive alien species. In: Cavender-Bares J, Gamon JA, Townsend PA (eds) Remote sensing of plant biodiversity. Springer International Publishing, pp 267–307. https://doi.org/10.1007/978-3-030-33157-3_12

Carlton JT (1994) Exotic phytoplankton from ships' ballast waters: risk of potential spread to Mariculture sites on Canada's East Coast. Can Data Fish Aquatic Sci 937:1–51

Casella E, Collin A, Harris D, Ferse S, Bejarano S, Parravicini V, Hench JLJL, Rovere A (2017) Mapping coral reefs using consumer-grade drones and structure from motion photogrammetry techniques. Coral Reefs 36(1):269–275. https://doi.org/10.1007/s00338-016-1522-0

Catford JA, Vesk PA, Richardson DM, Pyšek P (2012) Quantifying levels of biological invasion: towards the objective classification of invaded and invasible ecosystems. Glob Chang Biol 18(1):44–62. https://doi.org/10.1111/j.1365-2486.2011.02549.x

Chen Z, Hu C, Muller-Karger FE, Luther ME (2010) Short-term variability of suspended sediment and phytoplankton in Tampa Bay, Florida: observations from a coastal oceanographic tower and ocean color satellites. Estuar Coast Shelf Sci 89(1):62–72

Chong WS, Zaki NHM, Hossain MS, Muslim AM, Pour AB (2021) Introducing Theil-Sen estimator for sun glint correction of UAV data for coral mapping. Geocarto Int 37:4527. https://doi.org/10.1080/10106049.2021.1892206

Colomina I, Molina P (2014) Unmanned aerial systems for photogrammetry and remote sensing: A review. ISPRS J Photogramm Remote Sens 92:79–97. https://doi.org/10.1016/j.isprsjprs.2014.02.013

Crusiol LGT, Nanni MR, Silva GFC, Furlanetto RH, da Silva Gualberto AA, Gasparotto ADC, De Paula MN (2017) Semi professional digital camera calibration techniques for Vis/NIR spectral data acquisition from an unmanned aerial vehicle. Int J Remote Sens 38:2717. https://doi.org/10.1080/01431161.2016.1264032

Dehnen-Schmutz K, Boivin T, Essl F, Groom QJ, Harrison L, Touza JM, Bayliss H (2018) Alien futures: what is on the horizon for biological invasions? Divers Distrib 24(8):1149–1157

Dierssen H, McManus GB, Chlus A, Qiu D, Gao B-C, Lin S (2015) Space station image captures a red tide ciliate bloom at high spectral and spatial resolution. Proc Natl Acad Sci 112(48): 14783–14787

Dinguirard M, Slater PN (1999) Calibration of space-multispectral imaging sensors: A review. Remote Sens Environ 68:194. https://doi.org/10.1016/S0034-4257(98)00111-4

Essl F, Hulme PE, Jeschke JM, Keller R, Pyšek P, Richardson DM, Saul W-C, Bacher S, Dullinger S, Estévez RA et al (2017) Scientific and normative foundations for the valuation of alien-species impacts: thirteen core principles. Bioscience 67(2):166–178

Feng Q, Liu J, Gong J (2015) UAV remote sensing for urban vegetation mapping using random forest and texture analysis. Remote Sens 7(1):1074–1094. https://doi.org/10.3390/rs70101074

Gallardo B, Clavero M, Sánchez MI, Vilà M (2016) Global ecological impacts of invasive species in aquatic ecosystems. Glob Chang Biol 22(1):151–163. https://doi.org/10.1111/gcb.13004

Gorelick N, Hancher M, Dixon M, Ilyushchenko S, Thau D, Moore R (2017) Google earth engine: planetary-scale geospatial analysis for everyone. Remote Sens Environ 202:18–27

Hamada Y, Stow DA, Coulter LL, Jafolla JC, Hendricks LW (2007) Detecting tamarisk species (Tamarix spp.) in riparian habitats of Southern California using high spatial resolution hyperspectral imagery. Remote Sens Environ 109(2):237–248

Hamylton SM (2017) Mapping coral reef environments. Prog Phys Geogr 41(6):803–833. https://doi.org/10.1177/0309133317744998

Hedley JD, Harborne ARR, Mumby PJJ (2005a) Simple and robust removal of sun glint for mapping shallow-water benthos. Int J Remote Sens 26(10):2107–2112. https://doi.org/10.1080/01431160500034086

Hedley JD, Harborne AR, Mumby PJ (2005b) Technical note: simple and robust removal of sun glint for mapping shallow-water benthos. Int J Remote Sens 26(10):2107–2112. https://doi.org/10.1080/01431160500034086

Hedley JD, Roelfsema CM, Chollett I, Harborne AR, Heron SF, Weeks SJ, Skirving WJ, Strong AE, Mark Eakin C, Christensen TRL, Ticzon V, Bejarano S, Mumby PJ (2016) Remote sensing of coral reefs for monitoring and management: A review. Remote Sens 8(2). https://doi.org/10.3390/rs8020118

Hedley JD, Roelfsema CM, Phinn SR, Mumby PJ (2012) Environmental and sensor limitations in optical remote sensing of coral reefs: implications for monitoring and sensor design. Remote Sens 4(1):271–302. https://doi.org/10.3390/rs4010271

Hellmann C, Große-Stoltenberg A, Thiele J, Oldeland J, Werner C (2017) Heterogeneous environments shape invader impacts: integrating environmental, structural and functional effects by isoscapes and remote sensing. Sci Rep 7(1):1–11

Hestir EL, Greenberg JA, Ustin SL (2012) Classification trees for aquatic vegetation community prediction from imaging spectroscopy. IEEE J Selected Topics Appl Earth Observ Remote Sensing 5(5):1572–1584

Higgins SI, Richardson DM, Cowling RM (2001) Validation of a spatial simulation model of a spreading alien plant population. J Appl Ecol 38(3):571–584

Hossain MS, Bujang JS, Zakaria MH, Hashim M (2015) The application of remote sensing to seagrass ecosystems: an overview and future research prospects. Int J Remote Sens 36(1): 61–114. https://doi.org/10.1080/01431161.2014.990649

Huang Y, Ding W, Li H (2016) Haze removal for UAV reconnaissance images using layered scattering model. Chin J Aeronaut 29:502. https://doi.org/10.1016/j.cja.2016.01.012

Ishii J, Washitani I (2013) Early detection of the invasive alien plant Solidago altissima in moist tall grassland using hyperspectral imagery. Int J Remote Sens 34(16):5926–5936. https://doi.org/10.1080/01431161.2013.799790

Issaris Y, Katsanevakis S, Salomidi M, Tsiamis K, Katsiaras N, Verriopoulos G (2012) Occupancy estimation of marine species: dealing with imperfect detectability. Mar Ecol Prog Ser 453:95–106

Jung S, Cho H, Kim D, Kim K, Han J-I, Myung H (2017) Development of algal bloom removal system using unmanned aerial vehicle and surface vehicle. IEEE Access 5:22166–22176

Katsanevakis S, Coll M, Piroddi C, Steenbeek J, Ben Rais Lasram F, Zenetos A, Cardoso AC (2014b) Invading the Mediterranean Sea: biodiversity patterns shaped by human activities. Front Mar Sci 1. https://doi.org/10.3389/fmars.2014.00032

Katsanevakis S, Deriu I, D'Amico F, Nunes AL, Sanchez SP, Crocetta F, Arianoutsou M, Bazos I, Christopoulou A, Curto G, Delipetrou P, Kokkoris Y, Panov VE, Rabitsch W, Roques A, Scalera R, Shirley SM, Tricarico E, Vannini A et al (2015) European alien species information network (EASIN): supporting european policies and scientific research. Manag Biol Invasions 6(2):147–157. https://doi.org/10.3391/mbi.2015.6.2.05

Katsanevakis S, Wallentinus I, Zenetos A, Leppäkoski E, Çinar ME, Oztürk B, Grabowski M, Golani D, Cardoso AC (2014a) Impacts of invasive alien marine species on ecosystem services and biodiversity: A pan-European review. Aquat Invasions 9(4):391–423. https://doi.org/10.3391/ai.2014.9.4.01

Katsanevakis S, Weber A, Pipitone C, Leopold M, Cronin M, Scheidat M et al (2012) Monitoring marine populations and communities: methods dealing with imperfect detectability. Aquat Biol 16(1):31–52

Kay S, Hedley JDJ, Lavender S (2009) Sun glint correction of high and low spatial resolution images of aquatic scenes: a review of methods for visible and near-infrared wavelengths. Remote Sens 1(4):697–730. https://doi.org/10.3390/rs1040697

Kislik C, Dronova I, Kelly M (2018) UAVs in support of algal bloom research: A review of current applications and future opportunities. Drones 2(4). https://doi.org/10.3390/drones2040035

Klein J, Verlaque M (2008) The Caulerpa racemosa invasion: A critical review. Mar Pollut Bull 56(2):205–225. https://doi.org/10.1016/j.marpolbul.2007.09.043

Kudela RM, Palacios SL, Austerberry DC, Accorsi EK, Guild LS, Torres-Perez J (2015) Application of hyperspectral remote sensing to cyanobacterial blooms in inland waters. Remote Sens Environ 167:196–205

Kutser T, Vahtmäe E, Praks J (2009) A sun glint correction method for hyperspectral imagery containing areas with non-negligible water leaving NIR signal. Remote Sens Environ 113(10): 2267–2274. https://doi.org/10.1016/j.rse.2009.06.016

Laba M, Downs R, Smith S, Welsh S, Neider C, White S, Richmond M, Philpot W, Baveye P (2008) Mapping invasive wetland plants in the Hudson River National Estuarine Research Reserve using quickbird satellite imagery. Remote Sens Environ 112(1):286–300

Langdon B, Pauchard A, Aguayo M (2010) Pinus contorta invasion in the Chilean Patagonia: local patterns in a global context. Biol Invasions 12(12):3961–3971. https://doi.org/10.1007/s10530-010-9817-5

Le Louarn M, Clergeau P, Briche E, Deschamps-Cottin M (2017) "Kill two birds with one stone": urban tree species classification using bi-temporal Pléiades images to study nesting preferences of an invasive bird. Remote Sens 9(9):916

Lim A, Hedley JD, LeDrew E, Mumby PJ, Roelfsema C (2009) The effects of ecologically determined spatial complexity on the classification accuracy of simulated coral reef images. Remote Sens Environ 113(5):965–978. https://doi.org/10.1016/j.rse.2009.01.011

Lin B, Ross SD, Prussin AJ, Schmale DG (2014) Seasonal associations and atmospheric transport distances of fungi in the genus fusarium collected with unmanned aerial vehicles and ground-based sampling devices. Atmos Environ 94:385–391. https://doi.org/10.1016/j.atmosenv.2014.05.043

Mannino AM, Borfecchia F, Micheli C (2021) Tracking marine alien macroalgae in the Mediterranean Sea: the contribution of citizen science and remote sensing. J Marine Sci Eng 9(3). https://doi.org/10.3390/jmse9030288

Marino A, Sanjuan-Ferrer MJ, Hajnsek I, Ouchi K (2015) Ship detection with spectral analysis of synthetic aperture radar: A comparison of new and well-known algorithms. Remote Sens 7(5): 5416–5439. https://doi.org/10.3390/rs70505416

Matijević S, Bogner D, Bojanić N, Žuljević A, Despalatović M, Antolić B, Nikolić V, Bilić J et al (2013) Biogeochemical characteristics of sediments under the canopy of invasive alga Caulerpa racemosa var. cylindracea (Pelješac peninsula, Adriatic Sea). Fresenius Environ Bull 22:3030–3040

Mazor T, Levin N, Possingham HP, Levy Y, Rocchini D, Richardson AJ, Kark S (2013) Can satellite-based night lights be used for conservation? The case of nesting sea turtles in the Mediterranean. Biol Conserv 159:63–72. https://doi.org/10.1016/j.biocon.2012.11.004

Minařík R, Langhammer J, Hanuš J (2019) Radiometric and atmospheric corrections of multispectral μMCA camera for UAV spectroscopy. Remote Sens 11. https://doi.org/10.3390/rs11202428

Mitchell JJ, Glenn NF (2009) Leafy spurge (Euphorbia esula) classification performance using hyperspectral and multispectral sensors. Rangel Ecol Manag 62(1):16–27. https://doi.org/10.2111/08-100

Muller-Karger FE, Hestir E, Ade C, Turpie K, Roberts DA, Siegel D, Miller RJ, Humm D, Izenberg N, Keller M et al (2018) Satellite sensor requirements for monitoring essential biodiversity variables of coastal ecosystems. Ecol Appl 28(3):749–760

Müllerová J, Bartaloš T, Bruuna J, Dvovrák P, Vitková M (2017) Unmanned aircraft in nature conservation: an example from plant invasions. Int J Remote Sens 38(8–10):2177–2198

Mumby PJ, Green EP, Edwards AJ, Clark CD (1997) Coral reef habitat-mapping: how much detail can remote sensing provide? Mar Biol 130(2):193–202. https://doi.org/10.1007/s002270050238

Mumby PJJ, Clark CDD, Green EPP, Edwards AJJ (1998) Benefits of water column correction and contextual editing for mapping coral reefs. Int J Remote Sens 19(1):203–210. https://doi.org/10.1080/014311698216521

Mumby PJPJ, Skirving W, Strong AEAE, Hardy JTJT, LeDrew EFEF, Hochberg EJEJ, Stumpf RPRP, David LTLT (2004) Remote sensing of coral reefs and their physical environment. Mar Pollut Bull 48:219–228

Muslim AM, Chong WS, Safuan CDM, Khalil I, Hossain MS (2019) Coral reef mapping of UAV: A comparison of sun glint correction methods. Remote Sens 11(20):2422. https://doi.org/10.3390/rs11202422

Olenin S, Gollasch S, Jonušas S, Rimkutė I (2000) En-route investigations of plankton in ballast water on a Ship's voyage from the Baltic Sea to the open Atlantic Coast of Europe. Int Rev Hydrobiol 85(596):577. https://doi.org/10.1002/1522-2632(200011)85:5/6<577::AID-IROH577>3.0.CO;2-C

Olenina I, Wasmund N, Hajdu S, Jurgensone I, Gromisz S, Kownacka J, Toming K, Vaiciute D, Olenin S (2010) Assessing impacts of invasive phytoplankton: the Baltic Sea case. Mar Pollut Bull 60(10):1691–1700

Petrocelli A, Cecere E, Portacci G, Micheli C, De Cecco L, Martini S, Borfecchia F (2015) Preliminary mapping of the alien seaweed Hypnea cornuta (Rhodophyta, Gigartinales) in the mar piccolo of Taranto (southern Italy, Mediterranean Sea). Biol Mar Medit 22:44–45

Pierrot Deseilligny M, Clery I (2012) Apero, an open source bundle Adjusment software for automatic calibration and orientation of set of images. ISPRS Int Arch Photogr Remote Sensing Spatial Inform Sci XXXVIII-5:269–276. https://doi.org/10.5194/isprsarchives-xxxviii-5-w16-269-2011

Purkis SJJ, Pasterkamp R (2004) Integrating in situ reef-top reflectance spectra with Landsat TM imagery to aid shallow-tropical benthic habitat mapping. Coral Reefs 23(1):5–20. https://doi.org/10.1007/s00338-003-0351-0

Purkis SJSJ (2005) A "reef-up" approach to classifying coral habitats from IKONOS imagery. IEEE Trans Geosci Remote Sens 43(6):1375–1390. https://doi.org/10.1109/TGRS.2005.845646

Ricciardi A, Blackburn TM, Carlton JT, Dick JTA, Hulme PE, Iacarella JC, Jeschke JM, Liebhold AM, Lockwood JL, MacIsaac HJ et al (2017) Invasion science: a horizon scan of emerging challenges and opportunities. Trends Ecol Evol 32(6):464–474

Rocchini D, Andreo V, Förster M, Garzon-Lopez CX, Gutierrez AP, Gillespie TW, Hauffe HC, He KS, Kleinschmit B, Mairota P et al (2015) Potential of remote sensing to predict species invasions: A modelling perspective. Prog Phys Geogr 39(3):283–309

Rouget M, Richardson DM (2003) Inferring process from pattern in plant invasions: a semimechanistic model incorporating propagule pressure and environmental factors. Am Nat 162(6):713–724

Roy H, Groom Q, Adriaens T, Agnello G, Antic M, Archambeau AS et al (2018) Increasing understanding of alien species through citizen science (Alien-CSI). Res Ideas Outcomes 4: e31412

Russell JC, Blackburn TM (2017) Invasive alien species: denialism, disagreement, definitions, and dialogue. Trends Ecol Evol 32(5):312–314. https://doi.org/10.1016/j.tree.2017.02.005

Safonova A, Tabik S, Alcaraz-Segura D, Rubtsov A, Maglinets Y, Herrera F (2019) Detection of fir trees (Abies sibirica) damaged by the bark beetle in unmanned aerial vehicle images with deep learning. Remote Sens 11(6):643

Safuan CDM, Ismail K, Khalil I, Ali A, Chong WS, Chan AAA, Ismail MN, Repin IIM, Bachok Z, Che Din MS, Ismail K, Khalil I, Ali A, Wei Sheng C, Chan AAA, Ismail N, Repin IIM, Bachok Z (2018) Quantification of coral reef benthos for coral health assessment in Labuan Marine Park, Malaysia. J Sustain Sci Manag 13:101–112

Smith GM, Milton EJ (1999) The use of the empirical line method to calibrate remotely sensed data to reflectance. Int J Remote Sens 20:2653. https://doi.org/10.1080/014311699211994

Techy L, Schmale DG III, Woolsey CA (2010) Coordinated aerobiological sampling of a plant pathogen in the lower atmosphere using two autonomous unmanned aerial vehicles. J Field Robotics 27(3):335–343. https://doi.org/10.1002/rob.20335

Toro FG, Tsourdos A (2018) UAV or drones for remote sensing applications (books volume 1 & 2). In MDPI (Vols. 1 & 2). https://doi.org/10.3390/books978-3-03897-092-7

Vaz AS, Alcaraz-Segura D, Campos JC, Vicente JR, Honrado JP (2018) Managing plant invasions through the lens of remote sensing: A review of progress and the way forward. Sci Total Environ 642:1328–1339. https://doi.org/10.1016/j.scitotenv.2018.06.134

Vaz AS, Alcaraz-Segura D, Vicente JR, Honrado JP (2019) The many roles of remote sensing in invasion science. Front Ecol Evol 7. https://doi.org/10.3389/fevo.2019.00370

Vaz AS, Kueffer C, Kull CA, Richardson DM, Schindler S, Muñoz-Pajares AJ, Vicente JR, Martins J, Hui C, Kühn I et al (2017) The progress of interdisciplinarity in invasion science. Ambio 46(4):428–442

Vergés A, Doropoulos C, Malcolm HA, Skye M, Garcia-Pizá M, Marzinelli EM, Campbell AH, Ballesteros E, Hoey AS, Vila-Concejo A, Bozec Y-M, Steinberg PD (2016) Long-term empirical evidence of ocean warming leading to tropicalization of fish communities, increased herbivory, and loss of kelp. Proc Natl Acad Sci 113(48):13791–13796. https://doi.org/10.1073/pnas.1610725113

Vicente JR, Alagador D, Guerra C, Alonso JM, Kueffer C, Vaz AS et al (2016) Cost-effective monitoring of biological invasions under global change: a model-based framework. J Appl Ecol 53(5):1317–1329

Vicente JR, Pinto AT, Araujo MB, Verburg PH, Lomba A, Randin CF et al (2013) Using life strategies to explore the vulnerability of ecosystem services to invasion by alien plants. Ecosystems 16:678–693

Vilà M, Hulme PE (2017) Non-native species, ecosystem services, and human Well-being. In: Vilà M, Hulme PE (eds) Impact of biological invasions on ecosystem services. Springer International Publishing, pp 1–14. https://doi.org/10.1007/978-3-319-45121-3_1

Visser V, Langdon B, Pauchard A, Richardson DM (2014) Unlocking the potential of Google Earth as a tool in invasion science. Biol Invasions 16:513–534

Wang H, Zhong G, Yan H, Liu H, Wang Y, Zhang C (2012) Growth control of cyanobacteria by three submerged Macrophytes. Environ Eng Sci 29(6):420–425. https://doi.org/10.1089/ees.2010.0286

West AM, Evangelista PH, Jarnevich CS, Kumar S, Swallow A, Luizza MW, Chignell SM (2017) Using multi-date satellite imagery to monitor invasive grass species distribution in post-wildfire landscapes: an iterative, adaptable approach that employs open-source data and software. Int J Appl Earth Obs Geoinf 59:135–146

Wilson JRU, Caplat P, Dickie IA, Hui C, Maxwell BD, Nuñez MA, Pauchard A, Rejmánek M, Richardson DM, Robertson MP, Spear D, Webber BL, van Wilgen BW, Zenni RD (2014) A standardized set of metrics to assess and monitor tree invasions. Biol Invasions 16(3):535–551. https://doi.org/10.1007/s10530-013-0605-x

Xiang T-Z, Xia G-S, Zhang L (2019) Mini-unmanned aerial vehicle-based remote sensing: techniques, applications, and prospects. IEEE Geosci Remote Sensing Magazine 7(3):29–63. https://doi.org/10.1109/mgrs.2019.2918840

Xu F, Gao Z, Jiang X, Shang W, Ning J, Song D, Ai J (2018) A UAV and S2A data-based estimation of the initial biomass of green algae in the South Yellow Sea. Mar Pollut Bull 128: 408–414. https://doi.org/10.1016/j.marpolbul.2018.01.061

Yao H, Qin R, Chen X (2019) Unmanned aerial vehicle for remote sensing applications - A review. Remote Sensing 11:1443. https://doi.org/10.3390/rs11121443

Zaki NHM, Chong WS, Muslim AM, Reba MNM, Hossain MS (2022) Assessing optimal UAV-data pre-processing workflows for quality ortho-image generation to support coral reef mapping. Geocarto Int 1–25:1. https://doi.org/10.1080/10106049.2022.2037732

Zarco-Tejada PJ, Morales A, Testi L, Villalobos FJ (2013) Spatio-temporal patterns of chlorophyll fluorescence and physiological and structural indices acquired from hyperspectral imagery as compared with carbon fluxes measured with eddy covariance. Remote Sens Environ 133:102. https://doi.org/10.1016/j.rse.2013.02.003

The Utilization of Agro-Based Wastes by Marine Phototrophic Microbes

15

Sujjat Al Azad and Mohammad Tamrin Bin Mohamad Lal

Abstract

Purple Non-Sulphur Bacteria (PNSB) is known as photosynthetic bacteria that are able to tolerate various habitats including marine sediment and mangrove environment. Species of PNSB are efficient in conversion of agro-based waste into nutritionally value-added products with reduction of Chemical Oxygen Demand (COD). Different levels of inoculum were used to evaluate the growth profiles of *Rv. sulfidophilum* in settled sardine processing wastewater. *Rv. sulfidophilum* grew well in unsterilized diluted and undiluted SPW under both anaerobic light and aerobic dark culture conditions. The growth in term of bacterial biomass and production of total carotenoids, as well reduction of chemical oxygen demand were observed better while cultured in aerobic dark condition at 300-rpm agitation speed. Sardine processing wastewater could be one of the best options in the production of Purple Non-Sulphur Bacteria (PNSB) in addition to the reduction of organic loading load. The produced biomass could be useful as feed additive in aquafeed industry. Dual benefits could be derived by culturing phototrophic bacteria in agro-based waste to produce bacterial biomass with the reduction of pollution load of the wastewater.

15.1 Introduction

Marine microbes are unique and diverse that live in a variety of complex oceanic ecosystems. In the marine environment, microbes are involved in the process of nutrient recycling. Marine microbes with diversified metabolic pathways feed on

S. Al Azad (✉) · M. T. B. Mohamad Lal
Borneo Marine Research Institute, Universiti Malaysia Sabah, Kota Kinabalu, Sabah, Malaysia
e-mail: sujjat@ums.edu.my

M. D. Shah et al. (eds.), *Marine Biotechnology: Applications in Food, Drugs and Energy*, https://doi.org/10.1007/978-981-99-0624-6_15

decomposed waste materials through the microbial loop and subsequently enter the ocean food chain. Feeding on waste not only accelerates microbial growth, but also eliminates ocean waste restoring the ecosystem's health. Large numbers of marine microbes have been discovered and identified. In this chapter, the roles of a few marine phototrophic bacteria are investigated with a special focus on the utilization of agro-based waste. Anaerobic phototrophic bacteria are mainly purple sulphur and purple non-sulphur bacteria (PNSB). Among them, PNSB, which utilize organic compounds are widely used for biotechnological applications. On the other hand, purple sulphur bacteria rely on sulphide and hydrogen as electron donors. Potential PNSB isolated from the marine environment are, *Afifella* sp. (Urdiain et al. 2008), *Rhodobium* sp. (Hiraishi et al. 1995), *Rhodovulum* sp. (Kompantseva 1986), *Rhodovibrio* sp. (Nissen and Dundas 1984), *Roseospira* sp. (Kompantseva and Gorlenko 1984) and *Rhodothalassium* sp. (Drews and Imhoff 1991).

15.2 Phototrophic Microbes

Phototrophic bacteria, previously known as photosynthetic bacteria (PSB), were described first by Engelmann in 1883. He characterized them based on pigmentation with a well define absorption spectrum similar to that of a green plant. Classical studies on photosynthetic bacteria (PSB) were started in 1931 and classified phototrophic bacteria as photosynthetic bacteria with the isolation of two types of PSB, one purple and the other green (Van Niel 1957). Van Niel defined four conditions for their isolation and growth as (a) illumination, (b) mineral medium, (c) a light alkaline reaction and (d) strict anaerobiosis. He also mentioned the overall metabolic pathway as:

(a) Green bacteria $2H_2S + CO_2 \xrightarrow[\text{Energy}]{\text{Light}} (CH_2O) + H_2O + 2S$

(b) Purple bacteria, $H_2S + 2H_2O + 2CO_2 \xrightarrow[\text{Energy}]{\text{Light}} 2(CH_2O) + H_2SO_4$

Green bacteria oxidize sulphide to sulphur with the assimilation of carbon dioxide, but in purple bacteria oxidation of sulphide proceeded through sulphate (Kobayashi and Kobayashi 2001). Dow (1982) reported three distinct groups of photosynthetic bacteria (PSB). They are cyanobacteria, purple PSB and green PSB. The major distinction is that phototrophic (with only one photosystem) bacteria can synthesize in anoxygenic conditions, whereas the cyanobacteria with two photosystems operating in series display oxygenic photosynthesis. Phototrophic bacteria do not use water as an electron donor, but cyanobacteria do so. Phototrophic bacteria unlike cyanobacteria in the presence of light can also use a variety of simple organic compounds as electron donors (Pfenning 1977; Imhoff 1992).

However, three major groups of photosynthetic bacteria are now recognized:

1. the purple and brown non-sulphur bacteria (Athiorhodaceae),
2. the purple sulphur bacteria (Thiorhodaceae) and
3. the green sulphur bacteria (Chlorobacteriaceae).

Phototrophic bacteria are widely distributed in nature, predominantly in aquatic and terrestrial habitats and even under the extreme conditions of Antarctica (Madigan 1999). In aquatic habitats, phototrophic bacteria can be found in freshwater, estuarine water and seawater, sulphur-containing hot springs (Pfenning 1977; Imhoff 1992) and in agro-industrial wastewater (Sasaki et al. 1991; Prasertsan et al. 1993).

15.3 The Potential of Agro-Based Wastes

The agriculture industry in Malaysia contributed 8.1% or RM89.5 billion to the 2018 Gross Domestic Product (GDP) (Department of Statistics, Malaysia 2019). The overall economic growth globally, especially in Malaysia, is influenced by the significant role of the agriculture industry (Dardak 2015). The suitability of Malaysia as a contributor to agricultural production could be factorized by the high humidity and temperature in Malaysia, which enable optimal conditions for agricultural activities to be carried out, such as plantation and aquaculture activities. Unfortunately, increased production of agriculture activities has also increased the waste generation due to agricultural activities and programs involving crop plantation, horticulture, livestock and aquaculture farming. A total of 998 million tons of agricultural waste were generated globally, while in Malaysia, the disposal of 1.2 million tons of agricultural waste into landfills was recorded annually. Agricultural waste or agro-based waste could be defined as the production of waste that originated from various agriculture activities that are deemed useless and often discarded when produced in farming activities. Agro-based waste produced by agricultural industry's activities contains various constituents. The main constituents that make up the composition of agro-based waste are soluble constituents such as organic acids, amino acids and sugar, while insoluble chemical constituents consist of cellulose and lignin (Caprara et al. 2011). In addition, resins, oil waxes, protein, minerals and fats make up the rest of the agro-based waste constituents. Although agro-industrial residues are becoming a main environmental concern due to their composition, quality control has been considered for agro-based waste and categorized as agro-industrial by-products due to its high nutritional perspective (Graminha et al. 2008). Agro-based waste or commonly known as agricultural waste, is considered the primary residue which uses valuable agricultural resources through the process of intensive conversion of the agricultural resources and could potentially produce high sustainability products such as materials, molecules, fertilizer and energy to contribute to the economic growth and preventing over exploitation of land usage (UNEP 2011). The agro-based waste consists of agricultural residues and farmed animal wastes. Agricultural residues could be defined as animal or plant residues as a result of co-products and by-products of the waste generation (Harris et al. 2001). The agricultural activity involving farmed animals, such as livestock or aquaculture contributes significantly to the generation of waste due to farmed animals' continuous production throughout the years. Farmed animal waste could be divided into aquaculture animal waste and livestock waste. Aquaculture hatcheries and farms are one of the main producers of aquaculture waste that plays an important role in seedling production and raising of fry and fingerlings used in aquaculture farms and fishing activities (Thorpe and Cho 1995). Aquaculture hatcheries consist of fish, crustaceans, bivalves and other related aquaculture animal rearing facilities. Livestock domesticated animals that are raised in an agricultural settings produce commodities mainly as food supply and sustainable agricultural production. Livestock animals could be categorized as domesticated animals which are reared in a controlled environment in the agricultural setting to produce profits. Livestock animals consist of various animals such as sheep, goats, cattle and poultry animals such as chickens, ducks and swine. Livestock produces waste in the form of manure with 55–90% of nitrogen and phosphorus content found in animal feed excreted as waste products in the form of faeces and urine (Tamminga et al. 2000). Production of ammonia by livestock waste leads to the eutrophication of rivers and lakes due to its potential as a source of pollutant, causing an ecological imbalance in the water system due to the high nutrient concentration produced (FAO 2008). Waste generated from aquaculture activities mainly consists of soluble and solid waste with both organic and inorganic components, where the source of wastewater comes from uneaten feed and fish faeces (Tlusty et al. 2001). Solid wastes were known as the type of waste that could clog the gills of aquatic animals, mainly fish which could lead to death due to anoxia (Akinwole et al. 2016). Dissolved waste produced from food metabolism in fish due to uneaten decomposed feed has two major components that are dangerous and harmful to aquaculture species at a high concentration, which are nitrogen and phosphorus-based products (Boyd and Massaut 1999). The inability to utilize a substantial percentage of nitrogen and phosphorus products by fish produces industrial waste has the potential to cause environmental pollution in aquaculture hatchery systems (Lazzari and Baldisserotto 2008).

Canned fish are popular and the cheapest of fisheries products. Fish canning industries, which produce huge quantities of processed wastewater, are located in a number of tropical countries like Malaysia and Thailand. These industries are required to set up treatment plants to treat their highly polluting wastewater. This further increases the operating costs. Treatment of wastewater in industries is considered an unproductive expenditure, which in turn increases the cost of the canned product. One approach to recycled wastewater is to convert it into value-added products such as the production of bacterial biomass or its products, which could be profitably used. Purple non-sulphur bacteria can be grown in both synthetic medium and agro-industrial waste substrates, which are rich in carbon or nitrogen sources. Because of their high tolerance to environmental conditions and wide nutrient utilization patterns, purple non-sulphur bacteria could be cultured in various types of wastewater (Kobayashi and Kobayashi 2001). Generally, industries considered wastewater treatment as an added expense. However, wastewater treatment before disposal to nature is mandatory by law. The Department of Environment of Malaysia as mandated under the Environmental Quality Act, of 1974 (DOE 2010), functions by preventing and controlling pollution of any portion of the environment by limiting the volumes, types and constituents of pollutants discharged, emitted or deposited into the environment. The National Economic Recovery Plan (NERP) is built to protect and preserve the environment (IMPAK 1998). This is a moral obligation, too. Fish processing factories in Malaysia have wastewater treatment plants. The treatment plants are designed only to treat wastewater by sieving, skimming and oxidation processes. The generated wastewater is not being utilized economically despite having a good potential to be beneficially utilized for microbial growth. In Malaysia, the sardine canning factory (Mafipro Sd. Bhd.) discharges about 60,000 litres of effluent per day. Effluents from fish processing factories are rich in nutrients and contain large amounts of organic matter that increase pollution loads in the environment. The wastewater from a sardine canning factory, if allowed to settle, would yield about 6000 L of a slurry of solid particles. Thus, the remaining 54,000 L of effluent could be used as a substrate for bacterial biomass production. Under optimal conditions the biomass of Purple non-sulphur bacterium *Rhodovulum sulfidophilum* (an isolate from Marine Mangrove mud) produced in sardine processing wastewater was 6.97 g/L (dry weight basis). Thus, total production of bacterial biomass per day will be 376 kg. The bacterial biomass is rich in proteins, vitamins and amino acids. The biomass of phototrophic bacteria is reported to be highly nutritive (Kobayashi and Kobayashi 2001). The nutritional value of phototrophic bacteria biomass could be used as a potential protein supplement in the aquaculture feed industry. There are advantages to using phototrophic bacterial cells incorporated in aquaculture diets. The survival rate of carp increases by up to 96.5% when PSB used as 0.1% is supplemented with commercial feed (Kobayashi and Kobayashi 2001). Phototrophic bacterial cells incorporated in a feed of juvenile fish have several advantages, such as (1) prevention of fish diseases; (2) increase in fish growth rates; (3) improvement in the quality of fish meat and (4) maintenance of water quality in ponds (Noparatnaraporn and Nagai 1986). Phototrophic bacteria cells have some anti-viral compounds that suppress viral diseases of shrimp as well as shellfish (Hirotani et al. 1991). In Japan, the gill disease of the prawn was prevented completely by adding anoxygenic phototrophic bacteria to the tank (Kobayashi and Kobayashi 2001). To improve fish growth, a fish growth hormone gene was introduced into Rhodobacter sp. strain NKPB0021, and bacterial biomass was used as a dietary supplement. The bacteria containing active recombinant growth hormone can be easily grown in seawater and used directly as a marine fish feed supplement (Burgess et al. 1993). The bacterium *Rhodospirillum rubrun* is used with fish meals and fed to juvenile freshwater prawns. The highest growth was obtained with a diet containing mixed other bacterial strains (*Pseudomonas*, *Moraxella*, *Vibrio* and *Micrococcus*), but the food conversion efficiency in a mixed diet of phototrophic bacteria was observed to be better (Manju and Dhevendaran 1997). The purple non-sulphur bacterium *Rv. sulfidophilum* was combined with commercial tilapia feed. In 119 days of growth, 14% higher survival and 30% higher body weight were obtained in tilapia fed with *Rv. sulfidophilum* biomass diet as compared to those without phototrophic bacterial biomass (Banerjee et al. 2000). More studies are needed to determine the optimum level of bacterial biomass to be incorporated into the diet, the nutritional composition of the selected strain and a suitable and cheap substrate for producing nutritionally richer microbial biomass. Phototrophic bacteria contained carotenoids, and these can contribute to the improvement of the skin pigmentation of culture organisms by the addition of bacterial biomass into diets (Noparatnaraporn and Nagai 1986). The pond ecosystem could be improved by adding phototrophic bacterial biomass. Phototrophic bacteria can convert sulphite to sulphate and have a high sulphur tolerance capacity. Phototrophic bacteria also have the strong nitrogen-fixing abilities. Anoxygenic phototrophic bacteria are important in aquaculture because of their photosynthetic capability, fixing atmospheric CO_2 and thus could be used as primary producers in the pond ecosystem. Bacteria could also enrich the pond with nitrogen because they could fix N_2 and convert various organic/inorganic sulphides in waste into useful SCP along (Sasikala and Ramana 1995). Various species of phototrophic bacteria have the efficacy to multiply cells in wastewater, thus increasing the single-cell protein parallel to reduce chemical oxygen demand. The addition of 1% (w/w) *Rv. sulfidophilum* biomass in the *Penaeus monodon* larval diet enhanced the growth and survival of larvae. In addition, a faster metamorphosis was observed in the shrimp larval life cycle. Further, with 1% *Rv. sulfidophilum* biomass, the NH_3-N level and overall water quality parameters were improved and ultimately led to higher growth and survival of *P. monodon* larvae. It can be concluded that the phototrophic bacterial biomass possesses growth co-factors or there could be probiotic effects while used as feed in aquaculture species. Inclusion of purple non-sulphur bacteria biomass in formulated feed to promote growth, feed conversion ratio and survival of Asian sea bass, *Lates calcarifer* juveniles. The addition of 0.3% bacterial biomass in feed shows the highest growth, survival rate and better feed conversion ratio (Shapawi et al. 2012). Bacterial biomass is also used for the replacement of fishmeal in formulated diets. The waste-grown phototrophic bacterial biomass as an ingredient capable to replace 33% and 66% of fishmeal without any negative effects on the growth and survival of Asian sea bass, *Lates calcarifer* (Delamare-Deboutteville et al. 2019). Purple non-sulphur bacteria are one of the potential candidates to be utilized as an aquaculture feed supplement.

15.4 PNSB in the Utilization of Agro-Based Waste

The purple phototrophic bacteria are classified into two broad distinct groups, that is, the purple non-sulphur bacteria (PNSB) and purple sulphur bacteria (PSB). The purple non-sulphur bacteria belong to the Athiorhodaceae and are photoheterotrophic, using organic compounds as both electron donors and carbon sources. These bacteria can reduce CO_2 but derive most of their cellular material from organic nutrients. Purple non-sulphur bacteria (PNSB) use diverse metabolic pathways and have the proven capability to grow in anaerobic light as well as in aerobic dark conditions. On the other hand, due to wide distribution in nature, they favour opportunistic growth in various substrates and are thus widely used for various purposes.

Purple Non-sulphur Bacteria (PNSB) are considered phototrophic microorganisms which could be characterized by their ability to possess diverse physiological and morphological characteristics. PNSB are considered proteobacteria with phylogenetic trees indicating close relation between various PNSB species to non-phototrophic species (Imhoff and Pfennig 1984). PNSB can grow in various cultivation modes and conditions due to its unique metabolism characteristics. PNSB are also able to tolerate a wide range of habitats and can grow under various culturing conditions such as photoautotrophic, photoheterotrophic and chemoheterotrophic (Imhoff and Hiraishi 2005). PNSB distribution can be found mainly in areas such as waste lagoons, coastal lagoons, wastewater ponds and mangrove swamps (Drews and Imhoff 1991) where a high concentration of nutrients can be found. PNSB are the largest group of phototrophs that can convert agro-based industrial waste through a bio-conversion process to produce nutritionally value-added products (Sasikala and Ramana 1995). Besides, PNSB possess the ability to execute various functions such as phosphate solubilization (Rana et al. 2016), nitrogen fixation (Kantha et al. 2015) and methane emission mitigation (Sakpirom et al. 2017). The advantages of using PNSB are possession of diverse metabolic pathways and the ability to grow on a substrate such as carbon or nitrogenous-based substrate. PNSB utilize carbon and energy sources from a broad range of organic compounds (Kantachote et al. 2005). PNSB can adapt to a variety of natural or synthetic substrates and are able to grow in various conditions such as in light or dark conditions and aerobic or anaerobic conditions (Imhoff and Hiraishi 2005). PNSB *Afifella marina* is used in the utilization of vegetable waste and bioprocess product is used as additives in aquafeed (Azad et al. 2018). The bioconversion into aquaculture feed supplement from leafy vegetable waste content a high calorific value for microbial production as a supportive substrate that is rich in nutritive values (Prakash and Singh 2013).

The cultivation of Purple Non-Sulphur Bacteria (PNSB) in agricultural waste, especially in aquaculture activities, has seen an increase in research purposes to gain valuable information on the utilization of agro-based waste through cultivation with various purple non-sulphur bacteria species to introduce efficient and effective management of aquaculture wastes to prevent the degradation of the environment and aquatic organisms due to impacts of pollution. The cultivation of PNSB species using various agro-based waste as substrate by utilizing microbes contained in the substrate provides a better understanding of the variety of PNSB yield production. PNSB yield production provides dual benefits in terms of the value-added products and at the same time reduces COD levels from the waste which could be considered an eco-friendly approach. The marine phototrophic bacterium *Rv. sulfidophilum* was successfully isolated from the mangrove mud of Malaysia and evaluated for the potential to utilize an agro-based waste, Sardine Processing Wastewater (SPW), in the production of bacterium biomass.

15.4.1 Assessment of the SPW as a Substrate in the Production of Bacterial Biomass

Purple non-sulphur bacteria isolated from the mangrove mud of Port Dickson, Malaysia, grow easily in synthetic Glutamate-malate Medium (GMM). The following factors are used to select isolates:

1. The bright reddish colour of a single colony
2. Larger in size compared to other colonies
3. Contamination free from other microbes
4. The latest isolation plate during the culture period
5. Salt-tolerant strain (marine origin)

The selected isolate starts to grow just after inoculation in synthetic Glutamate-malate media (GMM). The change of colour in culture media indicates the commencement of the growth of microbes, and a 48-h culture turns pink to reddish. The colour of the cell suspension turns yellow brown at the stationary phase. The cells are well distributed in the culture bottles as they show no sedimentation after a 120-h culture. Streaking on an agar plate containing GMM media, the bacterium forms round red colonies. Careful observation of agar plates reveals that the bacterium did not produce any slime in the growth medium. The details of morphometric and meristic characters are shown in Table 15.1.

Table 15.1 Morphometric, meristic and physiochemical characteristics of *Rhodovulum sulfidophilum*

Morphometric and meristic	
Cell shape	Rod
Cell size (μm):	
Length:	0.8–0.9
Wide:	0.3–0.5
Physiochemical	
Culture colour:	Reddish-brown (anaerobic light in SPW)
(after 48 h)	Reddish-pink (aerobic dark in SPW)
	Dark yellow-brown (anaerobic light in GMM)
Slime formation	No
Clumping	Yes
Gram strain	Negative
Bacteriochlorophyll	a, c
Major carotenoids	Spheroidenone and spheroidene

The absorption spectrum of the isolate indicated the dominance of carotenoids and bacteriochlorophyll. The maximum absorption spectrum for bacteriochlorophyll can be observed in the range of 756, 590, 483 and 377 nm. The selected strain was confirmed to be *Rv. sulfidophilum* by the International Biotechnological Institute of Japan, based on the 16 s rRNA sequence.

15.4.2 The Characteristics of SPW Use as a Media in the Production of *Rv. sulfidophilum* Bacterium Biomass

The nutrients available in the substrate are essentially important, as they must support the growth of microbes. The biochemical compositions of SPW are different after it allows the solid particles to settle. The amount of organic nitrogen, pH and salinity did not show any differences between the settled and non-settled wastewater. The nutritional values in SPW did not show major variations in non-settled and settled wastewater. However, the reduction of COD, removal of total solids, NH_3-N and phosphorus from wastewater in settled SPW were reduced by 50%, 58%, 70% and 45%, respectively. The physiochemical properties and proximate compositions of non-settled and settled SPW are presented in Tables 15.2 and 15.3 respectively.

The removal of solid particles from SPW did not affect the essential nutrients for the growth of phototrophic bacteria. The organic nitrogen content in both settled and non-settled wastewater did not change after the suspended particles were removed.

Table 15.2 Physiochemical parameters of non-settled and settled sardine processing wastewater (mean of triplicates ± standard deviation)

Parameters	Non-settled SPW	Settled SPW
pH	6.2 ± 0.164	6.3 ± 0.06
Salinity (ppt)	38.6 ± 0.115	38.5 ± 0.10
Chemical oxygen demand (mg/L)	63,000 ± 25.51	32,000 ± 13.65
Total solids (g/L)	201 ± 2.082	55.3 ± 1.34
Total suspended solids (mg/L)	16.9 ± 0.950	0.63 ± 0.02
NH_3-N (mg/L)	2.25 ± 0.087	0.629 ± 0.01
Total Kjeldahl nitrogen (mg/L)	685 ± 0.122	635 ± 0.09
Total phosphorus (mg/L)	225 ± 3.215	125 ± 3.06

Table 15.3 Proximate composition (% of dry weight) of sardine processing wastewater (mean of triplicates ± standard deviation)

(Percentage of composition)	Non-settled SPW	Settled SPW
Moisture	5 ± 0.081	4 ± 0.09
Crude protein	68 ± 0.176	63 ± 0.55
Crude lipid	9.5 ± 0.058	8.5 ± 0.40
Ash	15 ± 0.140	16 ± 0.21
Fibre	1.2 ± 0.006	0.7 ± 0.10
Nitrogen free extract	0.3 ± 0.228	7.8 ± 1.15

The additional benefit of using settled SPW was that the COD level was reduced by 50% and suspended particles were reduced by 27%. In general, the overall proximate compositions of SPW did not show any marked differences between non-settled and settled effluents.

15.4.3 *Rv. sulfidophilum* in the Utilization of SPW

15.4.3.1 Culture of *Rv. sulfidophilum* in Unsettled SPW

The inoculum of *Rv. sulfidophilum* developed in synthetic GMM media can be used in the utilization of unsettled non-sterilized SPW (Fig. 15.1). Typically, 5% (v/v basis) inoculum sizes are quite efficient to accelerate the growth of *Rv. sulfidophilum* in unsettled SPW. The growth of the bacterium is noticeable within 24 h, as the changes of colour from light pink to reddish pink. SPW are presented in Table 15.4.

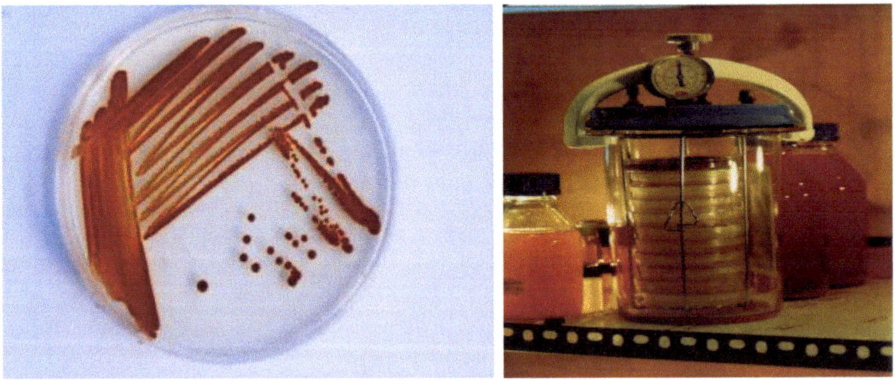

Fig. 15.1 Pure colonies of *Rv. sulfidophilum* (GMM agar plate (left) grown under anaerobic light culture conditions in an incubation jar (right), illuminated at 2500 lux at a temperature of 30 ± 2 °C

Table 15.4 Growth characteristics of *Rv. sulfidophilum* in unsettled sterilized undiluted and 50% diluted (v/v, with distilled water) SPW under anaerobic light (mean ± standard deviation)

	Parameters				
	X_{max}	μ_{max}	$Y_{x/s}$	Carotenoid	Reduction of COD
Experimental conditions	(g/L)	(per h)	(g cell/g COD)	(mg/g dry cell)	(%)
Undiluted SPW	2.45 ± 0.34	0.01 ± 0.00	0.11 ± 0.02	1.044 ± 0.09	70
50% diluted SPW	3.34 ± 0.20	0.026 ± 0.1	0.25 ± 0.02	0.816 ± 0.09	66

X_{max} = maximum dry cell weight (g/L) after 96 h
μ_{max} = maximum specific growth rate per h after 48 h
Yx/s = maximum cell yield (g cell/g of Chemical Oxygen Demand) after 96 h
Carotenoids = maximum production after 72 h
Reduction of COD (%) = maximum after 120 h

The pattern of growth in terms of dry cell biomass (g/L) is always higher when the bacterium culture is in 50% undiluted SPW. Of the substrate tested, undiluted SPW gave the highest carotenoid production of 1.044 (mg/g dry cell weight) after 72-h incubation. The reduction of COD is comparable if diluted or not diluted.

A maximum of 70% COD reduction can be achieved in the utilization of bacterium in undiluted SPW, but a mean cell yield of 0.25 g cell/g COD, was observed the highest during culture in 50% diluted SPW, which is 56% higher than the production of mean cell yield from undiluted SPW.

Marine bacteria will not grow or will grow poorly in a medium without salt. Further, the addition of 30 ppt NaCl in the media is a must for the growth of marine purple non-sulphur bacteria (Imhoff and Truper 1991). Purple non-sulphur bacterium *Rhodovulum* sp., from marine sources, required salinities from 0.5% to 7.5% for optimal growth (Hiraishi and Ueda 1994). The salinity of SPW was within the range for the growth of purple non-sulphur bacteria. Further, SPW contained a higher salinity of 3.8% than the optimum level of salinity required for the growth of marine purple non-sulphur bacteria.

Growth characteristics of *Rv. sulfidophilum* in the utilization of sardine processing wastewater are limited. Bacterium species of *Rhodovulum* can grew efficiently in aerobic dark condition, but bacterium growth profiling or pattern were not clear (Watanabe et al. 1998). Purple non-sulphur bacterium *Rhodocyclus gelatinosus*, isolated from seafood processing wastewater was grown in tuna condensate, where dilution of tuna condensate mixed with shrimp blanching in a ratio of 1:10 (v/v) was observed as best combination in the production of *R. gelatinosus* (Prasertsan et al. 1993). *Rhodocyclus gelatinosus* in diluted tuna condensate wastewater produces 5.6 g/L of dry cell biomass with an 86% reduction in COD. The composition of solid portion in tuna condensate were mainly blood, tissue and fish extract, whereas different types of composition like cooked sardine skin, tissues, soft bones, fish extracts and nitrogenous organic compounds found in SPW. However, the higher organic load was determined in tuna condensate when compared with the loading load that obtained in SPW. The COD of tuna condensate (73,617 mg/L) is comparatively higher than that of SPW (63,000 mg/L). But the total solids in SPW (201 g/L) are much higher compared to the total solids determined in tuna condensate (6.54 g/L).

15.4.4 Utilization of Settled SPW by *Rv. sulfidophilum* with Two Different Levels (10% and 15%) of Inoculum

The total solids contribute to the high loading of COD. Nutritional values that changes during solids removal in SPW were enough to support the production of bacterial biomass. Settled solids from SPW could also be used as feed ingredients, suitable for aquaculture or poultry feed. Bacterium *Rv. sulfidophilum* efficiently utilize the settled SPW, although nutrients, like total nitrogen and total phosphorus are reduced after the settlement of SPW. The inoculum size of 5% may not be suitable for the utilization of unsettled SPW, as *Rv. sulfidophilum* is unable to

Table 15.5 Optimization of parameters for growth of *Rv. sulfidophilum* in settled undiluted and settled 50% diluted (v/v with distilled water) non-sterilized SPW under anaerobic light with two levels (10% and 15%) of inoculum

	Parameters					
	X_{max}	μ_{max}	$Y_{x/s}$	Carotenoid	Reduction of COD	Soluble protein
Experimental conditions	(g/L)	(per h)	(g cell/g COD)	(mg/ g dry cell)	(%)	(μg/mL)
Settled undiluted						
10% inoculum	6.76	0.053	0.683	1.116	70	398.97
15% inoculum	7.64	0.025	0.997	1.161	77	366.73
Settled diluted						
10% inoculum	4.86	0.023	1.154	1.541	87	209.83
15% inoculum	4.91	0.025	1.002	1.621	85	184.12

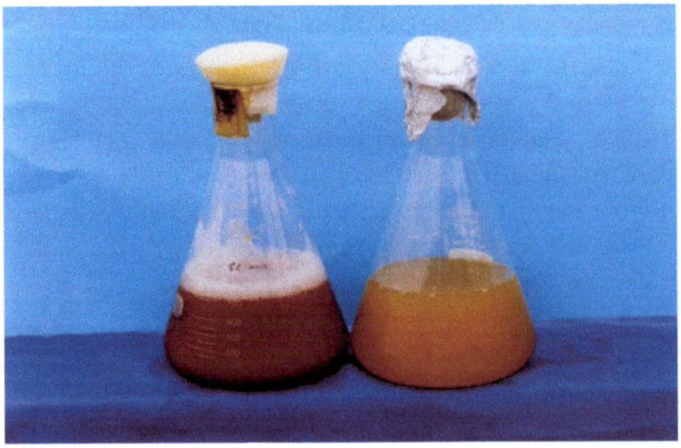

Fig. 15.2 Growth of *Rv. sulfidophilum* in settled undiluted non-sterilized SPW under anaerobic light with 15% (v/v) of inoculum

suppress the growth of opportunities bacteria that remain in that substrate. Production of biomass can be improved with the increase of inoculum sizes. The growth characteristics in settled diluted and undiluted SPM in terms of dry cell weight and production of total carotenoids can be improved as shown in Table 15.5. Inoculum sizes certainly affect the growth characteristics of *Rv. sulfidophilum* in settled and un-settled SPW (Fig. 15.2).

15.4.5 Utilization of Unsettled SPW by *Rv. sulfidophilum* with Two Different (15% and 20%) Levels of Inoculum

The growth characteristics of un-settled SPW with a higher level of inoculum are better in terms of production of dry cell weight and total carotenoids, as shown in Table 15.6.

Generally, it is best to prepare the inoculum (seed) in a similar medium as the production medium to enhance growth and reduce the log phase. The preparation of

Table 15.6 Growth characteristics of *Rv. sulfidophilum* in unsettled undiluted and 50% diluted (v/v, with distilled water) non-sterile SPW with two levels of inoculum under anaerobic light (values expressed as a mean of two replicates)

Experimental conditions		Parameters					
		X_{max}	μ_{max}	$Y_{x/s}$	Carotenoid	Reduction of COD	Soluble protein
SPW	Inoculum	(g/L)	(per h)	(g cell/g COD)	(mg/ g dry cell)	(%)	(μg/mL)
Undiluted	15%	8.76	0.044	0.493	0.747	71	466.56
Undiluted	20%	5.96	0.023	0.444	0.939	76	294.48
50% diluted	15%	7.90	0.030	0.370	0.872	62	439.42
50% diluted	20%	5.17	0.034	0.340	0.978	59	233.04

Rv. sulfidophilum inoculum is best in a synthetic media (GMM) rather than prepared in the supernatant sardine processing wastewater. The supernatant of SPW has limitations as a medium for inoculum (seed) preparation, including (1) shortage of nutrients, such as tryptophan and methionine; (2) higher percentage of suspended material as compared to synthetic GMM, which reduces transmission and uniform distribution of light in the culture bottles; and (3) incomplete utilization of the suspended particles in SPW (Azad et al. 2003). The bacterial numbers are likely to be lower in an inoculum prepared in SPW. In contrast, inoculum grown in GMM contains a higher number of bacterial cells and lower suspended particles. The higher number of bacterial cells thus speeds up the process of substrate utilization, suppresses the growth of other heterotrophic bacteria and hence produces more bacterial biomass within a short time. In addition, the supernatant SPW, however, could be considered as an inoculum medium, after the following steps are taken into account:

1. Complete removal of all suspended solid particles by centrifuging, although this procedure is expensive and laborious
2. Centrifuged SPW is diluted with water into various proportions (v/v) and
3. Inoculation of higher proportions of stock bacterial cells

Levels of inoculation have varying effects on the production of biomass produced, the carotenoid content and soluble protein *Rv. sulfidophilum* in SPW. The inoculum levels of 10% to 20% (v/v) are sufficient to produce *Rv. sulfidophilum* biomasses, with the production of carotenoids and soluble protein in a bacterial cell. On the other hand, a reduction of COD could be obtained in SPW by *Rv. sulfidophilum* with a 30% (v/v) level of inoculum. Therefore, the choice of inoculum level depends on the purpose. For bioremediation, the best inoculum (seed) level is 30% grown in GMM, while for the production of SCP, the optimum size is 10% to 20% (v/v) inoculum. On the other hand, for the production of biomass, carotenoid and soluble protein, a 15% (v/v) inoculum size will be better as compared to 10% and 20% (v/v) levels of inoculum when *Rv. sulfidophilum* utilizes SPW (Fig. 15.3).

Production of *Rv. sulfidophilum* biomass also varies with the diluted or undiluted, settled or unsettled SPW. The settled particles of sardine processing wastewater included cooked fish muscles, soft bones and skins. Major constraints on the use of unsettled and unsterilized wastewaters are:

1. Bacterial cells are difficult to be separated from settled particles.
2. Quantification of actual bacterial biomass will not be accurate.
3. Cultured biomass might contain other heterotrophic bacteria.

In fact, dilution of wastewater on an industrial scale may not be appropriate as this would incur extra costs. However, there are other advantages to the use of unsettled wastewater in culturing *Rv. sulfidophilum.* The mixed bacterial biomass and unutilized suspended particles could directly be used as a feed ingredient. In such a situation, the nutritional profiles of the produced bacterial biomass need to be

Fig. 15.3 Inoculum and growth of *Rv. sulfidophilum* in GMM (48 h). (Left to right: Sterilized GM media, inoculated with *Rv. sulfidophilum* (0 h) in GM-media, growth of *Rv. sulfidophilum* after 48 h)

assessed. Not only the nutritional profile, but the nature of other bacteria (if present other than phototrophic) must be evaluated for their toxic effect. The unused settleable suspended particles during culture in SPW may change the physiochemical properties of SPW. Therefore, the physiochemical and biological parameters of harvested biomass need to be evaluated throughout the culture period. However, the benefits of using settled SPW are:

1. A reduction of COD of 50% by the removal of suspended particles
2. Proximate composition in settled wastewater remains unchanged and suitable for the growth of *Rv. sulfidophilum*
3. Easier to handle by researchers as a high amount of particles has been removed
4. Facilitates better mixing of SPW in a culture system
5. Maximum light penetration during anaerobic culture conditions since it is less turbid

The use of SPW, settled or unsettled did not improve carotenoid production by *Rv. sulfidophilum* in an anaerobic light culture system. In unsettled SPW, suspended particles inhibited exposure to uniform light. Agitation under aerobic dark conditions can improve in carotenoid production. Generally, the production of carotenoids was observed to be at a maximum within a 36-h culture in anaerobic light conditions. The production of carotenoids did not directly relate to the production of biomass, but rather an inverse relationship in the production of carotenoids with the production of biomass. Although both the production of biomass and carotenoid concentration are the prime criteria used for the selection of the bacterium as a feed supplement, the main advantage of culturing *Rv. sulfidophilum* in SPW is that the wastewater can be used directly as a substrate for biomass production. The effluents of cow dung (Vrati and Verma 1983) and swine waste (Sasaki et al. 1991) must be digested anaerobically before being used as substrates for phototrophic bacteria. Another advantage is that *Rv. sulfidophilum* can be cultured under aerobic dark conditions, which reduces the energy cost. The culture could be similar to the production of bakery yeast.

15.4.6 Utilization of Non-sterilized, Undiluted, Settled SPW with *Rv. sulfidophilum* in Different Culture Conditions

Rv. sulfidophilum can also be grown in anaerobic light also in aerobic dark conditions (Fig. 15.4). Although dry cell weight production is higher in anaerobic light conditions, it is comparable with aerobic dark culture conditions (Table 15.7). To save energy, the best option is to use aerobic dark culture conditions, in which situation agitation speed plays a very important role. The scenario is different when using agitation speed. The total biomass production in *Rv. sulfidophilum* is rapid under aerobic dark culture conditions in SPW and remained high till 48 h of culture. In anaerobic light conditions, the highest bacterial biomass of 4.34 g/L can be obtained in 120-h culture. However, this value is 34% higher compared to the maximum biomass (4.34 g/L) obtained in aerobic dark culture after 48-h of incubation. Although, the highest biomass is attainable under anaerobic light culture conditions, the mean specific growth rate of 0.049 per h is faster under aerobic

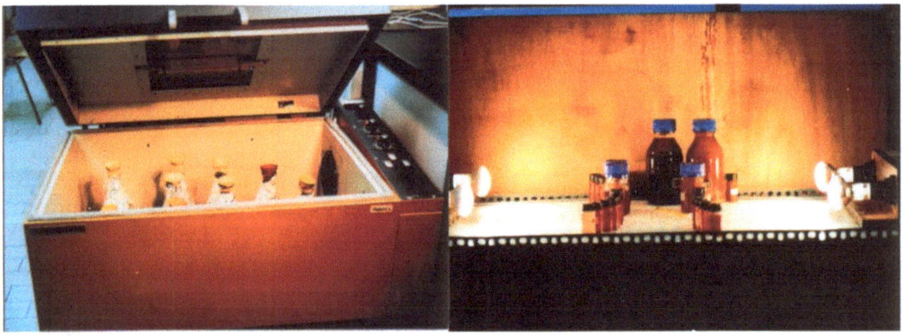

Fig. 15.4 Culture of *Rv. sulfidophilum* under aerobic dark conditions (left) and anaerobic light conditions (right) in non-sterilized undiluted settled SPW with 15% (v/v) inoculum

Table 15.7 Effect of culture conditions on growth of *Rv. sulfidophilum* in non-sterilized undiluted settled SPW with 15% (v/v) inoculum

	Parameters					
	X_{max}	μ_{max}	$Y_{x/s}$	Carotenoid		Soluble protein
Culture conditions	(g/L)	(per h)	(g cell/g COD)	(mg/ g dry cell)	COD reduction (%)	(μg/mL)
Anaerobic light	4.34	0.044	0.424	1.025	68	376
Aerobic dark	3.23	0.049	0.415	1.001	76	359

Table 15.8 Effect on agitation speeds (rpm) on the growth of *Rv. sulfidophilum* in non-sterilized undiluted settled SPW with 15% (v/v) inoculum level under aerobic dark condition

Culture conditions aerobic dark	Parameters					
	X_{max}	μ_{max}	$Y_{x/s}$ (g cell/g COD)	Carotenoid (mg/ g dry cell)	COD reduction (%)	Soluble protein (μg/ mL)
With 200 rpm	4.90	0.023	0.708	2.294	78	398.8
With 300 rpm	6.97	0.020	0.742	4.236	69	431.66
With 400 rpm	5.74	0.029	0.688	2.474	73	284.95

Fig. 15.5 Growth of *Rv. sulfidophilum* under anaerobic light conditions in non-sterilized settled SPW. (From left to right: Non-sterilized settled SPW without inoculation; Inoculated with *Rv. sulfidophilum* (0 h); growth after 24 h; growth after 48 h; growth after 72 h; growth after 96 h; growth after 120 h

dark conditions in a 24-h incubation. The trend in the removal of COD from sardine processing wastewater by *Rv. sulfidophilum* cultured both in anaerobic light and aerobic dark culture conditions is almost similar.

At 300-rpm agitation, initially, the growth of the cell (μ_{max} of 0.020 per h) and cell yield of bacterium biomass very slow in dark. Gradually, cells start to grow and the highest of 6.96 g/L can be obtained after 96 h at 300-rpm agitation speed. Parallel to the growth of bacterial biomass, exceptional improvement in the total carotenoids production were observed while cultured in aerobic dark condition. The highest of 4.2 (mg/g dry cell) of carotenoids and soluble protein of 431 μg/mL was also obtained under aerobic dark conditions and 300-rpm agitation speed (Table 15.8) (Fig. 15.5).

The higher reduction of COD in aerobic dark conditions may be due to the agitation of culture. Agitation is favourable to the culture of *Rv. sulfidophilum* in SPW because: (1) it prevented suspended particles from settling and even broke up large particles into fine particles that are more readily utilized by bacterial cells, (2) it circulated oxygen in the culture system and (3) the available oxygen accelerated oxidation and helped in the nitrification process, too (Fig. 15.6).

The unique property of *Rv. sulfidophilum* is its efficacy to multiply cells under anaerobic light and aerobic dark conditions. Bacterium belongs to the versatile group

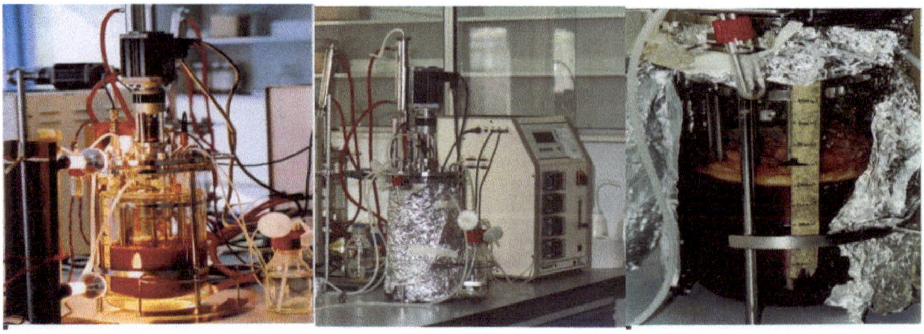

Fig. 15.6 Mass production of *Rv. sulfidophilum* in SPW with a bioreactor. (Left: in presence of light. Right: dark culture)

as capable of obtaining growth energy via alternative mechanisms, which facilitates dark fermentative growth (Madigan et al. 1980).

Agitation of the substrate, which normally distributes the nutrients play significant roles in microbial cell production and also in the production of carotenoids. On the other hand, in aerobic culture, concentration of oxygen suppressed the activity of carotenoid production, but needs to optimize to enhance the synthesis of the pigment. Bacterium *Rv. sulfidophilum* is in a group that links the facultative anaerobic and aerobic conditions, and the highest production of carotenoids can be obtained in aerobic dark conditions. Under anaerobic light conditions, the growth observed is comparable with the product obtained in dark conditions, but pigment synthesis is suppressed by light conditions (Shioi and Doi 1990). The production of higher biomass and carotenoids in bacterium cells was observed better with 300-rpm agitation speed under aerobic culture conditions. However, the reduction of COD was determined lower at 300 rpm, compared to the reduction in COD at 200- and 400-rpm agitation speeds. *Rhodovulum* sp. grew well in aerobic dark conditions with agitation, but vigorous aeration suppressed carotenoid synthesis (Shioi and Doi 1990). *Rv. sulfidophilum* growth in synthetic media has been reported to synthesize bacteriochlorophyll-a aerobically in dark conditions (Watanabe et al. 1998) and photosynthesize under anaerobic light conditions (Shioi and Doi 1990). The speciality of bacterium is the linkage between facultative anaerobic and aerobic conditions. The *Rv. sulfidophilum* can also grow in wastewater that contained high sulphide (Hiraishi and Ueda 1994). The capability of growing in anaerobic light and aerobic dark conditions and bioconversion of wastewater into the microbial cell of protein parallel to the reduction COD makes it a good candidate for the utilization of sardine processing wastewater into value-added products. The prime criteria of microbial feed in aquaculture include rapid growth and the highest production of biomass, good yields of carotenoids and high percentages of protein. In addition, that must ensure better nutritional quality and non-toxic properties of the microbial cell (Litchfield 1977).

The production of *Rv. sulfidophilum* biomass in SPW has several advantages: (1) it grows in both diluted and undiluted SPW, (2) it can grow in both settled and non-settled SPW and (3) grows in both anaerobic light and aerobic dark conditions. However, the production of bacterial biomass, production of carotenoids and reduction of COD and soluble protein by *Rv. sulfidophilum* cell are dependent on types of culture substrate and culture conditions. For the utilization of SPW by *Rv. sulfidophilum*, the choice should fulfil the following: protein-rich biomass potentially useful for aquaculture, high carotenoid content for enhancing pigmentation in fed animals and bioremediation of wastewater.

SPW could be utilized by *Rv. sulfidophilum*, producing valuable biomass and at the same time reducing the COD of the wastewater. The process of biomass production and treatment of wastewater go through a series of steps. After harvesting the bacterial cells, the COD level in sardine processing wastewater remains 7000 to 8000 mg/L. Thus, further processes for the reduction of COD must be considered. The reduction of COD could be further achieved by the cultivation of microalgae in SPW after harvesting bacterial biomass. Applications of micro-algae in the bioremediation of wastewater include agricultural and domestic wastes by *Anabaena* and *Chlorella* in treating rubber effluent (Phang et al. 2001).

The probability of contamination with the pathogens is high when SCP is obtained from such a waste substrate. In addition, the bacteria may accumulate harmful substances that exist in the substrate itself. Substrates containing harmful substances should not be considered for the production of SCP. Therefore, substrates from agro-based industries, such as SPW are less hazardous. The Malaysian sardine canning industry maintains high sanitation conditions. Strong precautionary and safety measures are maintained to avoid any microbial contamination. Canning waste has to be taken from the exhaust point, inside the canning factory, just before mixing with other waste stream ingredients (e.g. tomato, chilli, etc.). The proximate composition of *Rv. sulfidophilum* biomass indicated that the biomass, like other SCP, was suitable to be used as a feed additive. Comparatively, a higher percentage of bacterial protein is available when *Rv. sulfidophilum* is cultured in wastewater rather than cultured in synthetic media culture. The high content of protein only is not the best selection criterion for any microbial feed if the cells do not contain enough nutrients to stimulate the growth and survival of the organism. Microbial cells from SPW contain a significant amount of carotenoid, which when consumed can enhance pigment formation in shrimp and ornamental fish. It is also equally important to select the strain, for bioremediation of wastewater with simultaneous production of single-cell protein. Phototrophic bacteria can utilize agro-based waste like algae and yeast. However, the protein content of the bacterial cells (60%–70%) is relatively higher compared to that of the yeast, which contains 50% to 60% protein (Sasaki et al. 1991). Dried algal SCP has been found to contain a lower nutritional value for fish growth than either yeast SCP or bacterial SCP (Litchfield 1977). Information regarding the nucleic acid content in PNSB is lacking. It may be one of the important criteria for the selection of microbes to be used as microbial feed. Nucleic acids in phototrophic bacteria especially those grown in waste substrates, need to be taken into consideration before the commercialization of the strain as SCP.

Biotechnological applications of phototrophic bacteria are well documented (Sasikala and Ramana 1995). The purple non-sulphur bacterium strain *Rv. sulfidophilum* is not an exception. *Rv. sulfidophilum* possesses special characteristics that make it possible to culture the bacterium in a variety of cultural conditions. The strain of *Rv. sulfidophilum* can efficiently utilize sardine processing wastewater whether diluted or not, as well as settled or unsettled both in anaerobic light and aerobic dark conditions. The main advantage is that it can be grown in non-sterilized sardine processing wastewater. As a huge volume of waste is generated, sterilization is not convenient and economical. Finally, *Rv. sulfidophilum* is capable of reducing the organic load in sardine processing wastewater (Azad et al. 2003). Thus, like other strains of PNSB such as *Rhodobacter sphaeroides*, *Rhodocyclus gelatinosus* and *Rhodopseudomonas gelatinosa*, *Rv. sulfidophilum* also provides a dual benefit: treatment of wastewater and at the same time, production of single-cell protein. It is difficult to obtain pure bacterial biomass from non-sterilized wastewater. Bacterial biomass produced in SPW contains not only cells of *Rv. sulfidophilum*, but also other fine organic particles that are suspended in the wastewater. In this context, the phototrophic bacterium, SPW and the process involved in producing its biomass are equally important in its technical evaluation and economic feasibility. The produced biomass should not only be nutritionally rich but must also be non-toxic, free from pathogens, palatable and easily digestible. Further, several other factors may affect the cost of producing bacterial biomass in the utilization of SPW. These factors include the organisms, substrate, culture techniques, harvesting methods and production processes to be used in various applications. All these factors have to be assessed individually for economic consideration.

15.5 Conclusions

Agro-based wastewater, especially SPW, is a nutritionally enriched substrate and the bacterium for the *Rv. sulfidophilum* can be used in the production of bacterial biomass, but with 15% (v/v) inoculum under aerobic dark conditions at 300 rpm agitation speed. The bacterium produced the highest biomass of 6.97 g/L with the production of total carotenoids (4.24 mg/g dry cell weight) and the highest of 431 (μg/mL) of soluble protein in aerobic dark conditions compared to cultures grown in anaerobic light conditions. The ability of this agro-based waste to support bacterial growth shows the promising potential of this waste for future biotechnological applications. The high utilization rate of nutrient content in agro-based waste by various bacteria species would provide better information on the frequent usage of agro-based waste as a growth medium, especially in the cultivation of PNSB species. PNSB have many great characteristics, bringing various benefits as a producer of value-added products, bioremediation of wastewater and single-cell protein production that would also be able to function as an alternative in waste and wastewater treatment. Improper waste removal can be avoided, which could potentially increase pollution levels due to the high generation of waste annually.

References

Akinwole AO, Dauda AB, Ololade AO (2016) Hematological response of *Clarias gariepinus* juveniles reared in treated wastewater after waste solids removal using alum or *Moringa oleifera* seed powder. Int J Acarol 6(11):1–8

Azad SA, Vikineswary S, Chong VC, Ramachandran KB (2003) Rhodovulum sulfidophilum in the treatment and utilization of sardine processing wastewater. Lett Appl Microbiol 38(1):13–18

Azad SA, Jie WH, Lal MTBM (2018) Utilization of vegetable waste juice by purple non-sulfur bacterium *Afifella marina* strain ME for biomass production. J Geosci Environ Protect 6(5): 210–219

Banerjee S, Azad SA, Vikineswary S, Selvaraj OS, Mukherjee TK (2000) Phototrophic bacteria as fish feed supplement. Asian-Austr J Anim Sci 13(7):991–994

Boyd CE, Massaut L (1999) Risks associated with the use of chemicals in pond aquaculture. Aquac Eng 20(2):113–132

Burgess JG, Tsubaki K, Matsunaga T (1993) Expression of yellow tail (*Seriola quinqueradiata*) fish growth hormone cDNA in the marine photosynthetic bacterium *Rhodobacter* SP. NKPB0021. Biotechnol Lett 15(2):111–114

Caprara C, Colla L, Lorenzini G, Santarelli C, Stoppiello G, Zanella D (2011) Development of a model for technical-economic feasibility analysis of biomass and mud gasification plants. Int J Energy Technol 3:1–6

Dardak RA (2015) Transformation of agricultural sector in Malaysia through agricultural policy. Malaysian agricultural Research and Development Institute (MARDI), Selangor

Delamare-Deboutteville J, Batstone DJ, Kawasaki M, Stegman S, Michael S, Tabrett S, Smullen R, Barnes AC, Hülsen T (2019) Mixed culture purple phototrophic bacteria is an effective fishmeal replacement in aquaculture. Water Research X 4(100031):1–11

Department of Environment (2010) Environmental Requirements: A Guide For Investors, Ministry of Natural Resources and Environment. Wisma Sumber Asli, Precinct 4 Federal Government Administrative Centre 62574 PUTRAJAYA. 78pp

Department of Statistics, Malaysia (2019) Supply and utilization accounts selected agricultural commodities, Malaysia 2014–2018. Department of Statistics Malaysia and Official Portal

Dow CS (1982) Experiments with photosynthetic bacteria. In: Primrose SB, Wardlaw AC (eds) Source book of experiment for the teaching of microbiology. Academic Press, New York, pp 408–422

Drews G, Imhoff JF (1991) Phototrophic purple bacteria. In: Shively JM, Barton LL (eds) Variations in autotrophic life. Academic Press, London, pp 51–97

FAO (2008) Guidelines for sustainable manure Management in Asian Livestock Production Systems. Int Atom Energy Agency 158:1–6

Graminha E, Gonçalves A, Pirota R, Balsalobre M, Silva RD, Gomes E (2008) Enzyme production by solid-state fermentation: application to animal nutrition. Anim Feed Sci Technol 144(1–2): 1–22

Harris PJ, Allison M, Smith G, Kindness HM, Kelley J (2001) The potential use of waste-stream products for soil amelioration in peri-urban interface agricultural production systems. Closing the Rural-urban Nutrient Cycle in Sub-Saharan Africa, Waste Composting for Urban and Peri-urban Agriculture, pp 1–28

Hiraishi A, Ueda Y (1994) *Rhodoplanes* gen. Nov, a new genus of phototrophic bacteria including *Rhodopseudomonas rosea* as *Rhodoplanes roseus* camb. Nov. and *Rhodoplanes elegans* sp. nov. Int J Syst Bacteriol 44(4):665–673

Hiraishi A, Urata K, Satoh T (1995) A new genus of marine budding phototrophic bacteria, *Rhodobium* gen. Nov., which includes *Rhodobium orientis* sp. nov. and *Rhodobium marinum* comb. nov. Int J Syst Bacteriol 45:226–234

Hirotani H, Ohigashi H, Kobayashi M, Koshimizu K, Takashi E (1991) Inactivation of T5 phage by cis-vaccenic acid, and antivirus substances from *Rhodopseudomonas capsulata*, and by unsaturated fatty acids and related alcohols. FEMS Microbiol Lett 77:13–18

Imhoff JF (1992) Taxonomy, phylogeny and general ecology of anoxygenic phototrophic bacteria. In: Mann NH, Carr NG (eds) Photosynthetic prokaryotes. Plenum Press, New York, pp 53–92

Imhoff JF, Hiraishi A (2005) Genus I. *Rhodobium*. Bergey's Manual of Systematic Bacteriology 2(2):571–574

Imhoff JF, Truper HG (1991) The genus *Rhodospirillum* and related genera. In: Balows A, Truper HG, Dworkin M, Harder W, Schleifer KH (eds) The prokaryotes, vol III. Springer-Verlag, New York, pp 2141–2155

Imhoff JF, Truper HG, Pfennig N (1984) Rearrangement of the species and genera of the phototrophic purple on-sulfur bacteria. Int J Syst Bacteriol 34(3):340–343

IMPAK (1998) National economic recovery plan agenda for action: protect the environment 4:1–3. Bulletin Department of Environment (DOE), Malaysia

Kantha T, Kantachot D, Klongdee N (2015) Potential of biofertilizers from selected *Rhodopseudomonas palustris* strains to assist rice (*Oryza sativa* L. subsp. indica) growth under salt stress and to reduce greenhouse gas emissions. Ann Microbiol 65(4):2109–2118

Kantachote D, Torpee S, Umsakul K (2005) The potential use of anoxygenic phototrophic bacteria for treating latex rubber sheet wastewater. Electron J Biotechnol 8(3):314–323

Kompantseva EI, Gorlenko VM (1984) A new species of moderately halophilic purple bacterium *Rhodospirillum mediosalinum* sp. nov. Microbiology 53:954–961

Kompantseva EI (1986) New halophilic purple bacteria *Rhodobacter euryhalinus* sp. nov. Microbiology 54:974–982

Kobayashi M, Kobayashi M (2001) Roles of phototrophic bacteria and their utilization. In: Kojima H, Lee YK (eds) Photosynthetic microorganisms in environmental biotechnology. Springer-Verlag, Hong Kong, pp 11–26

Lazzari R, Baldisserotto B (2008) Nitrogen and phosphorus waste in fish farming. Bol Inst Pesca 34:591–600

Litchfield JH (1977) Comparative technical and economic aspects of single cell protein processes. In: Perlman D (ed) Advances in applied microbiology, vol 22. Academic Press, New York, USA, pp 267–305

Madigan MT (1999) Isolation and characterization of psychrophilic purple bacteria from Antarctica. In: Peschek GA, Loffelhardt W, Schmetterer G (eds) The phototrophic prokaryotes. Kluwer Academic/Plenum Publishers, New York, Boston, Dordrecht, London, Moscow, pp 699–706

Madigan MT, Cox JC, Gest H (1980) Physiology of dark fermentative growth of *Rhodopseudomonas capsulata*. J Bacteriol 142(3):908–915

Manju KG, Dhevendaran K (1997) Effects of bacteria and actinomycetes as single cell protein feed on growth of juveniles of *Macrobrachium idella* (Hilgendorf). Indian J Exp Biol 35:53–55

Nissen H, Dundas ID (1984) *Rhodospirillum salinarum* sp. nov., a halophilic photosynthetic bacterium isolated from a Portugese saltern. Arch Microbiol 138:251–256

Noparatnaraporn N, Nagai S (1986) Selection of *Rhodobacter sphaeroides* P47 as a useful source of single cell protein. J Gen Appl Microbiol 1(32):351–359

Pfenning N (1977) Photosynthetic green and purple bacteria: a comparative, systematic survey. Annu Rev Microbiol 31:275–290

Phang SM, Chui YY, Kumaran G, Jeyaratnam S, Hishim MA (2001) High rate algal ponds for treatment of wastewater: a case study for the rubber industry. In: Kojima H, Lee YK (eds) Photosynthetic microorganisms in environmental biotechnology. Springer-Verlag, Hong Kong, pp 51–76

Prakash EV, Singh LP (2013) Biomethanation of vegetable and fruit waste in co-digestion process. International Journal of Emerging Technology and Advanced Engineering 3:493–495

Prasertsan P, Choori W, Suwanno S (1993) Isolation, identification and growth condition of photosynthetic bacteria found in seafood processing wastewater. World J Microbiol Biotechnol 9:590–592

Rana G, Meikap S, Mondol M, Bose PP (2016) Green-fertilizer, *Rhodospirillum rubrum*, for agricultural development on fly-ash without any toxic metal ion release. Basic Research Journal of Agricultural Science and Review 5:109–117

Sakpirom J, Kantachote D, Nunkaew T, Khan E (2017) Characterizations of purple non-sulfur bacteria isolated from paddy fields, and identification of strains with potential for plant growth-promotion, greenhouse gas mitigation and heavy metal bioremediation. Res Microbiol 68(3): 266–275

Sasikala C, Ramana VC (1995) Biotechnological potentials of anoxygenic phototrophic bacteria. I. Production of single cell protein, vitamins, ubiquinones, hormones and enzymes and use in waste treatment. In: Neidleman SL, Laskin AI (eds) Advance in applied microbiology, vol 41. Academic Press, NY, pp 173–226

Sasaki K, Noparatnaraporn N, Nagai S (1991) Use of photosynthetic bacteria for the production of SCP and chemical from agro-industrial wastes. In: Martin AM (ed) Bio-conversion of waste materials to industrial products. Applied Science, London, pp 225–263

Shapawi R, Ting TE, Azad SA (2012) Inclusion of purple non-sulfur bacterial biomass in formulated feed to promote growth, feed conversion ratio and survival of Asian sea bass *Lates calcarifer* juveniles. J Fish Aquat Sci 7(6):475–480

Shioi Y, Doi M (1990) Aerobic and anaerobic photosynthesis and bacteriochlorophyll formation in *Rhodobacter sulfidophilus*. In: Baltscheffsky M (ed) Current research in photosynthesis, vol III. Kluwer Academic Publishers, Dordrecht, pp 853–856

Tamminga S, Jongbloed AW, Van Eerdt MM, Aarts HFM, Mandersloot F, Hoogervorst NJP, Westhoek H (2000) Standards for the excretion of nitrogen by farm animals. Institute for Animal Husbandry and Animal. Health 71

Thorpe JE, Cho CY (1995) Minimising waste through bioenergetically and behaviourally based feeding strategies. Water Sci Technol 31:29–40

Tlusty MF, Bengston DA, Halvarson HO, Oktay SD, Pearce JB, Rheault RB (2001) Marine aquaculture and the environment: a meeting for stakeholders in the northeast. Cape Cod Press, Falmouth

UNEP (2011) Towards a Green Economy: Pathways to Sustainable Development and Poverty Eradication. Retrieved from: www.unep.org/greeneconomy.

Urdiain M, Lopez-Lopez A, Gonzalo C, Busse HJ, Langser S, Kampfer P, Rossello-Mora R (2008) Reclassification of *Rhodobium marinum* and *Rhodobium pfennigii* as *Afifella marina* gen. Nov. comb. nov. and *Afifella pfennigii* comb. nov., a new genus of photoheterotrophic Alphaproteobacteria and emended descriptions of *Rhodobium*, *Rhodobium orientis* and *Rhodobium gokarnense*. Syst Appl Microbiol 31:339–351

Van Niel CB (1957) Bergey's manual of determinative bacteriology, 7th edn. William and Wilkins Co. USA, Baltimore, pp 35–67

Vrati S, Verma J (1983) Production of molecular hydrogen and single cell protein by *Rhodopseudomonas capsulata* from cow dung. J Ferment Technol 61(2):157–162

Watanabe M, Sasaki K, Nakashimada Y, Nishio N (1998) High-density cell culture of marine photosynthetic bacterium *Rhodovulum* sp. with self flocculated cells. Biotechnol Lett 20(12): 1113–1117

Artificial Intelligence Methods in Marine Biotechnology

16

Muthu Subash Kavitha, Aurelia Jackson, V. Manikandarajan, Muralindran Mariappan, Anand Paul, and Balu Alagar Venmathi Maran

Abstract

Artificial Intelligence (AI) approaches allow users to apply algorithms and evaluate data and apply them in various industries, including marine biotechnology. In the field of oceanography, AI techniques are used in remote sensing, maritime transportation, data collection and management, ocean monitoring and predicting the occurrences of various oceanic phenomena. One of the subsets of AI techniques is Machine Learning (ML), and Deep Learning (DL) is a subset of ML that is inspired by the functionality of neurons, and its development has led to the concept of a computational model called the Artificial Neural Network (ANN). It has been well discussed in this chapter. Drug discovery is another untapped technique that can be developed to look into detecting and discovering the secondary metabolites of existing marine organisms. With the help of AI, knowledge gaps regarding the ocean's resources can be overcome, and this review compiles the current uses of this intelligent technique in the marine science discipline.

M. S. Kavitha
School of Information and Data Sciences, Nagasaki University, Nagasaki, Japan

A. Jackson · B. A. Venmathi Maran (✉)
Borneo Marine Research Institute, Universiti Malaysia Sabah, Kota Kinabalu, Sabah, Malaysia
e-mail: bavmaran@ums.edu.my

V. Manikandarajan
Kinabalu International School, Bukit Padang, Kota Kinabalu, Sabah, Malaysia

M. Mariappan
Faculty of Engineering, Universiti Malaysia Sabah, Kota Kinabalu, Sabah, Malaysia

A. Paul
The School of Computer Science and Engineering, Kyungpook National University, Daegu, South Korea

M. D. Shah et al. (eds.), *Marine Biotechnology: Applications in Food, Drugs and Energy*, https://doi.org/10.1007/978-981-99-0624-6_16

16.1 Introduction

In this era of globalization, humans are no longer strangers to the term "Artificial Intelligence (AI)", a term coined as intelligence generated by machines. It is an area in the field of information technology that works by combining a lot of data with iterative, fast processing and intelligent algorithms ranging from simple to complex. AI allows the software to learn the patterns from the sets of data provided automatically. Agarwala (2021) provided a detailed chronology of the evolution of AI-based learning approaches. While the concept of AI was proposed well before 1943, its development was difficult to accomplish due to the lack of resources and available information during that period. These issues were further exacerbated by periods of funding drops in the field, also known as "AI winter". However, decades of research and developments have conveniently caused AI to make its way into our lives thanks to the development of technology that has made a way for high-end computational power. It can store extremely large datasets for computational analysis to reveal patterns, trends and associations concerning human behaviour and interactions. From face recognition to navigation, the implementation of AI has undoubtedly and significantly improved our quality of life. Whereas in the case of marine biotechnology, the introduction of AI can fill research and knowledge gaps to further enhance the development of the field.

How AI is helpful in our daily lives and its working principles are interesting to beginners. As known to the public, AI is the algorithm for streaming services to access a movie or series applied to online shopping sites. It is also a part of the phone's security system – to bypass it, a thumb or facial print is required to "recognize" the owner. Simply put, AI is a technique that was developed to enable machines to mimic human behaviour and essentially duplicate their natural responses. The machines can perform the functions after being trained to accomplish specific tasks by processing enormous amounts of data into the systems, and then operating and adjusting themselves based on the input. This technology is sought after by large corporations because it is cost and time-efficient compared to conducting the workload manually (Davenport and Ronanki 2018).

16.1.1 Different Types of AI

One of the subsets of AI techniques is Machine Learning (ML) (Majumdar 1985; Badillo et al. 2020), which utilizes statistical methods to enable machines to improve the validity of their data as more information becomes available. In the field of marine science, ML has been used to create species distribution models and predict environmental variables based on pre-processed data, among many others (e Silva et al. 2022). On the other hand, Deep Learning (DL) (LeCun et al. 2015) is a subset of ML (Kubat 2017; Aggarwal and Murty 2021) that is inspired by the functionality of neurons, and its development has led to the concept of a computational model called the Artificial Neural Network (ANN). It can be tasked to run several checklists while at the same time filtering available data to get the most accurate prediction. In

Table 16.1 Comparison between Deep Learning and Machine Learning

Deep Learning	Machine Learning
Data-dependent; requires a large amount of data to understand the algorithm perfectly	Not data-dependent; can function consistently well with smaller datasets
Heavily dependent on high-end hardware; including GPUs, which are integral for optimization, to conduct large matrix multiplication operations	Can work on low-end machines
Feature engineering: Process of putting domain knowledge to reduce the complexity of data & make patterns more visible to learning algorithms Difficult & expensive process in terms of time & expertise	Features need to be identified by experts & hand coded per domain & data type
Approaches problem by solving it from beginning to end without requiring third party	A problem-solving approach is done by breaking down each problem into subparts, solving them individually and combining altogether for the desired result. Emphasizes detection more than recognition of the problems
Execution time takes longer and too many parameters	Shorter time to execute
Interpretability: Hard	Interpretability: Easy

marine science, DL has been used to classify, detect and segment objects onto fish images or audio files (Goodwin et al. 2022).

Using DL to run a task compared to ML can reduce costs and labour and improve accuracy, among other things (Table 16.1). DL is also more reliable as it can solve more complex tasks, such as detecting patterns in visual and acoustic data that are difficult to detect or discriminate (Malde et al. 2020). However, further advancements in terms of unsupervised or self-learning are required in this field as it is still in its growing phase (Gupta et al. 2021). The basics of the workflow used in AI-based strategies start with gathering, evaluating and pre-processing the collected data. The data will then be used to train in an AI method, and this dataset is called the training data, which can then be evaluated and calibrated. The next step is to use validation data to improve the model, ensure the input data quality and accuracy and develop models and insights. Finally, testing data will be used to validate AI further to improve its accuracy (Goodwin et al. 2022).

16.2 Applications of AI in the Marine Field

The marine environment is an integral part of the earth's climate. Aside from mediating temperature and weather, the oceans are a crucial component in determining rainfall, droughts and floods, on top of circulating 83% of the world's carbon through its waters. The ecosystems within the environment provide half of the global biological production and essential ecosystem services; hence, there is a need to

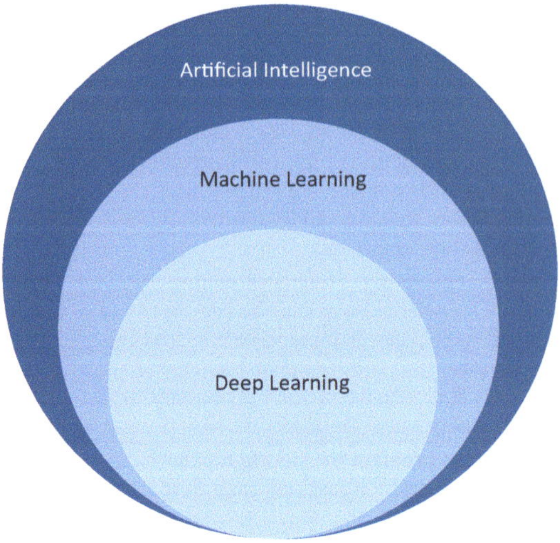

Fig. 16.1 Diagram of AI subsets

utilize marine science to obtain information on the sustainable use of the oceans. Despite that, huge data volumes from the vast oceans impose a challenge for the marine science disciplines. Issues such as data complexity and reduced data quality in the marine field should be acknowledged. The development of AI techniques can be used to address these issues and challenges. Novel ideas borne out of these solutions can be utilized to improve the efforts being done on the sustainable use of the marine environment without exhausting human labour via manual work. AI also plays a vital role in the exploration of the marine environment and also aids in its development (Fig. 16.1). Several tasks by AI in the marine field include monitoring, detection, identification, segmentation, classification and prediction (Fig. 16.2). It may help to speed up the acquisition of knowledge of the blue ocean, especially what lies deep within.

16.3 Artificial Intelligence in Marine Remote Sensing

One of the uses of AI techniques in the marine field can be found in remote sensors. Remote sensing is the process of detecting, monitoring and acquiring information regarding an object or phenomenon by measuring reflected and emitted radiation from a distance. Starting from the year 1972, satellites were utilized to monitor terrestrial terrain, but the technology was not used to monitor marine terrains until the late 1970s (Clark 1993). The development of various sectors has prompted technological advancements to bring forth remote sensing to detect, track, assess and monitor pollution in the marine environment (Loughland & Saji 2008). Aside from marine pollution, other ocean variables, such as temperature, pH, salinity, dissolved oxygen (DO), turbidity, conductivity, fluorescence due to chlorophyll and colour-dissolved organic matter (CDOM), can also be detected by remote

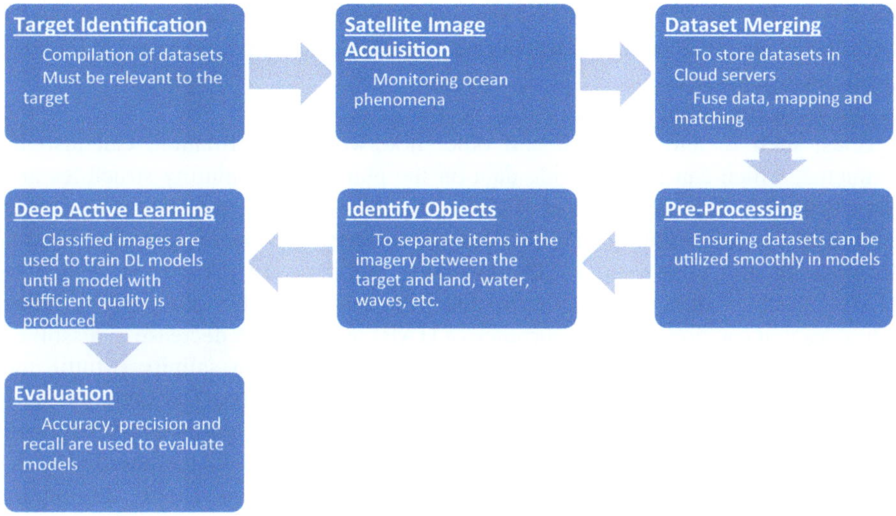

Fig. 16.2 Simplified framework in AI models for processing marine data (adapted from Sagi et al. 2020; Goodwin et al. 2022; Northeast Fisheries Science Center 2022)

sensors and are installed into satellites, ships and ocean observatories (Fig. 16.2). Various types of sensors with different capabilities have been developed with time, despite each sensor having its limitations (Hafeez et al. 2018).

To thoroughly understand the sustainability of human activities in an environment as vast and deep as the marine environment, AI machines and models are deployed to process and analyse the data collected using remote sensing. It is not limited to satellites, drones and probes. The dispatch of AI models combined with consistent remote sensing can further simplify the process of monitoring. It can help save time and reduce efforts without disrupting the marine environment (Tabak et al. 2019; Costello 2020). For example, Project Natick, the world's first deployed underwater data centre, implements AI to track changes in its surrounding environment. CoralNet is another model used to analyse corals and rocky reefs by processing images in the benthic seabed, while projects like the Allen Coral Atlas, Shell Ocean Discovery XPrize and LarvalBot are AI models that are programmed to reseed endangered reefs (Agarwala 2021). With enough resources, these models can be utilized to monitor and mitigate the effects of marine life due to climate change (Zhu 2018).

16.3.1 Harmful Algal Blooms (HABs) Monitoring

Poisoning incidents caused by harmful algal blooms have caused more human illness than any other toxin in seafood, and early detection of impeding HABs is necessary to reduce their harmful impacts. One of the ways to accomplish this is by installing early-warning systems by training AI models to predict and detect HABs (Recknagel et al. 1997). AI models in remote sensing can be programmed to identify plankton species in a selected area, especially species that can excrete neurotoxins. Aside from predicting the taxonomy of the detected species, the features of the species, such as length, width and other note-worthy information, can also be extracted, which can then provide data on the plankton community structures and their functions. The prediction of HAB occurrences using AI models can be further improved by incorporating satellite data and imagery into the system. Sea surface temperature and photosynthetically available radiation are parameters related to HABs, and their data are easily extracted from satellites (Shin et al. 2019). Other variables that can predict the occurrence of HABs are a sudden decrease in dissolved oxygen, higher chlorophyll-a conditions and fluctuations in salinity (Muttil and Chau 2007; Zhang et al. 2016; Shehhi & Chau 2020; Yñiguez and Ottong 2020). Further improvisations to the models will be the inclusion of autonomous platforms such as genomics, acoustics and pigments, that combine image-based data to conduct coastal plankton monitoring of the variables mentioned more thoroughly (Gorsky et al. 2019; Lombard et al. 2019).

16.3.2 Artificial Intelligence in Marine Monitoring

Climate change threatens marine ecosystems as it affects the physicochemical parameters of the ocean, such as pH, temperature and salinity, which ultimately lead to the degradation of marine life. This will cause habitat-forming species and ocean productivity to decline, and the changes in the geographical distribution of marine organisms will also occur (Wooldridge and Done 2009; Rizzi et al. 2016). Hence, it is essential to constantly monitor variables or parameters that may disrupt the health of marine ecosystems, and data collection plays a massive role in keeping the oceans healthy. A healthy aquatic environment would require consistent efforts to prevent pollution, protect marine species and habitats, mitigate the impacts of climate change and practice sustainable fishing (Agarwala 2021). With a data volume as vast as the ocean, the analysis of available data would also require efforts just as extensive, or even more, to ensure the longevity of the ocean resources for the future benefits of generations to come (Visbeck 2018).

In topics related to pollution, AI models are trained to identify floating objects on the sea surface, such as plastics, oil, algae, seaweed and other marine debris, using satellite imagery (Biermann et al. 2020). However, the same methods cannot be used to detect pollutants within the ocean water body. Thus, Autonomous Underwater Vehicles (AUVs) or Remotely Operated Vehicles (ROVs), which are examples of marine robots, can be dispatched to collect data deep into the oceans, where satellite imagery cannot reach due to the capacity of satellite sensor bands to penetrate the ocean floors. The data provided by the marine robots can then be analysed using ML (Fulton et al. 2018; Watanabe et al. 2019) to assess the extent of pollution.

Aside from pollutants, similar efforts are also being done to monitor available ocean resources, including food, minerals, energy, sediment and other properties.

Consistent monitoring and mapping of the ocean are vital for us to understand various phenomena, which include the ocean currents and circulation, geomorphology, benthic organisms and their processes, the impact of human activities on these phenomena and the effects of climate change on marine life (Danovaro et al. 2020). The information provided is also helpful to the researchers in devising strategies to conserve marine ecosystems.

16.4 Blue Technology and Its Significance

Blue biotechnology involves living marine and freshwater organisms in various applications of technology to produce goods, services and knowledge for humans (OECD 2016). Till now, there is an increasing interest in the potential of ocean biotechnology, as currently, the global market is estimated at USD 6 billion and is predicted to increase to USD 8.4 billion by 2026 (https://www.researchandmarkets.com/reports/4911755/marine-biotechnology-global-market-trajectory). The unique environment present in the oceans imposes a challenge for marine organisms to adapt and survive under high pressures, temperatures, salinities and pH, hence giving rise to animals with various specializations. To date, there are 194,000 species of bacteria, plants and animals inhabiting the world's oceans (Mora et al. 2011). Still, only a few of them are exploited for their active compounds and secondary metabolites compared to terrestrial organisms (Hu et al. 2011). Marine algae are one of the non-conventional sources of food (Gouveia et al. 2006; Lum et al. 2013), cosmetics (Wang et al. 2015), medicines (Hamed et al. 2015), biofuel production (Lum et al. 2013), packaging (Adli et al. 2018), dyeing and pigmentation (Azeem et al. 2019), bioremediation (Neveux et al. 2018) and use for wastewater treatment (Uysal et al. 2016; Rashid et al. 2018) due to their high growth rate and ability to make use of CO_2 and seawater (Griffiths et al. 2011; Goswami et al. 2015). Despite that, traditional algae processing techniques require improvement in terms of efficiency, energy cost and quality, and the use of AI can be implemented to speed up these methods, especially in food and biofuel production (Wang et al. 2019; Schenk 2021; Zhang et al. 2021). Studies in identifying algae have also been conducted to allow real-time microalgae identification (Teng et al. 2020). As interest, applications and cultivation of this product continue to skyrocket, the same can be said regarding their value, as it is estimated that the value of seaweed in worldwide markets will reach USD 9.98 billion by 2024 (Statista-The Statistic Portal 2011). By sustainably farming and harvesting algae, it will simultaneously solve the limitation of resources along with environmental problems the world is currently facing, such as water purification and emission of greenhouse gases, which may lead to global warming and ocean acidification (Wallington et al. 2009).

Other unconventional sources of marine products can be derived from jellyfish. Compared to other organisms such as fungi, Porifera and bacteria, jellyfish have been under-exploited when obtaining natural products. And even then, most of the derived products are from those of benthic jellyfish (Merquiol et al. 2019). Jellyfish under the class Scyphozoa are known to possess three kinds of compounds, including collagen and fatty acids, which have boomed in various industries due to their importance and uses as nutraceuticals (Hsieh et al. 2001), cosmeceuticals (Leone et al. 2015, 2019; Kim et al. 2016) and biomedical applications due to their collagen (Ovchinnikova et al. 2006). Additionally, the Green Fluorescent Protein (GFP) found in the crystal jellyfish, *Aequorea* sp., was also found to be usable in organic structures for DNA studies as a marker protein (Shimomura et al. 1962). The presence of crude venom in jellyfish, which is found to have anti-tumoral activities, is toxic to colon cancer and hepatoma cells in humans (Li et al. 2012; Lee et al. 2017). As jellyfish blooms are random and unpredictable, the implementation of AI to monitor jellyfish in real-time can observe and predict impending blooms (Martin-Abadal et al. 2020). However, some of the challenges faced when attempting to successfully market products of blue biotechnology include the small market size, production costs and constraints regarding specifications of quality, safety assurance, extraction techniques, minimization of environmental impact and palatability issues (Gellenbeck 2012; Xiong et al. 2018; Camacho et al. 2019). Hence, further studies are required to make use of these marine products safely and sustainably thoroughly.

16.5 AI in the Development of Drug Discovery in Marine Organisms

The discovery of drugs is a process of identifying active compounds for new medications, whether by identifying active compounds or ingredients in currently known remedies or through accidental discoveries (Aliper et al. 2016). It is a complex process that is developed based on the primary concern that it heavily consumes both time and resources. As the efficiency of research and development continues to dwindle (Scannell et al. 2012), computational approaches such as *in silico* drug discovery (Loging et al. 2007; Kirchmair et al. 2015) are now being studied due to their ability to handle large volumes of data and process them automatically (Miles and Walker 2006). As a result, many pharmaceutical companies have been exploring the applications of AI in drug discovery (Zarringhalam et al. 2018; Zhou et al. 2019; Davies et al. 2020; Kotsias et al. 2020; Montanari et al. 2020; Rohall et al. 2020). It includes the identification of a drug target, screening of available drugs, peptide synthesis, physiochemical activity and toxicity (Gupta et al. 2021). One of the first companies to apply AI to drug discovery was Atomwise, which currently has 550 projects utilizing AI to deal with issues such as discovery, potency optimization, selectivity optimization and off-target toxicity testing. Other companies with the same approaches to AI applications in drug discovery include Benevolent AI, Recursion and Standigm (Elbadawi et al. 2021).

The capabilities of AI in biotechnology are not just limited to drug target identification but also drug screening, image screening and predictive modelling. The way AI works in the aforementioned tasks was by testing molecules with molecular descriptions. It can also be provided in binary codes to represent the absence or presence of a molecular feature, in this case, the active compounds (Lorena et al. 2008). The AI models identify the cells and their features (Tripathy et al. 2014). Furthermore, the data from the testing molecules are then known and used to train an AI model to detect or predict whether or not a novel molecule is the targeted molecular feature. Various studies have made use of AI to discover novel drugs. However, the same approaches have not yet been available in the field of marine science, even though the ocean houses numerous species with all sorts of active compounds (Hu et al. 2011). Efforts in implementing AI in marine drug discovery should be focused on, as the economic value of anticancer marine drugs continues to increase, valued at USD 563 billion to 5.7 trillion (Erwinn et al. 2010; Abdelmohsen et al. 2017).

If companies like Atomwise were to incorporate their drug discovery approaches with AI models that can identify marine organisms down to the species level automatically, the process of discovery of marine-derived drugs using AI can be studied (Watson et al. 2018; Zhavoronkov et al. 2020), especially when dealing with data volumes as large as the oceans. Aside from that, this approach can also lessen the harm and disturbance caused to the marine environment and the target species to preserve the active compounds. For instance, some compounds' presence depends directly on their hosts' activities, if both organisms are in a symbiotic association. With the inclusion of drug discovery from marine organisms, it is expected that further advancements in medication can be personalized and catered according to the needs of individual patients worldwide.

16.6 Deep Learning Methods in Marine Science Applications

Deep learning is used in varying applications in the field of marine science (Table 16.2). For instance, deep learning is used to study the development of effective fish sampling techniques to estimate and monitor fish biomass and assemblages. However, there is a gap in poor luminosity, fish shape, texture, orientation, moving seaweed or seagrass in the background and seabed structures (Salman et al. 2020). A regional-based convolutional neural network was used to solve generic object detection and localization problems and was trained using a new method to make use of motion from fish in videos through background subtraction and optical flow. It merged the results with raw image data to create regions that are fish dependent. Another study looked at lobster detection by generating synthetic parts of lobsters such as the antennae, instead of their whole body to generate higher detection accuracy, as there is a lack of annotated training datasets (Mahmood et al. 2020). The trained model is then tasked to detect wild western rock lobsters that are partially visible and occluded in the west Australian seas. Additionally, recent developments of artificial intelligence in fisheries science are providing a promising opportunity relating to a massive sampling of fish catches. To illustrate this, a study tackled the statistics of fish catch for marine resource management, whereby images of common fish caught on decks of fishing vessels are automatically identified using a machine vision technique, called VGG-16 (Lu et al. 2020). Álvarez-Ellacularía

Table 16.2 Methods and modalities of AI applications in marine science

Source	Methods	Image/video modalities
Salman et al. (2020)	Gaussian mixture modelling, optical flow and a residual neural network (ResNet-152)	Datasets for fish detection were obtained from Fish4Knowledge, including videos of fish under different environmental conditions and LCF-15 fish tasks.
Mahmood et al. (2020)	You only look once (YOLOv3)	237 ground annotated images, collected from the stereo camera at a resolution of 1360 X 1024 pixels. A poisson blending scheme is used to aid in adding synthetic parts to images.
Lu et al. (2020)	Visual geometry group (VGG-16)	16,517 images collected from Taiwan's Council of Agriculture and resized to 330 X 250 pixels.
Álvarez-Ellacuría et al. (2020)	Mask regional convolutional neural network (R-CNN)	Webcam images were obtained at a resolution of 1280 X 760, and manually annotated using LABELBOX
French et al. (2020)	Mask R-CNN	Image data were extracted from CCTV footage in 800p HD in MPEG-4 format.
Garcia et al. (2020)	Mask R-CNN	Deep vision is used to acquire stereo image pairs of fish in JPG format at a resolution of 1392 X 1040 pixels.
Tseng and Kuo (2020)	Mask R-CNN	700 videos acquired from SeaTube at 30 fps at a resolution of 1280 X 720 pixels.
Proud et al. (2020)	Random forest	Acoustic data collected using hull-mounted Kongsberg Sinrad EK60 scientific echo sounders
Brautaset et al. (2020)	U-net	Simrad EKM60 echosounder systems operating transducers attached on four trawlers,
Li et al. (2020)	Faster R-CNN, feature pyramid network (FPN), single shot MultiBox detector (SSD), YOLOv3, & RetinaNet	Olympus BX53 was used to collect microscopic images of phytoplankton with an overall magnification of x200 and a resolution of 2040 x 1536
Campbell et al. (2020)	Inception v3	PWS plankton camera with onboard computer segments each image and stores it in TIFF format.

et al. (2020) studied the unsupervised estimation of European hake length using Mask R-CNN. A different study reported the development of a computer vision system that analyses surveillance footage for monitoring and quantifying discarded fish catch, which is impacted by the various conditions on board fishing trawlers

(French et al. 2020). Another recently developed tool for the detection of bycatch as a result of overfishing is echo sounders, which are normally used to detect schools of fish and qualitatively estimated the species present or not and estimated the amount. Garcia et al. (2020) studied the image-based method for individual fish detection targeted at a juvenile fish catch that is undersized, whereby the deep vision imaging system acquires the images directly from the trawl, which are then pre-processed for a Mask R-CNN model to localize and segment each fish in the images. The refining of the segmentation using local gradients is used to accurately estimate the boundary regions of every individual fish, enabling this technique to be useful when detecting cluttered images with overlapping fish. Aside from monitoring fish populations in the wild, AI applications have also been used to detect and recognize phytoplankton, which plays an important role in the marine environment, and aquaculture. However, current efforts in doing so mostly rely on manual labour and operations. The study by Li et al. (2020) presented the phytoplankton microscopic dataset PMID2019 for the autonomous detection of phytoplankton, which includes 24 different categories. The dataset was developed for phytoplankton microscopic vision technology in the future, and can also be used to assess the performance of existing AI models to detect phytoplankton within the datasets. Campbell et al. (2020) developed a novel plankton imager to be deployed on board for profiling. The images are manually identified into classes, and a hybrid CNN was trained to identify the images. While the accuracy of classification varied among the classes, applying thresholds to the output of the CNN improved its classification accuracy.

Datasets are extracted from a large Fish4Knowledge underwater video repository, Complex Scenes dataset and the LifeCLEF 2015 fish dataset to validate the effectiveness of the hybrid approach and achieved a detection accuracy (F-Score) of 87.44% and 80.02% respectively on these datasets. It advocated the utilization of a deep-learning approach for fish detection tasks (Salman et al. 2020). Deep convolutional neural network models are developed to identify the species from the images and achieved an accuracy of 96.24% (Lu et al. 2020). The estimated mean of fish lengths at the box level is accurate; for average lengths ranging from 20 to 40 cm, the root-mean-square deviation was 1.9 cm and the maximum deviation between the estimated and the measured mean body length was 4.0 cm. Álvarez-Ellacularía et al. (2020) found the classification accuracy varied among the different classes, and applying thresholds to the output of the neural network (interpretable as probabilities or classifier confidence), improved classification accuracy in non-ambiguous groups to between 80% and 100% (Campbell et al. 2020).

16.7 Conclusion

As the data volume in marine science continues to increase, new challenges are also introduced to researchers and scientists alike. But with the help of AI, the burden of manually collecting data from the vast oceans can be significantly reduced. Drug discovery techniques with the help of DL have been studied successfully, and with further development of this technology, scientists can investigate detecting and discovering secondary metabolites from marine organisms residing in the oceans. As the use of AI in the marine field is currently focused on data processing, efforts to ensure a more integrated approach should be implemented, especially in the collection and management of data. Seeing how limited available data is in marine science, archives of marine datasets should be created to preserve and share the knowledge and expertise among other marine researchers and scientists to evaluate AI models.

Acknowledgement Financial support received from the International Grant: LPA2007 to BAVM.

References

Abdelmohsen UR, Balasubramanian S, Oelschlaeger TA, Grkovic T, Pham NB, Quinn RJ (2017) Potential of marine natural products against drug-resistant fungal, viral, and parasitic infections. Lancet Infect Dis 17:e30–e41

Adli SA, Ali F, Azmi AS (2018) Preparation and characterization of PLA/algae nanocomposite for packaging industry. J Adv Res Mat Sci 50(1):19–25

Agarwala N (2021) Managing marine environmental pollution using artificial intelligence. Maritime Technol Res 3(2):120–136

Aggarwal M, Murty MN (2021) Deep learning. In: SpringerBriefs in applied sciences and technology. Springer, Singapore, pp 1–6

Aliper A, Plis S, Artemov A, Ulloa A, Mamoshina P, Zhavoronkov A (2016) Deep learning applications for predicting pharmacological properties of drugs and drug repurposing using transcriptomic data. Mol Pharm 13:2524–2530

Álvarez-Ellacuría A, Palmer M, Catalán IA, Lisani J-L (2020) Image-based, unsupervised estimation of fish size from commercial landings using deep learning. ICES J Mar Sci 77(4): 1330–1339

Azeem M, Iqbal N, Mir RA, Adeel S, Batool F, Khan AA, Gul S (2019) Harnessing natural colorants from algal species for fabric dyeing: A sustainable eco-friendly approach for textile processing. J Appl Phycol 31:3941–3948

Badillo S, Banfai B, Birzele F (2020) An introduction to machine learning. Clin Pharmacol Therapeut 107(4):871–885

Biermann L, Clewley D, Martinez-Vicente V, Topouzelis K (2020) Finding plastic patches in coastal waters using optical satellite data. Sci Rep 10:5364

Brautaset O, Waldeland AU, Johnsen E, Malde K, Eikvil L, Salberg A-B, Handegard NO (2020) Acoustic classification in multifrequency echosounder data using deep convolutional neural networks. ICES J Mar Sci 77(4):1391–1400

Camacho F, Macedo A, Malcata F (2019) Potential industrial applications and commercialization of microalgae in the functional food and feed industries: A short review. Mar Drugs 17(6):312

Campbell RW, Roberts PL, Jaffe J (2020) The Prince William sound plankton camera: A profiling *in situ* observatory of plankton and particulates. ICES J Mar Sci 77(4):1440–1455

Costello M (2020) AI is a promising new tool for monitoring marine biodiversity. University of Auckland. Retrieved from: https://www.oceansofbiodiversity.auckland.ac.nz/2020/06/22/ai-is-a-promising-new-tool-for-monitoring-marine-biodiversity/

Danovaro R, Fanelli E, Aguzzi J, Billett D, Carugati L, Corinaldesi C, Dell'Anno A, Gjerde K, Jamieson AJ, Kark S, McClain C, Levin L, Levin N, Ramirez-Llodra E, Ruhl H, Smith CR, Snelgrove PVR, Thomsen L, Dover CLV, Yasuhara M (2020) Ecological variables for developing a global deep-ocean monitoring and conservation strategy. Nat Ecol Evol 4:181–192

Davenport TH, Ronanki R (2018) Artificial intelligence for the real world. Harv Bus Rev. Retrieved from: https://hbr.org/webinar/2018/02/artificial-intelligence-for-the-real-world

Davies M, Jones RDO, Grime K, Jansson-Löfmark R, Fretland AJ, Winiwarter S, Morgan P, McGinnity DF (2020) Improving the accuracy of predicted human pharmacokinetics: lessons learned from the AstraZeneca drug pipeline over two decades. Trends Pharmacol Sci 41:390–408

e Silva LO, Resende M, Galhardas H, Manquinho V, Lynce I (2022) DeepData : machine learning in the marine ecosystems. Expert Syst Appl 206:117841

Elbadawi M, Gaisford S, Basit AW (2021) Advanced machine-learning techniques in drug discovery. Drug Discov Today 26(3):769–777

Erwinn PM, Lopez-Legentil S, Schuhmann PW (2010) The pharmaceutical value of marine biodiversity for anti-cancer drug discovery. Ecol Econ 70:445–451

French G, Mackiewicz M, Fisher M, Holah H, Kilburn R, Campbell N, Needle C (2020) Deep neural networks for analysis of fisheries surveillance video and automated monitoring of fish discards. ICES J Mar Sci 77(4):1340–1353

Fulton M, Jungseok H, Islam MJ, Sattar J (2018) Robotic detection of marine litter using deep visual detection models, computer science and engineering. In: Proceedings of the international conference on robotics and automation. Montreal, Canada. arXiv:1804.01079

Garcia R, Prados R, Quintana J, Tempelaar A, Gracias N, Rosen S, Vågstøl H, Løvall K (2020) Automatic segmentation of fish using deep learning with application to fish size measurement. ICES J Mar Sci 77(4):1354–1366

Gellenbeck KW (2012) Utilization of algal materials for nutraceutical and cosmeceutical applications-what do manufacturers need to know? J Appl Phycol 24:309–313

Goodwin M, Halvorsen KT, Jiao L, Knausgård KM, Martin AH, Moyano M, Oomen RA, Rasmussen JH, Sørdalen TK, Thorbjørnsen SH (2022) Unlocking the potential of deep learning for marine ecology: overview, applications, and outlook. ICES J Mar Sci 72(2):319–336

Gorsky G, Bourdin G, Lombard F, Pedrotti ML, Audrain S, Bin N, Boss E (2019) Expanding Tara oceans protocols for underway, ecosystemic sampling of the ocean-atmosphere interface during Tara Pacific expedition (2016–2018). Front Mar Sci 6:750

Goswami G, Bang V, Agarwal S (2015) Diverse applications of algae. Int J Adv Res Sci Eng 4(1): 1102–1109

Gouveia L, Raymundo A, Batista AP, Sousa I, Empis J (2006) *Chlorella vulgaris* and *Haematococcus pluvialis* biomass as colouring and antioxidant in food emulsions. Eur Food Res Technol 222:362–367

Griffiths MJ, Dicks RG, Richardson C, Harrison STL (2011) Advantages and challenges of microalgae as a source of oil for biodiesel. In: Biodiesel – feedstocks and processing technologies. IntechOpen, pp 178–200

Gupta R, Srivastava D, Sahu M, Tiwari S, Ambasta RK, Kumar P (2021) Artificial intelligence to deep learning: machine intelligence approach for drug discovery. Mol Divers 25:1315–1360

Hamed I, Özogul F, Özogul Y, Regenstein JM (2015) Marine bioactive compounds and their health benefits: A review. Compr Rev Food Sci Food Saf 14:446–465

Hsieh PYH, Leong FM, Rudloe J (2001) Jellyfish as food. Hydrobiologia 451:11–17

Hu G-P, Yuan J, Sun L, She Z-G, Wu J-H, Lan X-J, Zhu X, Lin Y-C, Chen S-P (2011) Statistical research on marine natural prodcuts based on data obtained between 1985 to 2008. Mar Drugs 9: 514–525

Kim DW, Park TS, Kim JY, Choi SG, Lee DW (2016) Moisturising effect of jellyfish collagen extract. J Soc Cosmet Sci Korea 42(2):153–162

Kirchmair J, Göller AH, Lang D, Kunze J, Testa B, Wilson ID, Glen RC, Schneider G (2015) Predicting drug metabolism: experiment and/or computation? Nat Rev Drug Discov 14:387

Kotsias P-C, Arús-Pous J, Chen H, Engkvist O, Tyrchan C, Bjerrum EJ (2020) Direct steering of de novo molecular generation with descriptor conditional recurrent neural networks. Nat Mach Intell 2:254–265

Kubat M (2017) An introduction to machine learning. Springer International Publishing, Cham, pp 1–291

LeCun Y, Bengio Y, Hinton G (2015) Deep learning. Nature 521(7553):436–444

Lee H, Bae SK, Kim M, Pyo MJ, Kim M, Yang S, Won C-K, Yoon WD, Han CH, Kang C (2017) Anticancer effect of *Nemopilema nomurai* jellyfish venom on hepg2 cells and a tumor xenograft animal model. Evid Based Complement Alternat Med 2017:1–12
Leone A, Lecci RM, Durante M, Meli F, Piraino S (2015) The bright side of gelatinous blooms: nutraceutical value and antioxidant properties of three Mediterranean jellyfish (Scyphozoa). Mar Drugs 13:4654–4681
Leone A, Lecci RM, Milisenda G, Piraino S (2019) Mediterranean jellyfish as novel food: effects of thermal processing on antioxidant, phenolic, and protein contents. Eur Food Res Technol 245: 1611–1627
Li C, Li P, Feng J, Li I, Yu H (2012) Cytotoxicity of the venom from the nematocysts of jellyfish *Cyanea nozakii* Kishinouye. Toxicol Ind Health 28:186–192
Li Q, Sun X, Dong J, Song S, Zhang T, Liu D, Zhang H, Han S (2020) Developing a microscopic image dataset in support of intelligent phytoplankton detection using deep learning. ICES J Mar Sci 77(4):1427–1439
Loging W, Harland L, Williams-Jones B (2007) High-throughput electronic biology: mining information for drug discovery. Nat Rev Drug Discov 6:220–230
Lombard F, Boss E, Waite AM, Vogt M, Uitz J, Stemmann L, Sosik HM (2019) Globally consistent quantitative observations of planktonic ecosystems. Front Mar Sci 6:196
Lorena AC, de Carvalho ACPLF, Gama JMP (2008) A review on the combination of binary classifiers in multiclass problems. Artif Intell Rev 30:19–37
Lu Y-C, Tung C, Kuo Y-F (2020) Identifying the species of harvested tuna and billfish using deep convolutional neural networks. ICES J Mar Sci 77(4):1318–1329
Lum KK, Kim J, Lei XG (2013) Dual potential of microalgae as a sustainable biofuel feedstock and animal feed. J Anim Sci Biotechnol 4:1–7
Mahmood A, Bennamoun M, An S, Sohel F, Boussaid F, Hovey R, Kendrick G (2020) Automatic detection of Western rock lobster using synthetic data. ICES J Mar Sci 77(4):1308–1317
Majumdar DD (1985) Trends in pattern recognition and machine learning. Def Sci J 35(3):6027
Malde K, Handegard NO, Eikvil L, Salberg A (2020) Machine intelligence and the data-driven future of marine science. ICES J Mar Sci 77(4):1274–1285
Martin-Abadal M, Ruiz-Frau A, Hinz H, Gonzalez-Cid Y (2020) Jellytoring: real-time jellyfish monitoring based on deep learning object detection. Sensors 20:1708
Merquiol L, Romano G, Ianora A, D'Ambra I (2019) Biotechnological applications of scyphomedusae. Mar Drugs 17:604
Miles J, Walker A (2006) The potential application of artificial intelligence in transport. IEE Proceedings – Intelligent Transport Systems 153:183–198
Montanari F, Kuhnke L, Ter Laak A, Clevert DA (2020) Modeling physico-chemical ADMET endpoints with multitask graph convolutional networks. Molecules 25:44–56
Mora C, Tittensor DP, Adl S, Simpson AGB, Worm B (2011) How many species are there on earth and in the ocean? PLoS Biol 9(8):e1001127
Muttil N, Chau K-W (2007) Machine-learning paradigms for selecting ecologically significant input variables. Eng Appl Artif Intell 20(6):735–744
Neveux N, Bolton JJ, Bruhn A, Roberts DA, Ras M (2018) The bioremediation potential of seaweeds: recycling nitrogen, phosphorus and other waste products. In: *Blue biotechnology: production and use of marine molecules*. Wiley-VCH Verlag GmbH & Co. KGaA, pp 217–241
Northeast Fisheries Science Center (2022) Artificial intelligence: detecting marine animals with satellites. Noaa Fisheries. Retrieved from: https://www.fisheries.noaa.gov/new-england-mid-atlantic/science-data/artificial-intelligence-detecting-marine-animals-satellites
OECD (2016) Marine Biotechnology: Definitions, infrastructures and directions for innovations. Retrieved from: https://www.oecd.org/health/biotech/marine-biotechnology-ocean-productivity-sustainability.html
Ovchinnikova TV, Balandin SV, Aleshina GM, Tagaev AA, Leonova YF, Krasnodembsky ED, Men'Shenin AV, Kokryakov VN (2006) Aurelin, a novel antimicrobial peptide from jellyfish

Aurelia aurita with structural features of defensins and channel-blocking toxins. Biochem Biophys Res Commun 348:514–523

Proud R, Mangeni-Sande R, Kayanda RJ, Cox MJ, Nyamweya C, Ongore C, Natugonza V, Everson I, Elison M, Hobbs L, Kashindye BB, Mlaponi EW, Taabu-Munyaho A, Mwainge VM, Kagoya E, Pegado A, Nduwayesu E, Brierley AS (2020) Automated classification of schools of the silver cyprinid *Rastrineobola argentea* in Lake Victoria acoustic survey data using random forests. ICES J Mar Sci 77(4):1379–1390

Rashid N, Park WK, Selvaratnam T (2018) Binary culture of microalgae as an integrated approach for enhanced biomass and metabolites productivity, wastewater treatment, and bioflocculation. Chemosphere 194:67–75

Recknagel F, French M, Harkonen P, Yabunaka K-I (1997) Artificial neural network approach for modelling and prediction of algal blooms. Ecol Model 96:11–28

Rizzi J, Torresan S, Critto A, Zabeo A, Brigolin D, Carniel S, Pastres R, Marcomini A (2016) Climate change impacts on marine water quality: the case study of the northern Adriatic Sea. Mar Pollut Bull 102(2):271–282

Rohall SL, Auch L, Gable J, Gora J, Jansen J, Lu Y, Martin E, Pancost-Heidebrecht M, Shirley B, Stiefl N, Lindvall M (2020) An artificial intelligence approach to proactively inspire drug discovery with recommendations. J Med Chem 63:8824–8834

Sagi T, Lehahn Y, Bar K (2020) Artificial intelligence for ocean science data integration: current state, gaps, and way forward. Elementa Science of the Anthropocene 8:21

Salman A, Siddiqui SA, Shafait F, Mian A, Shortis MR, Khurshid K, Ulges A, Schwanecke U (2020) Automatic fish detection in underwater videos by a deep neural network-based hybrid motion learning system. ICES J Mar Sci 77(4):1295–1307

Scannell JW, Blanckley A, Boldon H, Warrington B (2012) Diagnosing the decline in pharmaceutical R&D efficiency. Nat Rev Drug Discov 11:191–200

Schenk PM (2021) Phycology: algae for food, feed, fuel and the planet. Phycology 1:76–78

Shimomura O, Johnson FH, Saiga Y (1962) Extraction, purification and properties of aequorin, a bioluminescent protein from the luminous hydromedusan, *Aequorea*. J Cell Comp Physiol 59(3):223–239

Shin J, Kim SM, Son YB, Kim K, Ryu JH (2019) Early prediction of *Margalefidinium polykrikoides* bloom using a LSTM neural network model in the South Sea of Korea. J Coast Res 90:236–242

Statista-The Statistic Portal (2011) The statistic portal value of the seaweed market worldwide in 2017 and 2024 (in billion U.S. Dollars). Retrieved from: https://www.statista.com/

Tabak MA, Norouzzadeh MS, Wolfson DW, Sweeney SJ, VerCauteren KC, Snow NP, Halseth JM, Di Salvo PA, Lewis JS, White MD, Teton B, Beasley JC, Schlichting PE, Boughton RK, Wight B, Newkirk ES, Ivan JS, Odell EA, Brook RK, Lukacs PM, Moeller AK, Mandeville EG, Clune J, Miller RS (2019) Machine learning to classify animal species in camera trap images: applications in ecology. Methods Ecol Evol 10(4):585–590. https://doi.org/10.1101/346809

Teng SY, Yew GY, Sukačová K, Show PL, Máša V, Chang J-S (2020) Microalgae with artificial intelligence: A digitalized perspective on genetics, systems and products. Biotechnol Adv 44: 107631

Tripathy RK, Mahanta S, Paul S (2014) Artifcial intelligencebased classifcation of breast cancer using cellular images. RSC Adv 4:9349–9355

Tseng C-H, Kuo Y-F (2020) Detecting and counting harvested fish and identifying in fish types electronic monitoring system videos using deep convolutional neural networks. ICES J Mar Sci 77(4):1367–1378

Uysal O, Uysal O, Ek K (2016) Determination of fertilizing characteristics of three different microalgae cultivated in raceways in greenhouse conditions could increase soil fertility and product yield. Agric Sci Res 59:15–19

Visbeck M (2018) Ocean science research is key for a sustainable future. Nat Commun 6:690

Wallington TJ, Srinivasan J, Nielsen OJ, Highwood EJ (2009) Greenhouse gases and global warming. Environ Ecol Chem 1:36–63

Wang HMD, Chen CC, Huynh P, Chang JS (2015) Exploring the potential of using algae in cosmetics. Bioresour Technol 184:355–362
Wang J, Zhang M, Fang Z (2019) Recent development in efficient processing technology for edible algae: A review. Trends Food Sci Technol 88:251–259
Watanabe JI, Yang S, Miura N (2019) Underwater and airborne monitoring of marine ecosystems and debris. J Appl Remote Sens 13(4):1–10
Watson O, Cortes-Ciriano I, Taylor A, Watson JA (2018) A decision theoretic approach to model evaluation in computational drug discovery. arXi:1807.08926
Wooldridge SA, Done TJ (2009) Improved water quality can ameliorate effects of climate change on corals. Ecol Appl 19(6):1492–1499
Xiong JQ, Kurade MB, Jeon BH (2018) Can microalgae remove pharmaceutical contaminants from water? Trends Biotechnol 36:30–44
Yñiguez AT, Ottong ZJ (2020) Predicting fish kills and toxic blooms in an intensive mariculture site in The Philippines using a machine learning model. Sci Total Environ 707:136173
Zarringhalam K, Degras D, Brockel C, Ziemek D (2018) Robust phenotype prediction from gene expression data using differential shrinkage of co-regulated genes. Sci Rep 8:1237
Zhang F, Wang Y, Cao M, Sun X, Du Z, Liu R, Ye X (2016) Deep learning-based approach for prediction of algal blooms. Sustainability 8(10):1060
Zhang W, Li J, Liu T, Leng S, Yang L, Peng H, Jiang S, Zhou W, Leng L, Li H (2021) Machine learning prediction and optimization of bio-oil production from hydrothermal liquefaction of algae. Bioresour Technol 342:126011
Zhavoronkov A, Vanhaelen Q, Oprea TI (2020) Will artificial intelligence for drug discovery impact clinical pharmacology? Clin Pharmacol Ther 107(4):780–785
Zhou H, Xi L, Ziemek D, O'Neil S, Lee J, Stewart Z, Zhan J, Zhao S, Zhang Y, Page K, Huang A, Maciejewski M, Zhang B, Gorelick KJ, Fitz L, Pradhan V, Cataldi F, Vincent M, Von Schack D, Hung K, Hassan-Zahraee M (2019) Molecular profiling of ulcerative colitis subjects from the TURANDOT trial reveals novel pharmacodynamic/efficacy biomarkers. J Crohn's Colitis 13: 702–713
Zhu X (2018) Monitoring environmental conditions near underwater datacenters using deep learning. Microsoft Azure. Retrieved from https://azure.microsoft.com/enin/blog/monitoring-environmental-conditions-near-underwater-datacenters-using-deeplearning

MIX
Papier aus verantwortungsvollen Quellen
Paper from responsible sources
FSC® C105338

If you have any concerns about our products,
you can contact us on
ProductSafety@springernature.com

In case Publisher is established outside the EU,
the EU authorized representative is:
Springer Nature Customer Service Center GmbH
Europaplatz 3, 69115 Heidelberg, Germany

Printed by Libri Plureos GmbH
in Hamburg, Germany